CLOSED CIRCUIT TRICKLE IRRIGATION DESIGN

Theory and Applications

Research Advances in Sustainable Micro Irrigation

VOLUME 7

CLOSED CIRCUIT TRICKLE IRRIGATION DESIGN

Theory and Applications

Edited by

Megh R. Goyal, PhD, PE, Senior Editor-in-Chief
Hani A. A. Mansour, PhD, Co-Editor

APPLE
ACADEMIC
PRESS

Apple Academic Press Inc. | Apple Academic Press Inc.
3333 Mistwell Crescent | 9 Spinnaker Way
Oakville, ON L6L 0A2 | Waretown, NJ 08758
Canada | USA

First issued in paperback 2021

Exclusive worldwide distribution by CRC Press, a member of Taylor & Francis Group
No claim to original U.S. Government works

ISBN 13: 978-1-77463-538-4 (pbk)
ISBN 13: 978-1-77188-116-6 (hbk)

Library and Archives Canada Cataloguing in Publication

Closed circuit trickle irrigation design : theory and applications / edited by Megh R. Goyal, PhD, PE (senior editor-in-chief), Hani A.A. Mansour, PhD (co-editor).

(Research advances in sustainable micro irrigation volume 7)
Includes bibliographical references and index.
ISBN 978-1-77188-116-6 (bound)
1. Microirrigation. 2. Sustainable agriculture. I. Goyal, Megh Raj, editor II. Mansour, Hani A. A., author, editor III. Series: Research advances in sustainable micro irrigation; v. 7

S619.T74C66 2015   631.5'87   C2015-904258-5

Library of Congress Cataloging-in-Publication Data

Closed circuit trickle irrigation design : theory and applications / Megh R. Goyal, PhD, PE, senior editor-in-chief ; Hani A.A. Mansour, PhD, co-editor.

pages cm
Includes bibliographical references and index.
ISBN 978-1-77188-116-6 (alk. paper)
1. Microirrigation. 2. Irrigation farming. 3. Irrigation water. 4. Water-supply, Agricultural-
-Management. 5. Electricity in agriculture. I. Goyal, Megh Raj, editor. II. Mansour, Hani A. A., editor.

S619.T74C56 2015   631.5'87--dc23   2015022736

Apple Academic Press also publishes its books in a variety of electronic formats. Some content that appears in print may not be available in electronic format. For information about Apple Academic Press products, visit our website at **www.appleacademicpress.com** and the CRC Press website at **www.crc-press.com**

CONTENTS

LIST OF CONTRIBUTORS

Abdullah S. Aljughaiman, PhD
Department of ESNR, College of Agriculture and Food Science, King Faisal University, Saudi Arabia

Ebtisam E. Eldardiry, I., PhD
Water Relations and Field Irrigation Department, Agricultural Division and National Research Center, Cairo, Egypt. E-mail: ebtisameldardiry@gmail.com

M. Abd El-Hady, PhD
Water Relations and Field Irrigation Department, Agricultural Division and National Research Center, Giza, Cairo Egypt. hadymnrc60@gmail.com

Mohamed E. El-Hagarey, PhD
Researcher at Irrigation and Drainage Unit, Division of Water Resources and Desert Land, Desert Research Center (DRC), Ministry of Agriculture and Land Reclamation, 1 Mathaf El-Mataria Street, El-Mataria, Cairo-Egypt, B.O.P 11753. Mobile: +2 01063031920. E-mail: elhagarey@gmail.com

Yousif El-Melhem, PhD
Department of Environmental and Natural Resources, College of Agriculture and Food Science, King Faisal University, P.O. Box 420, Al-Hassa 31982, Saudi Arabia

Megh R. Goyal, PhD, PE
Retired Professor in Agricultural and Biomedical Engineering from General Engineering Department, University of Puerto Rico – Mayaguez Campus; and Senior Technical Editor-in-Chief in Agriculture Sciences and Biomedical Engineering, Apple Academic Press Inc., PO Box 86, Rincon – PR – 00677 – USA. E-mail: goyalmegh@gmail.com

Ahmed S. Hassan, PhD
Researcher at Irrigation Department, Agriculture Engineering Research Institute, ARC, Egypt

Farid Hellal, PhD
Plant Nutrition Department, National Research Center, Giza, Egypt

Marvin E. Jensen, PhD, PE
Retired Research Leader at USDA – ARS. 1207 Spring Wood Drive, Fort Collins, Colorado 80525. E-mail: mjensen419@aol.com

David A. Lightfoot, PhD
Soil and Plant and Agricultural Systems Department, Southern Illinois University, Carbondale, Illinois, USA

Hani A. Mansour, PhD
Post Doc Research Fellow, 225 S. Univ., Street, ABE Department, Purdue University West Lafayette, Indiana 47907, United States; Researcher, WRFI Department, Agriculture Division, National Research Center, El-Behouth St., Eldokki, Giza, Postal Code 12311, Cairo, Egypt. Tel.: +201-068989517 and +201-123355393; E-mail: mansourhani2011@gmail.com

V. M. Mayande, PhD
Vice Chancellor, Dr. Panjabrao Deshmukh Krishi Vidyapeeth, Akola, 444104, Maharashtra, India. Mobile: +91 9423174299, E-mail: vmmayande@yahoo.com

Hany M. Mehanna, PhD
Professor, Water Relations and Field Irrigation Department (Agricultural and Biological Division), National Research Centre, Cairo, Egypt. E-mail: mr.mehana@gmail.com

Miguel A. Muñoz-Muñoz, PhD
Ex-President of University of Puerto Rico, University of Puerto Rico, Mayaguez Campus, College of Agriculture Sciences, Call Box 9000, Mayagüez, PR. 00681-9000. Tel. 787-265-3871, E-mail: miguel.munoz3@upr.edu

B. J. Pandian, PhD
Dean and Director, Water Technology Center, College of Agricultural Engineering and Technology, Tamil Nadu Agricultural University (TNAU), Coimbatore – 641003, India

Sabreen Kh. Pibars, PhD
Water Relations and Field Irrigation Department (Agricultural and Biological Division), National Research Center, Cairo, Giza, Egypt. E-mail: saberennrc@yahoo.com

Gajendra Singh, PhD
Former Vice President, Asian Institute of Technology, Thailand. C-86, Millennium Apartments, Plot E-10A, Sector 61, NOIDA, U.P.–201301, India, Mobile: +91-9971087591, E-mail: prof.gsingh@gmail.com

R. K. Sivanappan, PhD
Former Professor and Dean, College of Agricultural Engineering and Technology, Tamil Nadu Agricultural University (TNAU), Coimbatore. Mailing address: Consultant, 14, Bharathi Park, 4th Cross Road, Coimbatore–641043, India. E-mail: sivanappanrk@hotmail.com

David M. Sumner, PhD
Hydrologist, Associate Director for Studies, United States Geological Survey, USGS Caribbean – Florida Water Science Center, 4446 Pet Lane, Suite 108, Lutz, FL, 33559, USA. Phone: 813-4985025, E-mail: dmsumner@usgs.gov

Mohamed Y. Tayel, PhD
Water Relation and Field Irrigation Department, National Research Center, Dokki, Giza, Egypt. E-mail: tayelmynrc@yahoo.com

LIST OF ABBREVIATIONS

ANOVA	analysis of variance
AVHRR	Advanced Very High Resolution Radiometer
AW	available water
BD	bulk density
BID	bubbler irrigation
BTU	British thermal units
BY	biomass yield
CM1DIS	closed circuits with one manifold for lateral lines
CM2DIS	closed circuits with two manifolds for lateral lines
CR	clogging ratio
CU	uniformity coefficient
CV	coefficient of variation
CVm	manufacturer's coefficient of emitter variation
CWR	wheat water requirement
CWUE	crop water use efficiency
DIC	drip irrigation closed circuits
DU	distribution uniformity
Epan	pan evaporation
EROS	Earth Resources Observation Systems
ET	evapotranspiration
ETc	evapotranspiration for corn
ETo	reference evapotranspiration
FAO	Food and Agriculture Organization
FC	field capacity
FL	friction loss
FV	flow velocity
FWUE	field water use efficiency
FYM	farm yard manure
HARC	Hofuf Agricultural Research Centre
HC	hydraulic conductivity
IWRg	gross irrigation water requirements

KSA	Kingdom of Saudi Arabia
LIS	localized irrigation systems
LLL	lateral line lengths
LSD	least significant difference
MODFLOW	modular finite-difference ground-water flow
MSIS	mini-sprinkler irrigation system
NCEP	National Centers for Environmental Prediction
NDVI	normalized difference vegetation index
NOAA	National Oceanic and Atmospheric Administration
NRC	National Research Center
NUE	nutrient use efficiency
OM	organic matter
PAR	photosynthetically active radiation
PE	polyethylene
PH	pressure head
PSIS	permanent sprinkler irrigation system
PSIS	permanent sprinkler irrigation system
PVEM	pressure value of effective more
SDI	surface drip irrigation
SI	seed index
SIUC	Southern Illinois University of Carbondale
SPSIS	semiportable sprinkler irrigation system
SSDI	subsurface drip irrigation
SY	Stover yield
TDIS	traditional drip irrigation system
TDR	time-domain reflectometry
UAE	United Arab Emirates
UC	uniformity coefficient
VCp	pressure variation coefficient
VCq	flow variation coefficient per plant
WCU	water consumptive use
WDE	water distribution efficiency
WP	water productivity
WP	wilting point
WUE	water use efficiency

LIST OF SYMBOLS

A	cross sectional flow area (L^2)
AW	available water (Θ_w, %)
B	Bowen ratio, equal to the ratio of sensible and latent heat fluxes
c	vapor density (g/m^3) or virtual temperature (in °C)
C_{1j}–C_{3j}	empirical parameters within evapotranspiration model for surface cover j
C_a	absolute water-budget closure, in mm/yr
C_p	specific heat capacity of air, in J/(g·°C)
C_r	relative water-budget closure, in percent
CV	coefficient of variation
D	accumulative intake rate (mm/min)
d	momentum displacement height of vegetation, in m
E	evapotranspiration rate, in g/(m^2·s)
e	vapor pressure (kPa)
e_a	actual vapor pressure (kPa)
E_{cp}	cumulative class A pan evaporation (mm)
eff	irrigation system efficiency
E_i	irrigation efficiency of drip system
E_p	pan evaporation (mm/day)
E_{pan}	class A pan evaporation
ER	cumulative effective rainfall (mm)
e_s	saturation vapor pressure (kPa)
Es	saturation vapor pressure, in kPa
e_s–e_a	vapor pressure deficit (kPa)
ET	evapotranspiration (mm/year)
ETa	reference ET
ET_c	crop evapotranspiration (mm/day)
ET_o	reference evapotranspiration (mm/day)
ET_{pan}	pan evaporation-derived evapotranspiration

EU	emission uniformity
F	factor used in krypton hygrometer correction
F	flow rate of the system (GPM)
F.C.	field capacity (v/v,%)
f_i	PAR-weighted fraction of the day
G	soil heat flux at land surface, in W/m^2
g_i	fractional contribution of burned area
gpm	gallons per minute
h	canopy height, in m
H	plant canopy height in meter
H	sensible heat flux, in W/m^2
h	soil water pressure head (L)
H_{cor}	sensible heat flux
h_{wt}	water-table depth below a reference level
i	an index for the burn zones (I to IV)
I	infiltration rate at time t (mm/min)
IR	injection rate, GPH
IRR	irrigation
K	the unsaturated hydraulic conductivity (LT^{-1})
K_c	crop coefficient
kg	kilograms
K_o	extinction coefficient of hygrometer for oxygen
K_p	pan coefficient
Kp	pan factor
K_w	extinction coefficient of hygrometer for water
L	leakage to the Upper Floridan aquifer, in mm/yr
lph	liter per hour
lps	liters per second
msl	mean sea level
n	number of emitters
NDVI	normalized-difference vegetation index, dimensionless
NIR	reflectance of near-infrared radiation, dimensionless
P	percentage of chlorine in the solution
P.W.P.	permanent wilting point ($\ominus_w$%)
P_a	atmospheric pressure, in P_a
Pa	atmospheric pressure, in Pa

PAR	photosynthetically active radiation, in moles/(m²·s)
PET	potential evapotranspiration, in mm/yr or W/m²
pH	acidity/alkalinity measurement scale
ppm	part per million
psi	pounds per square inch
Q	flow rate in gallons per minute
q	mean emitter discharges of each lateral (lh^{-1})
q	specific humidity, in g water/g moist air
R	rainfall
R	runoff, in mm/yr
r_a	aerodynamic resistance ($s\ m^{-1}$)
Ra	extraterrestrial radiation
R_d	gas constant for dry air, equal to 0.28704 J/°C/g
R_e	effective rainfall depth (mm)
r_h	aerodynamic resistance, in seconds per meter
R_i	individual rain gauge reading in mm
RMAX	maximum relative humidity
RMIN	minimum relative humidity
R_n	net radiation at the crop surface ($MJ\ m^{-2}day^{-1}$)
R_n	net radiation, in W/m²
R_{nb}	net radiation for burned areas, in W/m²
R_{nu}	net radiation for unburned areas, in W/m²
RO	surface runoff
r_s	bulk surface resistance ($s\ m^{-1}$)
Rs	incoming solar radiation
S	sink term accounting for root water uptake (T^{-1})
S	change in storage of energy in the biomass and air, in W/m²
Se	effective saturation
S_p	plant-to-plant spacing (m)
S_r	row-to-row spacing (m)
SU	statistical uniformity (%)
Sy	specific yield, in mm³ water/mm water-level change/mm²
S_ψ	water stress integral (MPa day)
T	time (hours)
t	time (min)
T_a	air temperature, in °C

TMAX	maximum temperature
TMIN	minimum temperature
T_s	sonic temperature, in °C
u	lateral wind speed along coordinate x-direction, in m/s
$u*$	friction velocity, in m/s
V	volume of water required (liter/day/plant)
v	lateral wind speed along coordinate y-direction, in m/s
V_{id}	irrigation volume applied in each irrigation (liter tree^{-1})
V_{pc}	the plant canopy volume (m^3)
W	canopy width
w	wind speed along coordinate z-direction, in m/s
w_b	fraction of the measured latent heat flux originating from burned areas, dimensionless
W_p	fractional wetted area
x	one of two orthogonal coordinate directions within a plane parallel to canopy surface
y	one of two orthogonal coordinate directions within a plane parallel to canopy surface
z	vertical coordinate (L)
z_m	roughness length of canopy for momentum, in m
z_s	height of sensors above land surface, in m

PREFACE

Due to increased agricultural production, irrigated land has increased in the arid and subhumid zones around the world. Agriculture has started to compete for water use with industries, municipalities and other sectors. This increasing demand along with increases in water and energy costs have made it necessary to develop new technologies for the adequate management of water. The intelligent use of water for crops requires understanding of evapotranspiration processes and use of efficient irrigation methods.

Every day, news on water scarcity appears throughout the world, indicating that government agencies at central/state/local levels, research and educational institutions, industries, sellers and others are aware of the urgent need to adopt micro irrigation technology that can have an irrigation efficiency up to 90% compared to 30–40% for the conventional gravity irrigation systems. I stress the urgent need to implement micro irrigation systems in water scarcity regions.

Micro irrigation is sustainable and is one of the best management practices. The water crisis is getting worse throughout the world, including Middle East and Puerto Rico, where I live. We can therefore conclude that the problem of water scarcity is rampant globally, creating the urgent need for water conservation. The use of micro irrigation systems is expected to result in water savings and increased crop yields in terms of volume and quality. The other important benefits of using micro irrigation systems include expansion in the area under irrigation, water conservation, optimum use of fertilizers and chemicals through water, and decreased labor costs, among others. The worldwide population is increasing at a rapid rate and it is imperative that the food supply keeps pace with this increasing population.

Micro irrigation, also known as trickle irrigation or drip irrigation or localized irrigation or high frequency or pressurized irrigation, is an irrigation method that saves water and fertilizer by allowing water to drip slowly to the roots of plants, either onto the soil surface or directly onto the root zone, through a network of valves, pipes, tubing, and emitters. It is done through narrow tubes that deliver water directly to the base of the plant. It supplies controlled delivery of water directly to individual plants and can be installed on the soil surface or subsurface. Micro irrigation systems are often used for farms and large gardens, but are equally effective in the home garden or even for houseplants or lawns.

The mission of this compendium is to serve as a reference manual for graduate and under graduate students of agricultural, biological and civil engineering; horticulture, soil science, crop science and agronomy. I hope that it will be a valuable reference for professionals that work with micro irrigation and water management; for professional training institutes, technical agricultural centers, irrigation centers, Agricultural Extension Service, and other agencies that work with micro irrigation programs.

After my first textbook, *Drip/Trickle or Micro Irrigation Management* by Apple Academic Press Inc., and response from international readers, I was motivated to bring out for the world community this ten-volume series on *Research Advances in Sustainable Micro Irrigation*. This book series will complement other books on micro irrigation that are currently available on the market, and my intention is not to replace any one of these. This book series is unique because it is complete and simple, a one stop manual, with worldwide applicability to irrigation management in agriculture. This series is a must for those interested in irrigation planning and management, namely, researchers, scientists, educators and students.

Among all irrigation systems, micro irrigation has the highest irrigation efficiency and is most efficient. *Fertigation* is the application of fertilizers, soil amendments, or other water-soluble products through an irrigation system. Chemigation, a related and sometimes interchangeable term, is the application of chemicals through an irrigation system. Fertigation is used extensively in commercial agriculture and horticulture. The irrigator must take into consideration suggestions, such as: (i) Fertigation is used to spoon-feed additional nutrients or correct nutrient deficiencies detected in plant tissue analysis. It is usually practiced on high-value crops such as vegetables, turf, fruit trees, and ornamentals; (ii) Injection during the middle one-third or the middle one-half of the irrigation is recommended for fertigation using micropropagation and drip irrigation; (iii) The water supply for fertigation is to be kept separate from the domestic water supply to avoid contamination; and (iv) The change of fertilizer during the growing season is important in order to adjust for fruit, flower, and root development.

The contribution by all cooperating authors to this book series has been most valuable in the compilation of this volume. Their names are mentioned in each chapter and in the list of contributors of each volume. This book would not have been written without the valuable cooperation of these investigators, many of whom are renowned scientists who have worked in the field of micro irrigation throughout their professional careers.

I am glad to introduce Dr. Hani A. Mansour, Distinguished Research Soil and Water Engineer Research Engineer at Water Relations Field Irrigation Department, Agricultural and Biological Division, National Research Center; and Visiting Post-Doc Research Fellow, Agricultural and Biological Engineering Department at Purdue University. He joins as co-editor for this volume. Without his support and extraordinary job, readers would not have this quality publication. Most of the research studies in this volume were conducted by Dr. Hani A. A. Mansour and his colleagues.

In this volume, we have included the chapter titled, "Evapotranspiration for cypress and pine forests: Florida, USA by Dr. David M Sumner." Theory, procedures, guidelines and applications in this chapter are equally applicable in tree crops under micro irrigation. Finally, this volume is unique as it includes a chapter titled "Drip irrigation in Rice" by Dr. R. K. Sivanappan (father of drip irrigation in India). Research studies confirm economical/physiological/crop/irrigation benefits in rice production under water scarcity conditions.

I will like to thank editorial staff, Sandy Jones Sickels, Vice President, and Ashish Kumar, Publisher and President at Apple Academic Press, Inc., for making every effort to publish the book when the diminishing water resources is a major issue worldwide. Special thanks are due to the AAP Production Staff for typesetting the entire manuscript and for the quality production of this book. We request that the reader sends us your constructive suggestions that may help to improve the next edition.

I express my deep admiration to my family for understanding and collaboration during the preparation of this ten volume book series. With my whole heart and best affection, I dedicate this volume to all researchers/

educators/engineers who work on micro irrigation technology and encourage them to come up with new ideas/developments, etc., in order alleviate problems of water scarcity and salinity. My salute to them for their devotion and vocation.

As an educator, I wish to offer this piece of advice to one and all in the world: "*Permit that our Almighty God, our Creator and excellent Teacher, irrigate the life with His Grace of rain trickle by trickle, because our life must continue trickling on...*"

—Megh R. Goyal, PhD, PE, Senior Editor-in-Chief

May 30, 2015

FOREWORD 1

With only a small portion of cultivated area under irrigation and the need to bring addition a land under irrigation, it is clear that the most critical input for agriculture today is water. It is important that all available supplies of water should be used intelligently to the best possible advantage. Recent research around the world has shown that the yields per unit quantity of water can be increased if fields are properly leveled, water requirements of the crops as well as the characteristics of the soil are known, and correct methods of irrigation are followed. Significant gains can also be made if the cropping patterns are changed so as to minimize storage during the hot summer months when evaporation losses are high, if seepage losses during conveyance are reduced, and if water is applied at critical times when it is most useful for plant growth.

Irrigation is mentioned in the Holy Bible and in the old documents of Syria, Persia, India, China, Java, and Italy. The importance of irrigation in our times has been defined appropriately by N.D. Gulati: "In many countries irrigation is an old art, as much as the civilization, but for humanity it is a science, the one to survive." The need for additional food for the world's population has spurred rapid development of irrigated land throughout the world. Vitally important in arid regions, irrigation is also an important improvement in many circumstances in humid regions. Unfortunately, often less than half the water applied is used by the crop-irrigation water may be lost through runoff, which may also cause damaging soil erosion, deep percolation beyond that required for leaching to maintain a favorable salt balance. New irrigation systems and design and selection techniques are continually being developed and examined in an effort to obtain increase efficiency of water application.

The main objective of irrigation is to provide plants with sufficient water to prevent stress that may reduce the yield. The frequency and quantity of water depends upon local climatic conditions, crop and stage of growth, and soil-moisture-plant characteristics. The need for irrigation can be determined in several ways that do not require knowledge of evapo-transpiration (ET) rates. One way is to observe crop indicators such as change of color or leaf angle, but this information may appear too late to avoid reduction in the crop yield or quality. Other similar methods of scheduling include determination of the plant water stress, soil moisture status, or soil water potential. Methods of estimating crop water require-ments using ET combined with soil characteristics have the advantage of not only being useful in determining when to irrigate, but also enables us to know the quantity of water needed. ET estimates have not been calcu-lated for the developing countries though basic information on weather data is available. This has contributed to one of the existing problems in which vegetable crops are over-irrigated and tree crops are under-irrigated.

Water supply in the world is dwindling because of luxury use of sources; competition for domestic, municipal, and industrial demands; declining water quality; and losses through seepage, runoff, and evapora-tion. Water rather than land is one of the limiting factors in our goal for self-sufficiency in agriculture. Intelligent use of water will avoid problems of seawater seeping into aquifers. Introduction of new irrigation methods has encouraged marginal farmers to adopt these methods without taking into consideration economic benefits of conventional, overhead, and drip irrigation systems. What is important is "net in the pocket" under limited available resources. Irrigation of crops in tropics requires appropriately tailored working principles for the effective use of all resources peculiar to the local conditions. Irrigation methods include border-, furrow-, subsur-face-, sprinkler-, sprinkler, micro, and drip/trickle, and xylem irrigation.

Drip irrigation is an application of water in combination with fertilizers within the vicinity of plant root in predetermined quantities at a specified time interval. The application of water is by means of drippers, which are located at desired spacing on a lateral line. The emitted water moves due to an unsaturated soil. Thus, favorable conditions of soil moisture in the root zone are maintained. This causes an optimum development of the crop. Drip/micro or trickle irrigation is convenient for vineyards, tree orchards, and row crops. The principal limitation is the high initial cost of the system for crops with very narrow planting distances. Forage crops may not be irrigated economically with drip irrigation. Drip irrigation is adaptable for almost all soils. In very fine textured soils, the intensity of water application can cause problems of aeration. In heavy soils, the lateral movement of the water is limited, thus more emitters per plant are needed to wet the desired area. With adequate design, use of pressure compensating drippers and pressure regulating valves, drip irrigation can be adapted to almost any topography. In some areas, drip irrigation is used successfully on steep slopes. In subsurface drip irrigation, laterals with drippers are buried at about 45 cm depth, with an objective to avoid the costs of transportation, installation, and dismantling of the system at the end of a crop. When it is located permanently, it does not harm the crop and solves the problem of installation and annual or periodic movement of the laterals. A carefully installed system can last for about 10 years.

The publication of this book series is an indication that things are beginning to change, that we are beginning to realize the importance of water conservation to minimize hunger. It is hoped that the publisher will produce similar materials in other languages.

In providing this book series on micro irrigation, Megh Raj Goyal, as well as the Apple Academic Press, has rendered an important service to farmers. Dr. Goyal, *Father of Irrigation Engineering in Puerto Rico*, has done an unselfish job in the presentation of this series that is simple, thorough, and informative. I have known Megh Raj since 1973 when we were working together at Haryana Agricultural University on an ICAR research project on "Cotton Mechanization and Acid Delinting in India."

Gajendra Singh, PhD,
Former Vice Chancellor, Doon University,
Dehradun, India; Adjunct Professor, Indian
Agricultural Research Institute, New Delhi;
Ex-President (2010–2012), Indian Society of
Agricultural Engineers; Former Deputy Director
General (Engineering), Indian Council of Agricultural
Research (ICAR), New Delhi; Former Vice-President/
Dean/Professor and Chairman, Asian Institute of
Technology, Thailand

New Delhi
May 30, 2015

FOREWORD 2

Monsoon failure in June of 2014 created shock waves once again across India. The Indian Meteorological Department reported a shortage of rains in major parts of India with a country average of 42%, Karnataka 35%, Konkan and Goa 56%, Kerala 24%, Gujarat 88% and Rajasthan 80%. India is still 62% agriculturally dependent on monsoon rain, and most of the 83% small and marginal farmers are living in these regions. Monsoon failure in 2014 affected food production and the livelihood of the majority of the population of India. The Government of India took timely and laudable initiatives to develop a contingency program. India has observed this type of monsoon situation 12 times during the last 113 years, meaning a huge deficit of rain once in 10 years. Although contingency plans provide some relief, there is a need to address fundamental issues of water management in India. India has 1896 km³ total renewable water resources; in addition only 5% of the total precipitation is harvestable. Improving water productivity is a major challenge. Improving irrigation efficiency, effective rainwater management, and recycling of industrial and sewage water will provide enough water available for agriculture in the state. Micro irrigation can mitigate abiotic stress situation by saving over 50% irrigation water and can be useful in a late monsoon situation for timely sowing.

Agricultural engineers across India have made several specific recommendations on water conservation practices, fertigation practices, ground water recharge, improving water productivity, land management practices, tillage/cultivation practices, and farm implements for moisture conservation. These technologies have potential to conserve water that can facilitate timely sowing of crops under the delayed monsoon situation that has occurred this year and provide solutions to monsoon worries. Agricultural engineers need to provide leadership opportunities in the water resources and water management sector, which includes departments of Command Area Development, Rural Development, Panchayat Raj, water resources, irrigation, soil conservation, watersheds, environment and energy for stability of agriculture, and in turn, stable growth of the Indian economy.

This book series on micro irrigation addresses the urgent need to adopt this water-saving technology not only in India but throughout the world. I would like to see more literature on micro irrigation for use by the irrigation fraternity. I appeal to all irrigation engineering fraternities to bring such issues to the forefront through research publications, symposiums, seminars and discussions with planners and policymakers at regional, state and national levels so that agricultural engineers will have a well-deserved space in the development process of the country.

V. M. Mayande, PhD
Former President (2012–2015), Indian Society
of Agricultural Engineers; Vice Chancellor,
Dr. Panjabrao Deshmukh Krishi Vidyapeeth, Akola,
Maharashtra–444104, India. Tel.: +91 9423174299.
E-mail: vmmayande@yahoo.com

May 30, 2015

FOREWORD 3

Irrigation has been a vital resource in farming since the evolution of humans. Due importance to irrigation was not accorded because of the fact that the availability of water has been persistent in the past. Sustained availability of water cannot be possible in the future, and there are several reports across the globe that severe water scarcity might hamper farm production. Hence, in modern-day farming, the most limiting input being water, much importance is needed for conservation and judicious use of the irrigation water for sustaining the productivity of food and other cash crops. Though the availability of information on micro irrigation is adequate, its application strategies must be expanded for the larger benefit of the water-saving technology by the clients.

In this context, under Indian conditions, the attempt made by Prof. R. K. Sivanappan, Former Dean, Agricultural Engineering College of TNAU, in collating all pertinent particulars and assembling them in the form a precious publication proves that the author is continuing his eminent service and support to the farming community by way of empowering them in adopting the micro irrigation technologies at ease, and the personnel involved in irrigation also enrich the knowledge on modern irrigation concepts.

While seeking the blessings of Dr. R. K. Sivanappan and Dr. Megh Raj Goyal (editor of this book series), I wish the publisher and authors success in all their endeavors, for helping the users of micro irrigation.

B. J. Pandian, PhD
Dean and Director, Water Technology Center,
College of Agricultural Engineering and Technology,
Tamil Nadu Agricultural University (TNAU),
Coimbatore – 641003, Tamil Nadu, India

May 30, 2015

FOREWORD 4

The micro irrigation system, more commonly known as the drip irrigation system, has been one of the greatest advancements in irrigation system technology developed over the past half century. The system delivers water directly to individual vines or to plant rows as needed for transpiration. The system tubing may be attached to vines, placed on or buried below the soil surface.

This book series, written by experienced system designers/scientists, describes various systems that are being used around the world; the principles of micro irrigation, chemigation, filtration systems, water movement in soils, soil-wetting patterns; and design principles, use of wastewater, crop water requirements and crop coefficients for a number of crops. The book series also includes chapters on hydraulic design, emitter discharge and variability, and water and fertigation management of micro irrigated vegetables, fruit trees, vines, and field crops. Irrigation engineers will find this book series to be a valuable reference.

Marvin E. Jensen, PhD, PE
Retired Research Program Leader at USDA-ARS;
and Irrigation Consultant, 1207 Spring Wood Drive,
Fort Collins, Colorado 80525, USA.
E-mail: mjensen419@aol.com

May 30, 2015

WARNING/DISCLAIMER

The goal of this compendium, *Closed Circuit Trickle Irrigation Design,* is to guide the world community on how to manage efficiently for economical crop production. The reader must be aware that dedication, commitment, honesty, and sincerity are most important factors in a dynamic manner for complete success. This reference is not intended for a one-time reading; we advise you to consult it frequently. To err is human. However, we must do our best. Always, there is a place for learning new experiences.

The editor, the contributing authors, the publisher, and the printer have made every effort to make this book as complete and as accurate as possible. However, there still may be grammatical errors or mistakes in the content or typography. Therefore, the contents in this book should be considered as a general guide and not a complete solution to address any specific situation in irrigation. For example, one size of irrigation pump does not fit all sizes of agricultural land and work for all crops.

The editor, the contributing authors, the publisher and the printer shall have neither liability nor responsibility to any person, organization, or entity with respect to any loss or damage caused, or alleged to have caused, directly or indirectly, by information or advice contained in this book. Therefore, the purchaser/reader must assume full responsibility for the use of the book or the information therein.

The mention of commercial brands and trade names are only for technical purposes and does not imply endorsement. The editor, contributing authors, educational institutions, and the publisher do not have any preference for a particular product.

All web links that are mentioned in this book were active on December 31, 2014. The editors, the contributing authors, the publisher, and the printing company shall have neither liability nor responsibility if any of the web links are inactive at the time of reading of this book.

ABOUT SENIOR EDITOR-IN-CHIEF

Megh R. Goyal, PhD, PE, is a Retired Professor in Agricultural and Biomedical Engineering from the General Engineering Department in the College of Engineering at University of Puerto Rico–Mayaguez Campus; and Senior Acquisitions Editor and Senior Technical Editor-in-Chief in Agriculture and Biomedical Engineering for Apple Academic Press Inc. He received his BSc degree in engineering in 1971 from Punjab Agricultural University, Ludhiana, India; his MSc degree in 1977 and PhD degree in 1979 from the Ohio State University, Columbus; and his Master of Divinity degree in 2001 from Puerto Rico Evangelical Seminary, Hato Rey, Puerto Rico, USA. He spent one-year sabbatical leave in 2002–2003 at the Biomedical Engineering Department at Florida International University in Miami, Florida, USA. Since 1971, he has worked as Soil Conservation Inspector (1971); Research Assistant at Haryana Agricultural University (1972–75) and Ohio State University (1975–79); Research Agricultural Engineer/Professor at the Department of Agricultural Engineering of UPRM (1979–1997); and Professor in Agricultural and Biomedical Engineering in the General Engineering Department of UPRM (1997–2012).

He was first agricultural engineer to receive the professional license in Agricultural Engineering in 1986 from College of Engineers and Surveyors of Puerto Rico. On September 16, 2005, he was proclaimed as "Father of Irrigation Engineering in Puerto Rico for the twentieth century" by the ASABE, Puerto Rico Section, for his pioneer work on micro irrigation, evapotranspiration, agroclimatology, and soil and water engineering. During his professional career of 45 years, he has received awards such as Scientist of the Year, Blue Ribbon Extension Award, Research Paper Award, Nolan Mitchell Young Extension Worker Award, Agricultural Engineer of the Year, Citations by Mayors of Juana Diaz and Ponce, Membership Grand Prize for ASAE Campaign, Felix Castro Rodriguez Academic Excellence, Rashtrya Ratan Award and Bharat Excellence Award and Gold Medal, Domingo Marrero Navarro Prize, Adopted Son of Moca, Irrigation Protagonist of UPRM, and Man of Drip Irrigation by Mayor of Municipalities of Mayaguez/Caguas/Ponce and Senate/Sec-retary of Agriculture of ELA, Puerto Rico.

He has authored more than 200 journal articles and textbooks, including *Elements of Agroclimatology* (Spanish) by UNISARC, Colombia, and two *Bibliographies on Drip Irrigation.* Apple Academic Press Inc. (AAP) has published his books, namely *Biofluid Dynamics of Human Body, Management of Drip/Trickle or Micro Irrigation, Evapotranspiration: Principles and Applications for Water Management, Sustainable Micro Irrigation Design Systems for Agricultural Crops: Practices and Theory, Biomechanics of Artificial Organs and Prostheses,* and *Scientific and Technical Terms in Bioengineering and Biotechnology.* During 2014–15, AAP is publishing his ten-volume set, Research Advances in Sustainable Micro Irrigation.

Readers may contact him at goyalmegh@gmail.com.

ABOUT CO-EDITOR

Hani A. Mansour, PhD, is a Distinguished Research Engineer in Soil and Water Engineering at the Water Relations Field Irrigation Department (Agricultural and Biological Division) of the National Research Center, Egypt. Now he was a Visiting Post-Doc Research Fellow in the Agricultural and Biological Engineering Department at Purdue University, until January 2015.

At Purdue University, he worked on using models and simulation programs in irrigation management under localized and developed irrigation systems. He has also been a Postdoc at SZIU, Godollo-Hungary, during October 2013 to April 2014. He has worked on development/design/management of drip irrigation systems, deficit irrigation systems, water and fertigation management, and treatment of low quality water in irrigation systems. He is an expert on closed circuits of drip irrigation system.

He obtained his BSc degree in Agricultural Engineering in 2000 from Monofiya University, Egypt; his MSc degree in Agricultural Engineering from Ain Shams University in 2006, Cairo, Egypt; and his PhD degree in 2012 through a scientific channel between the Plant and Soil and Agricultural System Department at Southern Illinois University at Carbondale (SIUC); Water Relations and Field Irrigation Department, NRC; and the Agricultural Engineering Department, Ain Shams University, Cairo, Egypt.

He is a critical reader, thinker, planner and fluent writer and has published more than 40 publications on micro irrigation technology in arid regions. Most of research studies in this volume are by Dr. Mansour and his colleagues in Egypt. Readers may contact him at: mansourhani2011@gmail.com.

BOOK REVIEWS

"I congratulate the editors on the completion and publication of this book volume on micro irrigation design. Water for food production is clearly one of the grand challenges of the twenty-first century. Hopefully this book will help irrigators and famers around the world to increase the adoption of water savings technology such as micro irrigation."

> —Vincent F. Bralts, PhD, PE, Professor and Ex-Associate Dean, Agricultural and Biological Engineering Department, Purdue University, West Lafayette, Indiana

"This book is user-friendly and is a must for all irrigation planners to minimize the problem of water scarcity worldwide. The *Father of Irrigation Engineering in Puerto Rico of twenty-first century and pioneer on micro irrigation in the Latin America*, Dr. Goyal (my longtime colleague) has done an extraordinary job in the presentation of this book."

> —Miguel A Muñoz, PhD, Ex-President of University of Puerto Rico; and Professor/ Soil Scientist

"I am moved by seeing the dedication of this textbook and recalling my association with Dr. Megh Raj Goyal while at Punjab Agricultural University in India. I congratulate him on his professional contributions and his distinction in irrigation. I believe that this innovative book will aid the irrigation fraternity throughout the world."

> —A. M. Michael, PhD, Former Professor/Director, Water Technology Centre – IARI; Ex-Vice-Chancellor, Kerala Agricultural University, Trichur, Kerala, India

OTHER BOOKS ON MICRO IRRIGATION TECHNOLOGY FROM APPLE ACADEMIC PRESS, INC.

Management of Drip/Trickle or Micro Irrigation
Editor: Megh R. Goyal

Principles and Management of Clogging in Micro Irrigation
Editors: Megh R. Goyal, Vishal K. Chavan, and Vinod K. Tripathi

Sustainable Micro Irrigation Design Systems for Agricultural Crops:
Methods and Practices
Editors: Megh R. Goyal, and P. Panigrahi

Book Series: Research Advances in Sustainable Micro Irrigation

Volume 1: Sustainable Micro Irrigation: Principles and Practices
Senior Editor-in-Chief: Megh R. Goyal

Volume 2: Sustainable Practices in Surface and Subsurface Micro Irrigation
Senior Editor-in-Chief: Megh R. Goyal

Volume 3: Sustainable Micro Irrigation Management for Trees and Vines
Senior Editor-in-Chief: Megh R. Goyal

Volume 4: Managementrformance, and Applications of Micro Irrigation Systems
Senior Editor-in-Chief: Megh R. Goyal

Volume 5: Applications of Furrow and Micro Irrigation in Arid and Semi-Arid
Regions
Senior Editor-in-Chief: Megh R. Goyal

Volume 6: Management Practices for Drip Irrigated Crop
Editors: Kamal Gurmit Singh, Megh R. Goyal, and Ramesh P. Rudra

Volume 7: Closed Circuit Micro irrigation Fesign: Theory and Applications
Senior Editor-in-Chief: Megh R. Goyal; Editor: Hani A. A. Mansour

Volume 8: Wastewater Management for Irrigation: Principles and Practices
Senior Editor-in-Chief: Megh R. Goyal

Volume 9: Water and Fertigation Management in Micro Irrigation
Editor-in-Chief: Megh R. Goyal

Volume 10: Innovations in Micro Irrigation Technology
Senior Editor-in-Chief: Megh R. Goyal;
Coeditors: Vishal K. Chavan and Vinod K. Tripathi

PART I

IRRIGATION METHODS

CHAPTER 1

SOIL MOISTURE AND SALINITY DISTRIBUTIONS UNDER MODIFIED SPRINKLER IRRIGATION

M. E. EL-HAGAREY, H. M. MEHANNA, and H. A. A. MANSOUR

CONTENTS

1.1 INTRODUCTION

Irrigation development is a gateway to increased agricultural/water/land productivity, increased farm-hold and national food security. However, irrigation development has been a major challenge in many developing countries, including Egypt. Hanson [8] mentioned that efficient furrow

Modified from *M. E. El-Hagarey, H. M. Mehanna, and H. A. A. Mansour, 2014. Movable Surface Irrigation System (MSIS) Impact on Spatial and Temporal Distribution of Soil Moisture and salinity. Journal of Agriculture and Veterinary Science, 7(6): 49–57. Open access at: www.iosrjournals.org.*

irrigation requires reducing deep percolation and surface runoff losses. Water that percolates below the root zone (deep percolation) is not available for crop production, though deep percolation may be necessary to control soil salinity [7, 13]. Deep percolation can be reduced by improving the irrigation uniformity and preventing over irrigation. Benham and Eisenhauer [3] reported that regardless of whether you dike or block the ends of the furrows, or if one irrigates using every or every-other furrow, soil texture, slope and surface conditions (whether the furrow is smooth or rough, wet or dry) all influence rate of water advance down the furrow. The speed of advance is directly related to uniformity of moisture distribution within the soil profile. The soil infiltration rate is also affected by surface conditions of soil.

Center-pivot irrigation (also called central pivot irrigation, waterwheel and circle irrigation) is a method of irrigation in which equipment rotates around a pivot and crops are watered with sprinklers. A circular area centered on the pivot is irrigated, often creating a circular pattern in crops when viewed from above (sometimes referred to as crop circles). Most center pivots were initially water-powered, and today most are propelled by electric motors. Center-pivot irrigation was first used in 1948 by a farmer Frank Zybach, native of Strasburg, Colorado. It was recognized as a method to improve water distribution to fields. Center pivot irrigation is a form of overhead sprinkler irrigation consisting of several segments of pipe (usually galvanized steel or aluminum) joined together and supported by trusses, mounted on wheeled towers with sprinklers positioned along its length. The machine moves in a circular pattern and is fed with water from the pivot point at the center of the circle. The outside set of wheels sets the master pace for the rotation (typically once every three days). The inner sets of wheels are mounted at hubs between two segments and use angle sensors to detect when the bend at the joint exceeds a certain threshold, and thus, the wheels should be rotated to keep the segments aligned. Center pivots are typically less than 500 meters in length (circle radius) with the most common size being the standard 400 m machine. To achieve uniform application, center pivots require an even emission flow rate across the radius of the machine. Since the outer-most spans (or towers) travel farther in a given time period than the innermost spans, nozzle sizes are smallest at the inner spans and increase with distance from the pivot point. Most center pivot systems now have drops hanging from a u-shaped pipe called a gooseneck attached at the top of the pipe with sprinkler heads that are positioned a few feet (at most) above the crop, thus limiting evaporative losses and wind drift. There are many different nozzle configurations available including static plate, moving plate and part circle. Pressure regulators are typically installed upstream of each nozzle to ensure each is operating at the correct design pressure. Drops can also be used with drag hoses or bubblers that deposit the water directly on the ground between crops. This type of system is known as LEPA (Low Energy Precision Application) and is often associated with the construction of small dams along the furrow length (termed furrow diking). Crops may be planted in straight rows or are sometimes planted in circles to conform to the travel of the irrigation system

Originally, most center pivots were water-powered. These were replaced by hydraulic systems and electric motor-driven systems. Most systems today are driven by an electric motor mounted at each tower. For a center pivot to be used, the terrain needs to be reasonably flat; but one major advantage of center pivots over alternative systems is the ability to function in undulating country. This advantage has resulted in increased irrigated acreage and water use in some areas. The system is in use, for example, in parts of the United States, Australia, New Zealand, Brazil and also in desert areas such as the Sahara and the Middle East.

The center-pivot irrigation system is considered to be a highly efficient system, which helps conserve water. Center pivot irrigation typically uses less water compared to many surface irrigation and furrow irrigation techniques, which reduces the expenditure of and conserves water. It also helps to reduce labor costs compared to some ground irrigation techniques, which are often more labor-intensive. Some ground irrigation techniques involve the digging of channels on the land for the water to flow, whereas the use of center-pivot irrigation can reduce the amount of soil tillage that occurs and helps to reduce water runoff and soil erosion that can occur with ground irrigation. Less tillage encourages more organic materials and crop residue to decompose back into the soil, and reduces soil compaction. Center pivot irrigation systems have experienced tremendous increase around the world in recent years due to: The potential for highly efficient and uniform water applications, the high degree of automation requiring less labor than most other irrigation methods, large coverage of areas, and the ability to economically apply water and water soluble nutrients over a wide range of soil, crop and topographic conditions. However, sprinkler pivot irrigation system requires high operating pressure, careful assessment of chemigation to reduce hazards. It also causes some soil compaction, and splash erosion. Systems are not flexible to irrigate tree orchards and gardens.

Low energy precision application (LEPA) irrigation technology has been developed to reduce water consumption and energy use in irrigated agriculture. Research studies have involved to minimize high spray evaporation losses common in Texas, USA. For instance, Clark and Finley [6] found that at a wind speed of 15 miles per hour (which is the annual average for the Texas High Plains) evaporative losses were 17%, and at speeds of 20 miles per hour losses were over 30 percent. In the Southern High Plains of Texas, losses on a linear-move sprinkler system were measured upto 94% at an average wind speed of 22 miles per hour with gusts of 34 miles per hour [11]. Another aspect involved designing a system to be used in conjunction with microbasin land preparation or furrow diking which prevents runoff and maximizes the use of rainfall and applied irrigation water. A double-ended sock was developed to accomplish both goals. No wind losses were observed, since water was discharged directly into the furrow. Also, the open ends help preserve the dikes. However, this method can only be used for irrigation [14].

Center pivot irrigation systems (CPIS) are used widely where most of the systems are low-pressure systems, including LEPA, low elevation spray application (LESA), mid-elevation spray application (MESA) and low-pressure in-canopy (LPIC). Low-pressure systems offer cost savings due to reduced energy requirements as compared to high-pressure systems. They also facilitate increased irrigation application efficiency, due to decreased evaporation losses during irrigation. Considering high energy costs in many areas of limited water capacities, high irrigation efficiency can help to lower overall pumping costs, or at least optimize crop yield/ quality return relative to water and energy inputs. ASAE [2] defined low energy precision application (LEPA) as a water, soil, and plant management regime where precision down-in-crop applications of water are made on the soil surface at the point of use. Application devices are located in the crop canopy on drop tubes mounted on low-pressure center pivot or linear move sprinkler irrigation systems.

FIGURE 1.1   Overview of center pivot irrigation system in the Egyptian desert.

In Saudi Arabian desert, an installed quarter-mile-long center pivot system costs about $70,000 to $120,000 on average. The pivot's speed is adjustable. It takes 36 h to make a full circle on 167 acres, putting down 12 mm of water (Fig. 1.1). In Egypt, first farm was at El-Salihia (Cairo – Ismielia desert road) that used center pivot irrigation. Later Dina farm (1978) and big farms at Cairo-Alexandria desert road also invested in pivot irrigation systems, because of low operating cost, low repairs, low maintenance requirements and promising results. Many farmers in Egypt also use this system to irrigation shrubs and vegetable crops.

This chapter discusses research results, on use of movable surface irrigation system (MSIS) in Egypt, to: (i) study spatial and temporal distribution of soil moisture and salinity; (ii) evaluate the reduction in irrigation losses; (iii) reduce chemigation hazard and operating costs; and (iv) evaluate the uniformity coefficient of MSIS system.

1.2. MATERIALS AND METHODS

The movable surface center pivot irrigation system was located at Farm about 70 km on Alexandria-Cairo desert road. Soil and irrigation water analyzes were conducted according to standard procedures. The MSIS consisted of following components:

Control head consisted of centrifugal pump 5"/5" (5"/5", 50 m lift and 80 m³/h discharge) driven by diesel engine, sand media filter (48" two tanks), back flow prevention devices, pressure gages, control valves, inflow gate valves and fertilizer injection pump.

Tower of center pivot: Two towers of center pivot irrigation system were 48 m in radius, 127 mm diameter of mainline (3 mm thickness of pipe), 75 cm spacing between holes, according to guidelines for center pivot irrigation system.

1.2.1. MODIFICATION OF CENTER PIVOT IRRIGATION SYSTEM

Sprinkler pivot irrigation system is operated at high operating pressure with high-energy consumptive use. The basic of modification of pivot system depended on replacing the sprinkler heads by polyethylene (PE) hoses, which can be operated at lower pressure. It can be observed that a span length of pivot line depends on pressure head because friction losses are related to the dynamic head. Continuity equation defines the relationship between liquid velocity and cross-sectional area at the exit pipe. The following equation by Abdel-Rahman et al. [1] can be used to calculate the inside diameter.

$$D = 536.3 \, [Q]^{0.5}/[\sqrt[4]{h}] \tag{1}$$

where: D = inside diameter of nozzle (mm), Q = discharge rate of nozzle (m³/s), and h = nozzle operating head (m). Generally nozzles with calculated inside diameter are not available in the market. Therefore the available inside diameter must be selected from the design table provided by the manufacturer so that calculated diameter matches to the closest higher value in the table.

1.2.2. DESIGN OF A NOZZLE

In this study, authors used five diameters of the outlets: 8.5, 9, 9.5, 10, and 10.5 mm. In Egypt, nozzles for MSIS are constructed from pierced cylinder delrin (a type of plastic that can be formed into a shape), with low friction and excellent dimensional stability. Basic components of modified nozzle are:

- Polyethylene hose (20 mm in diameter and 200 cm in length) with barbed ends.
- Cylinder of delrin stick (pierced type): 20 mm in diameter.
- Distribution of diameter category at pivot main line: There are five diameters and 45 holes, which refer to laterals' pivot. So, one diameter category was constructed at nine laterals from the beginning of center pivot main line at next.
- From results, the next type of arrangement of diameter category appear from the following, the pivot main line was raised 0.5 m from middle: this is change in elevation of pivot main line that affects the discharge (Fig. 1.2).

$$Hi = Ha + 0.75 \, Hfr + 0.5 \, \Delta Z + Hr + Hcv \qquad (2)$$

$$Hd = Hi - (Hf + \Delta Z + Hr) \qquad (3)$$

$$Q = AV = QL = QR \qquad (4)$$

where: Hi = pressure head of mainline beginning (m), Hd = pressure head of mainline end (m), Hfr = head losses at main line (m), $Havg$ = average pressure head, ΔZ = difference of mainline elevation (m), Hcv = total of secondary losses of connection parts (m), $g = 9.81$ m/s², Q = volume flow rate (m³/s), V = liquid velocity (m/s), and $A = \pi D^2/4$ = cross sectional area, and D = inside diameter. Energy equations are defined by Eqs. (2) and (3). Continuity equation for continuous flow is defined by Eq. (4). Subscripts L and R are for left and right side of tower. Applying continuity equation on two sides of a tower, we conclude that the discharge must be equal on both sides. For left hand side of tower (Fig. 1.2) and using Eqs. (2) and (3), we have:

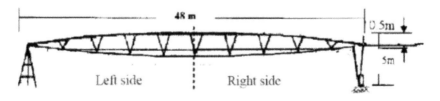

FIGURE 1.2   Two sides of a simple tower for a center pivot irrigation.

$Hi = 5 + 0.75(0.013) - 0.5(0.5) + (0.2) + (0.255) = 5.2$ m

$Hd = 5.2 - (0.013 + 0.5 + 0.2) = 4.5$ m

$Havg = (5.2 + 4.5)/2 = 4.9$ m

For right hand side of tower (Fig. 1.2) and using Eqs. (2) and (3), we have:

$Hi = 5 + 0.75(0.013) + 0.5(0.5) + (0.2) + (0.255) = 5.7$ m

$Hd = 5.7 - (0.013 + 0.5 + 0.2) = 5$ m

$Havg = (5 + 5.7)/2 = 5.3$ m

Applying continuity equation on both sides of a tower, we get:

$A_L V_L = A_R V_R$, and $V = [2 \ g \ Havg]^{0.5}$ or

$A_L [(2 \ g)(Havg)_L]^{0.5} = AR = A_R [(2 \ g)(Havg)_R]^{0.5}$ or

$\{A_L/A_R\} = \{[(2 \ g)(Havg)_R]^{0.5}\}/\{[(2 \ g)(Havg)_L]^{0.5}\}$ or

$\qquad = \{[(Havg)_R]^{0.5}\}/\{[(Havg)_L]^{0.5}\}$ or

$\qquad = \{[5.3]^{0.5}\}/\{[4.9]^{0.5}\} = [5.3/5.9]^{0.5} = 0.92$ or

$\{D_L/D_R\}^2 = 0.92$ \hfill (5)

Ratio of mean diameter on left side of a tower to mean diameter on right side of a tower is 0.92 according to the Eq. (5). The difference on area on both sides is only 8% that can be neglected for simplicity. However, we cannot neglect the new arrangement of diameters' category.

1.2.3. HYDRAULIC MEASUREMENTS

Water samples were taken by selecting 22 nozzles from 44 hoses. The time to collect same quantity of water sample was recorded. The water collected in each sample was measured with a graduated cylinder. Each observation was repeated four times and the average of four observations was calculated. Procedure described by Keller and Karmeli [10] was used to reduce the experimental error. Pump discharge, outlet pressure and nozzle discharge were calculated and tabulated. Uniformity coefficient was calculated according to Bralts et al. [4] as follows:

$$UC = 100[1 - \{\Sigma(Qd/Qavg)\}] \tag{6}$$

where: UC = uniformity coefficient (%), Qd = absolute deviation of each ample from the mean (lps), and Qavg = the mean of discharge from nozzle outlet (lps).

1.2.4. SOIL MEASUREMENTS

Soil samples were taken by a screw auger before and after each irrigation at three locations at a distance of 15 m from beginning of mainline of a center pivot (at beginning, middle and end of a main line), and at three soil depths (20, 40, and 60 cm). Samples were analyzed for soil moisture and salt accumulation. SURFER version-11 program was used to analyze the data to determine values of parameters. "Kriging" regression method was carried out for the regression analysis and to draw the contour maps.

1.3. RESULTS AND DISCUSSION

1.3.1. WATER APPLICATION UNIFORMITY OF MOVABLE SURFACE PIVOT IRRIGATION SYSTEM

Data in Figs. 1.3 and 1.4 shows the deviation of discharge and pressure from the mean values along pivot mainline. Also average pressure head of hose is equal to 5.25 m and it is nearly constant along pivot line. Besides, total dynamic pressure head is 15 m and deviates from 0.2 to 0.8 m. Average total discharge is 47.5 m^3/h and average discharge of MSIS nozzle is 0.3 lps at pressure head of pivot sprinklers of 50 m. Average total discharge for pivot sprinklers is 20 m^3/h, and average discharge of pivot sprinkler is 0.107 lps according to Broner [5].

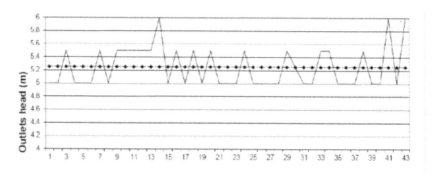

FIGURE 1.3 Deviation of outlet pressure head (m) from the mean pressure head (m). Solid line: Measured operating pressure head at outlets; and dotted line: Mean value of piezometric pressure head at the outlet

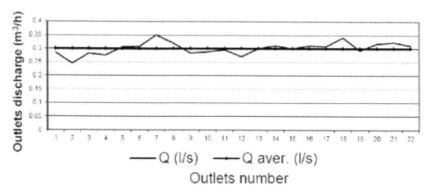

FIGURE 1.4 Deviation of outlet discharge (lps) from the mean discharge with long hoses (lps): 22 nozzles along the mail line.

Water distribution of nozzles is nearly constant for four replications of measurements. Uniformity coefficient is 90.7% and is excellent according to Merriam and Keller [12] and good according to IRYDA [9] for both hose lengths.

Regarding the mean total discharge for different replications, the mean total discharge was 47.5 m³/h and the mean discharge at outlets was 0.3 lps as shown in Fig. 1.4. The difference between discharge of each nozzle (deviation) for all of samples in the tower (total) and mean total discharge is due to experimental errors, which result in differences of discharge measurements for both sides of a tower. The discharge stability due to the pressure head follows an oscillating line. The deviation of pressure head from the mean ranged from 0.2 to 0.8 meter (Figs. 1.3 and 1.4).

The high uniformity coefficient of MSIS nozzle is a result of nozzle design by presenting the graduated diameters according to the changeable piezometeric head for nozzles. Changeable diameters were obtained by using Eq. (1). High water application is due to big size of nozzle diameter compared to size of sprinkler hole diameters, sequence irrigation cycle at low pressure head comparing to high pressure for sprinklers of the pivot system. Operating pressure for MSIS nozzle was low because of big size of outfit diameters. Pressure plot is of oscillating nature along mean pressure line. However, the deviation in pressure plot was very small due to a constant outlet pressure, mean of the total discharge, outlet discharge, total dynamic head, outlet piezometric head, and uniformity coefficient of water application.

1.3.2. SOIL MOISTURE DISTRIBUTION

Soil moisture distribution under MSIS is very important indicator of water application. Application efficiency of system was 90%. The amount of applied irrigation water was 4744 m^3/ha according to Abdel-Rahman et al. [1] and amount of applied irrigation water under sprinkler pivot was 5702 m^3/ha, according to El-Gindy, et al. [7]. This implies that mean water application under MSIS was 16.8% lower compared to applied water under sprinkler pivot. The ratios of water stored in the root zone to the water delivered to the field are thus influenced by:

- Evaporation losses from water flowing on the soil surface or in the air from sprinkler nozzle spray;
- Deep percolation below the root zone;
- Runoff; and
- Soil surface evaporation during irrigation.

Movable surface irrigation system in this chapter involved designing a system to be used in conjunction with microbasin land preparation or furrow diking which prevents runoff and maximizes the use of rainfall and applied irrigation water. Nozzles were developed to accomplish both goals (Fig. 1.5). No wind losses resulted, since water is discharged directly

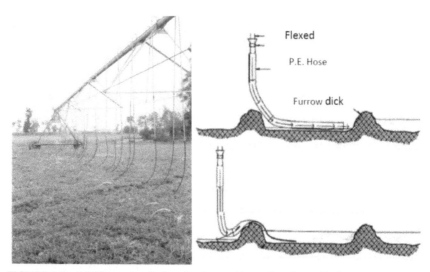

FIGURE 1.5   MSIS nozzles designed to be used in conjunction with furrow dikes.

into the furrow. Also, protected the crop from water out of nozzles, which causes fungal diseases. It also protects from hazard due to pesticide usage and generally chemigation through MSIS.

1.3.3   SOIL MOISTURE DISTRIBUTION PATTERNS

The system efficiency can be evaluated by measuring the moisture distribution in the soil profile (three locations from the center and three depths of the root zone). Soil moisture distribution under movable surface irrigation system followed the same moisture profile of modified surface irrigation. The soil moisture content increased with increment of soil profile depth.

This moisture distribution helps to reduce water losses by evaporation, because water was stored in the root zone. The vertical distribution of water was more difficult than the horizontal movement of water under sprinkler irrigation, where the greatest saved quantity of irrigation water was at the first layer of the soil profile. Using MSIS, the soil moisture was distributed uniformly and it supported salt leaching, as well as salt appearance according to Table 1.1 and Fig. 1.6.

TABLE 1.1   Soil Moisture Values Before and After Irrigation Process

Soil depth	Soil moisture, %					
	Before irrigation			After irrigation		
	Sample location from the upstream of main line, m					
	15	30	45	15	30	45
0–20	22.52%	20.60%	22.14%	28.58%	27.07%	28.41%
20–40	23.77%	22.53%	22.52%	26.84%	28.78%	27.48%
40–60	23.63%	23.73%	23.45%	29.59%	27.88%	29.52%

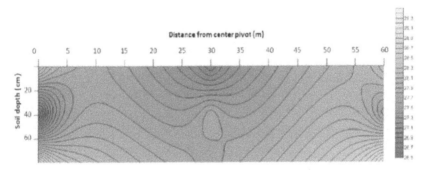

FIGURE 1.6   Spatial and temporal distribution of soil moisture after irrigation under MSIS.

The study reveals that furrow irrigation is not the efficient method of irrigation due to undesired percolation losses, which affect plant water uptake and the growth and yield of the cultivated crop (Fig. 1.7). Generally, there was a 4.1 cm of percolation loss in the case of furrow irrigation compared to no percolation loss under drip irrigation [7].

While under MSIS, it can achieve uniform water application of 90% and minimize deep percolation and runoff. According to moving of MSIS and applied water at all of land surface by had a 90% of uniformity distribution.

Contour map for soil moisture distribution before irrigation cycle depends on soil texture, slope, and climate. Contour map for soil moisture content after irrigation (Fig. 1.8) indicated the greatest amount of water saving at the third layer of soil (40–60 cm).

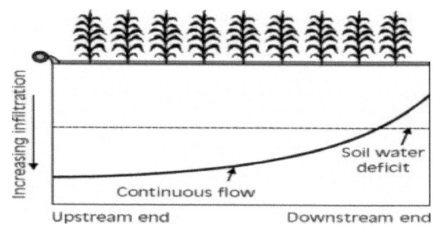

FIGURE 1.7   Poor uniformity and infiltration patterns for traditional furrow irrigation.

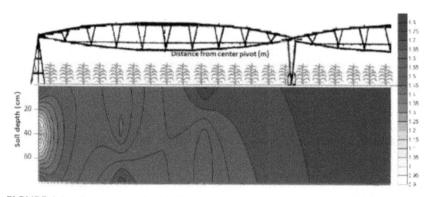

FIGURE 1.8   Contour map for depth of applied water under MSIS, after the irrigation.

MSIS can irrigate crops with the right amount of water, avoiding excess and runoff, and minimize foliar damage, which was common with saline water irrigation. Also, distribution of soil moisture was homogenous in the vertical and horizontal directions with respect to soil depth, as indicated in Figs. 1.6 and 1.8.

1.3.4. SALT CONCENTRATION DISTRIBUTION

Movable surface irrigation system showed advantages compared to both modified surface and pivot systems. Water application is at amount to prevent the salt appearance and the water application of 0.3 lps supports the leaching process. However, this enquires a good nutrient management of nutrient to minimize or avoid nutrient losses by leaching during irrigation.

Precision irrigation delivers water only on the soil surface so that water moves both vertically and laterally from the point of application. Only the water need of plant is supplied by the system thus avoiding losses. Plant roots will extract water from the moving soil solution. Salt concentration increases with distance away from the nozzle.

Any excess water applied through a dripper will leach salts primarily from the zone immediately around the dripper, but will have less impact on salts that have accumulated at greater horizontal distances away the lateral drip line [7, 13]. Rain, on the other hand, falls across the whole soil surface and is the major mechanism through which salts can leach downwards.

The potential for managing root zone salinity and the application of leaching fractions is increasingly important as precision irrigation is implemented [15]. Table 1.2 and Figs. 1.9 and 1.10 show the spatial and temporal distribution of salts before and after the irrigation.

Under irrigated conditions in arid and semiarid climates, the build-up of salinity in soils is inevitable. The severity and rapidity of build-up depends on a number of interacting factors such as the amount of dissolved salt in the irrigation water, chemical composition of soil and irrigation water, irrigation method and the local climate. However, soil salinity can be managed to prolong field productivity with proper management of: soil

TABLE 1.2   Soil Salt Concentration (ppm) Before and After the Irrigation

	Soil salt concentration (ppm)					
	Before irrigation			After irrigation		
	Sample location from the upstream of main line, m					
Soil depth	15	30	45	15	30	45
0–20	896	960	960	768	1024	768
20–40	1216	1024	896	896	768	768
40–60	1280	1280	896	832	896	960

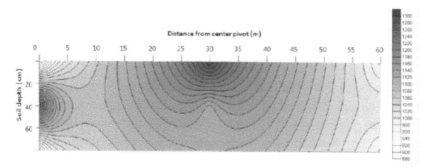

FIGURE 1.9   Spatial and temporal distribution of soil salts before irrigation under MSIS.

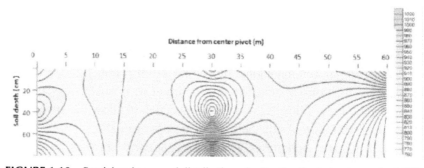

FIGURE 1.10   Spatial and temporal distribution of soil salts after irrigation under MSIS.

moisture, irrigation system uniformity and efficiency, local drainage, and the right choice of crops. Salt distribution in the soil profile under MSIS indicates that the application of adequate irrigation water (plus leaching requirements) leads to leached salts from the upper layers without salt appearance on the soil surface (Table 1.2 and Figs. 1.9 and 1.10).

1.4. CONCLUSIONS

For main nozzle flow of 0.3 lps, high uniformity coefficient for MSIS, total dynamic head (TDH) was reduced from 5 bars for sprinkler pivot system to 1.5 bars for MSIS. Also operating pressure head of nozzles in MSIS reduced energy requirement and irrigation costs.

Under MSIS, mean soil moisture content increased with increment of soil profile depth, thus reducing water losses by evaporation because of moisture storage in the root zone. Vertical dynamics of water under MSIS was more difficult than water dynamics under sprinkler irrigation. Soil salt distribution under MSIS before irrigation indicates that water inflow leached salts at soil layer, thus almost no appearance of salts at soil surface.

MSIS can irrigate crops with the right amount of water, avoiding excess runoff, and minimize foliar damage, which was common with saline water irrigation. MSIS is beneficial compared to both modified surface and pivot systems. Under MSIS, water application is at a proper amount to avoid salt accumulation at soil surface and to promote leaching to deeper layers.

1.5. SUMMARY

MSIS was based on the modification of center pivot irrigation system and depends on replacing the sprayers by polyethylene hoses with barbed ends. Field experiments were conducted at farm on Alexandria-Cairo desert road to assess moving surface irrigation system. This chapter discusses research results, on use of "MSIS in Egypt, to: (i) study spatial and temporal distribution of soil moisture and salinity; (ii) evaluate the reduction in irrigation losses; (iii) reduce chemigation hazard and operating costs; and (iv) evaluate the uniformity coefficient of MSIS system.

Operating the MSIS at low-pressure head of 1.5 bars, there was 16.8% water saving with a uniformity coefficient of 91%. Also, the movable irrigation surface system reduced the hazard of chemigation. Finally, high efficiency of applied water distribution reduced deep percolation and runoff.

KEYWORDS

- **Cairo desert**
- **center pivot**
- **chemigation**
- **deep percolation**
- **discharge**

- drip irrigation
- Egypt
- hydraulics
- leaching
- movable irrigation system
- nozzle
- polyethylene
- pressure head
- runoff
- salinity distribution
- soil moisture
- surface irrigation
- total dynamic head
- tower
- uniformity coefficient
- water saving

REFERENCES

1. Abdel-Rahman, M. E., A. M. El-Kot; A. A. Abdel-Aziz; and A. M. El-Gindy, 2005. Design and management of modified pivot irrigation systems. *Annals of Agricultural Science (Cairo), 50(2),* 415–432.
2. ASAE, (Soil and Water Steering Committee), 1995. *Soil and Water Terminology.* ASAE S526.1 March 95.
3. Benham, B. L., D. E. Eisenhauer., 2000. Management recommendations for blocked-end furrow irrigation. Nebraska Cooperative Extension G98–1372-A, USA.
4. Bralts, S. V., D. M. Edwards and I. P. Wu, 1987. Drip irrigation design and evaluation based on statistically uniformity concept. In: *Advances in Irrigation,* by D. Hillel. New York: Academic Press Inc., pp. 67–117.
5. Broner, I., 1991. *Center Pivot Irrigation Systems.* Cooperative Extension Crop Series Irrigation No. 4.704. Colorado State University, Fort Collins, CO.
6. Clark, R. N., and W. W. Finley, 1975. Sprinkler evaporation losses in the Southern Plains. *Am. Soc. Agric. Eng.* Paper No. 75–2573, St. Joseph, MI.
7. El-Gindy, A. M., M. F. Abd el-Salam, A. A. Abdel-Aziz and E. A. El-Sahar, 2003. Some engineering properties of maize plants, ears and kernels under different irrigation systems. *J. Agric., Sci., Mansura Univ., 28(6),* 4339–4360.

8. Hanson, B., 1993. Furrow irrigation. *Drought Tips* by University of California, Agriculture and Natural Resources, 300 Lakeside Drive, 6th Floor, Oakland, CA 94612–3560.

9. IRYDA (instituo de reforma *y* desarrollo agrario), 1983. Guidelines to prepare projects for localized irrigation (Spanish). Agric. Ministry of Fishing and Food, Madrid, Spain.

10. Keller, J. and D. Karmeli, 1975. *Trickle irrigation design.* Rain Bird Sprinkler Manufacturing Corp., Glendora, CA, USA.

11. Lyle M., and J. P. Bordovsky, 1981. Low energy precision application (LEPA) irrigation system. *Transactions of ASABE,* 24(5):1241–1245.

12. Merriam, J. L. and J. Keller, 1978. Farm irrigation systems evaluation: A guide for management. Utah State University, Logan, Utah, USA.

13. Panigrahi, B., D. P. Roy, and S. N. Panda, 2010. Water use and yield response of tomato as influenced by drip and furrow irrigation. *International Agricultural Engineering Journal,* 19(1):19.

14. Porter, D. O., and T. H. Marek, 2009. Center pivot sprinkler application depth and soil water holding capacity. Proceedings of the 21st Annual Central Plains Irrigation Conference, Colby, Kansas, February 24–25.

15. Stevens, R., 2002. Interactions between irrigation, salinity, leaching efficiency, salinity tolerance and sustainability. *The Australian & New Zealand Grape grower & Winemaker,* 466:71–76.

CHAPTER 2

PERFORMANCE OF SPRINKLER IRRIGATED WHEAT – PART I

M. Y. TAYEL, H. A. MANSOUR, and KH. PIBARS SABREEN

CONTENTS

2.1 INTRODUCTION

In the sprinkler method of irrigation, water is sprayed into the air and allowed to fall on the ground surface, thus simulating rainfall. The spray is developed by the flow of water under pressure through small orifices or

Modified and printed from *Tayel, M. Y., H. A. Mansour, and Sabreen, Kh. Pibars, 2014. Effect of two sprinkler irrigation types on wheat i-uniformity, vegetattve growth and yield. International Journal of Advanced Research, 2(1), 47–56. http://www.journalijar.com.*
In this chapter: 1 feddan = 0.42 hectares = 4200 m² = 1.038 acres = 24 kirat. A feddan (Arabic) is a unit of area. It is used in Egypt, Sudan, and Syria. The feddan is not an SI unit and in Classical Arabic, the word means 'a yoke of oxen': implying the area of ground that can be tilled in a certain time. In Egypt the feddan is the only nonmetric unit, which remained in use following the switch to the metric system. A feddan is divided into 24 kirats (175 m²). In Syria, the feddan ranges from 2295 square meters (m²) to 3443 square meters (m²).

nozzles. The pressure is usually obtained by pumping. With careful selection of nozzle sizes, operating pressure and sprinkler spacing, the amount of irrigation water required to refill the root zone can be applied uniformly at the rate to suit the soil infiltration rate [8].

One of the main challenges of world is water and food security. The increasing food demand and decreasing water allocation suggest that the agricultural sector must increase agricultural water productivity for producing more food with less water [4]. This challenge in arid regions of the world, such as Egypt, is more complicated. Agriculture sector is the main user of water in developing countries. In Egypt, 80% of the supplied water (70 billion cubic meters) belongs to this sector. Hence, increasing water productivity and water use efficiency (WUE) has very important role in reduce problems of water shortage. Water is a critical agricultural input in arid regions that affects crop yield and crop performance. In arid regions, agriculture is impossible without irrigation. Therefore, management and upgrading of irrigation systems have an important role in water productivity [8].

According to FAO [7], the total growing season of winter wheat ranges from 180 to 250 days. Mean daily temperature for optimum growth and tillering is between 15 and 20 °C. The crop is moderately tolerant to soil salinity. For high yields, water requirements of wheat vary from 450 to 650 mm depending on climatic conditions and length of the growing season.

Wheat (Triticum spp.) is a cereal grain, originally from the Levant region of the Near East but now cultivated worldwide. In 2010, world production of wheat was 651 million tons, making it the third most-produced cereal after maize (844 million tons) and rice (672 million tons). Wheat was the second most-produced cereal in 2009 with a world production in that year of 682 million tons. This grain is grown on more land area than any other commercial food. World trade in wheat is greater than for all other crops combined. Globally, wheat is the leading source of vegetable protein in human food, having a higher protein content than other major cereals, maize (corn) or rice. In terms of total production tonnages used for food, it is currently second to rice as the main human food crop and ahead of maize, after allowing for maize's more extensive use in animal feeds. Malr [16] reported that the wheat was cropped area on about 3,048,601 feddans during 2011–2012 in Egypt. Winter wheat can adjust its growth under soil water deficit conditions [7]. Winter wheat uses lower soil water content than many other crops [10], due to deep, fibrous and perfused nature of the root system. The rapid increase in population and the limited water resources in many parts of world led to an ever-increasing food problem. To narrow the gap between food production and consumption, we have to rationalize our limited natural resources of soil and water. FAO [7] stated that for irrigated wheat, the yield is 4 to 6 tons/ha (12–15% grain moisture) with a water utilization efficiency of 0.8 to 1 kg/m^3.

The evaluation of an irrigation system is based on the irrigation efficiency indices: uniformity coefficient, distribution uniformity and application efficiency [8]. These indices affect the irrigation system design and irrigation hydro-module determination. In addition, these indices are good indicators of the success of irrigation projects. In many modern research studies, water productivity (WP) has been introduced as a more comprehensive index for evaluation of water management in agriculture [11, 27]. Preference to use these indices in the decision and planning process depends on the choice by managers and experts [11, 25, 27]. However, determination of these indices and WP index can lead to an agreement among decision makers, engineers, researchers and water users in planning, designing and operating strategies. Water application uniformity in the field is expressed in terms of uniformity coefficient (CU). Increase of CU through improvement or upgrading of irrigation systems requires extra investment. Irrigation adequacy is another key management and operational parameter that is defined as the percentage of the field receiving the desired amount of water or more. This parameter affects the total applied water in the field and hence affects crop yield and WP. Various studies have been carried out to study effects of water distribution and irrigation system on crop yield. Spatial soil moisture variation and irrigation distribution uniformity have been analyzed theoretically by Warrick and Gardner [26]. Ayars et al. [3] studied effects of irrigation uniformity on sugar beet and cotton yield. Moteos et al. [18] studied uniformity sprinkler irrigation on cotton yield.

In a study on sprinkler irrigation system in semiarid region of Spain, Ortega Alvarez et al. [19] reported that economic benefits for barely, maize, garlic, onion crops were achieved with high uniformity coefficient of 90%. Based on water use efficiency (WUE) in sprinkler and trickle irrigation systems, Colaizzi et al. [4] reported that trickle irrigation was the best choice. Grassini et al. [9] studied impact of agronomic practices on maize yield and reported that applied irrigation water was 41 and 20% less under pivot and conservation tillage than under surface irrigation and conventional tillage, respectively.

Many research studies have shown that main changes in irrigation management are essential to optimize water use. For example, deficit irrigation is as an effective water planning strategy for improvement of WUE [6, 12, 25]. This strategy can be applied by decreasing the irrigation adequacy for design and management. Water-yield function has a key role in optimization of deficit irrigation. Most studies and methodologies use this function to optimize water allocation in drought conditions [27].

Water supply is a major constraint to crop production. Efficient use of irrigation water is becoming increasingly important, and alternative water application methods, such as sprinkler irrigation, may contribute substantially toward making the best use of water for agriculture and improving irrigation efficiency especially under cereal crop production [1, 20].

This chapter discusses effects of two types of sprinkler irrigation systems and different amounts of water on wheat vegetative growth and yield, WUE at Western Egyptian desert, during two growing seasons (2011–2012 and 2012–2013).

2.2 MATERIALS AND METHODS

During two successive season (2011–2012 and 2012–2013), field experiments were conducted at two sites El-Emam Malek village and NRC Farm (according to the cultivation periods), Nubaria, Behaira Governorate in Egypt. The study area is located to the west of the Nile Delta between latitudes 30°31'44" and 30°36'44"N and longitudes 30°20'19" and 30°26' 50"E. The experiments were conducted to study effects of two types of sprinkler irrigation system and different water amount on vegetative growth, WUE, and yield of wheat crop (*Triticum aestivum* L. cv. Gemmaiza-9).

2.2.1 UNIFORMITY OF IRRIGATION SYSTEM

There are numerous methods to measure the uniformity of sprinkler irrigation, and discussion of all of these is beyond the scope of this chapter.

Authors of this chapter have included following two commonly used methods in this chapter: The uniformity coefficient (UC) proposed by J.E. Christiansen in 1942 is defined below:

$$UC = 100 \times [1-(D/M)] \tag{1}$$

$$DU = 100 \times [1-(LQ/M)] \tag{2}$$

where: UC = uniformity coefficient (%) proposed by J.E. Christiansen in 1942, D = average absolute deviation of irrigation amounts, M = average of irrigation amount; DU = distribution uniformity (%), LQ = average of the lowest 1/4 of the irrigation amounts. These two uniformity methods are (approximately) related by the equations:

$$UC = (0.63)\,(DU) + 37 \tag{3}$$

$$DU = (1.59)\,(UC) - 59 \tag{4}$$

Christiansen developed UC to measure the uniformity of sprinkler systems, and it is most often applied in sprinkler irrigation situations. UC has been occasionally applied to other forms of irrigation, though DU has been applied to all types of irrigation systems.

2.2.2 STEPS TO EVALUATE TWO TYPES OF SPRINKLER IRRIGATION SYSTEMS

Each sprinkler irrigation system under study was evaluated at two different farms using 20 containers. The depth of water collected in each container is shown in Table 2.3.

2.2.3 MEASUREMENTS OF SOIL, WATER, AND PLANT PROPERTIES

Some physical, chemical properties of soil and water were determined [13, 14, 21]. Soil texture at both sites was sandy loam. Selected soil physical and chemical characteristics at two sites are shown in Table 2.1. Analysis

TABLE 2.1   Soil Properties at Two Sites

					Soil moisture, v/v		
	pH	EC	OM	CaCO$_3$	FC	WP	AW
Site	—	dS/m			%		
Emam Malek	8.1	2.3	0.5	5.6	9.5	3.6	5.9
NRC Farm	8.2	2.6	1.3	3.8	1 2.6	4.7	7.9

EC = electrical conductivity in the extracted soil paste, NRC = National Research Center, OM = organic matter, FC = field capacity, WP = wilting point, AW = available water = FC – WP.

of farmyard manure for this study consisted of:: 4.85 dS/m of EC (1:20), 7.77 pH (1:20), 11.2% of organic matter, 5.4, 0.85 and 1.12% of N-P-K, and 1:16.5 of C:N ratio.

2.2.4   WHEAT FARMING OPERATIONS

A randomized complete block design was used at two sites with three replications. The area of the experimental plot was 12 × 14 m^2 (=0.04 feddan). Farm yard manure (FYM) was added at the rate of 10 m^3/fed. The organic manure was thoroughly mixed in the 0–30 cm of the surface soil layer before planting. Fertilizers were 100 kg/fed. of superphosphate (15.5% P$_2$O$_5$) and 50 kg/fed. of potassium sulfate (48% K$_2$O) were added and mixed well before planting. Additional recommended dose of Nitrogen of 100 kg/fed. in two equal dosages was applied at 4 and 10 weeks after complete germination. Wheat seeds were broadcasted @ of 100 kg/fed. At the maturity stage, the plants were harvested and separated into grains and straw.

Soil samples at 0–20 cm depth were taken after harvesting to determine soil hydro-physical and chemical characteristics [21]. Soil hydraulic conductivity (HC, cm/unit time-unit area) was measured in the laboratory using a constant head method [14] with the following formula:

$$HC = (Q \times L)/(A \times t \times \Delta H) \qquad (5)$$

where: Q = volume of water rate flowing through saturated soil sample per unit time (L^3/T), A = cross sectional flow area (L^2), L = length of the soil core, t = time, and ΔH = hydraulic gradient (H/L). Planted area in semi-portable sprinkler irrigation system (SPSIS) was decreased by 10% relative to the permanent sprinkler irrigation system (PSIS). This was necessary because SPSIS requires more area for moving laterals from place to other compared to PSIM.

2.2.5 COMPONENTS OF SPRINKLER IRRIGATION SYSTEM

The following components were used in portable sprinkler irrigation system as shown in Fig. 2.1:

A pump unit (50 m³/h); Tubings: main/submains and laterals of inside diameter 150, 110, 90 mm, respectively; Couplers; Sprinkler head (lph); and other accessories such as valves, bends, plugs and risers.

Semi portable system is similar to the portable system, except that the location of water source and pumping unit is fixed. A fully permanent sprinkler irrigation system consists of permanently laid mains, submains and laterals and a stationery water source and pumping plant (Fig. 2.2).

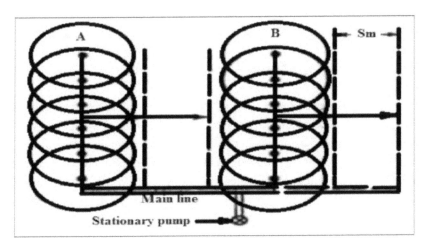

FIGURE 2.1   Semi-portable sprinkler irrigation system.

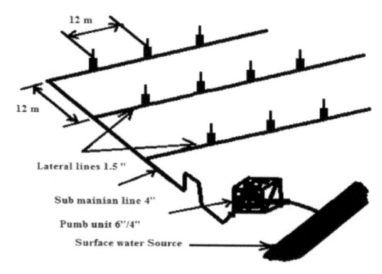

FIGURE 2.2   Permanent sprinkler irrigation system (PSIS).

2.2.6 *WATER REQUIREMENTS AND WATER USE EFFICIENCY*

Irrigation water was applied using sprinkler irrigation system through fixed lateral lines of 1.5″ in diameter with a spacing of 12 m between the laterals and 12 m between sprinkler heads down the lateral line. Number of sprinklers on every lateral line was 14. Riser height was 1.0 m. Submain and main lines were of 4″ and 6″ diameter, respectively. Mean of sprinkler discharge was 1.2 m³/h at a mean operating pressure of 2 bars. Mean wind speed was 1.5 m/s. Irrigation efficiency of the system was 78% for SPSIS and 83% for PSIS.

Irrigation water amounts were estimated using reference evapotranspiration and crop coefficients after Allen et al. [2], while the irrigation duration was determined using a water balance method. Irrigation interval was 4 days for all treatments. Estimated seasonal water requirement was 550 mm/season or 2310 m³/fed. Field water use efficiency (FWUE) was calculated using Eq. (6) [8, 17] and crop water use efficiency (CWUE) was calculated using Eq. (7) [8, 17].

$$FWUE = Y/WR \qquad\qquad (6)$$

TABLE 2.2   Water Requirements For Wheat Crop at Nubaria Sites, Egypt

Item	Month					
	Dec	**Jan**	**Feb**	**March**	**April**	**May**
Days	31	31	29	31	30	31
ET (mm/day)	2.8	6.3	5.9	4.2	7.4	2.0
Kc	0.4	0.4	0.8	1.3	0.5	0.4
ETc (mm/day)	1.1	2.5	4.7	5.4	3.7	0.8
Growth stage	**Planting establish- ment**	**Rapid vegeta- tive growth**		**Flowering and seed filling**	**Maturity and harvesting**	
IRn (mm/month)	33.1	77.6	136.4	167.5	111.1	24.9
IRg (mm/month)	36.4	85.3	150.0	184.4	122.0	27.4

Rn = Net irrigation requirements, IRg = Gross irrigation requirements, ET = evapo-transpiration, Kc = crop coefficient, ETc = crop evapotranspiration.

$$CWUE = Y/WCU \qquad (7)$$

where: FWUE = field water use efficiency, Y = Seed yield (kg/fed), WR = total amount of water applied in the field (m^3/fed.), CWUE = crop water use efficiency, and WCU = actual water consumptive use (m^3/fed.). All data for water requirements of wheat are presented in Table 2.2.

Data were subjected to analysis of variance in randomized complete block design and means were separated according to LSD test. Correlation and multiple regression analysis were conducted using a SAS computer program [22].

2.3   RESULTS AND DISCUSSION

2.3.1   *UNIFORMITY COEFFICIENT (% UC) AND DISTRIBUTION UNIFORMITY (% DU) FOR PERMANENT SPRINKLER IRRIGATION SYSTEM (PSIS)*

The data in Table 2.3 and Eqs. (1) and (4) were used to calculate % UC and % DU for permanent sprinkler irrigation system (PSIS):

d_{LQ} (Number of quarter cans ranked) = average of cans (1 to 5) of ascending order = 3.40

M = average absolute deviation of irrigation amounts in cans (1 to 20) = 4.46 = 4.5

D = average absolute, |di-dz| = 0.6

Substituting these values in Eq. (1), we get: UC = 100 (1–[D/M])

UC (%) = 100 × [1–(0.6/4.5)] = 100 × 0.867 = 86.7%

Substituting value of UC in in Eq. (4), we get distribution uniformity (DU, %):

DU = (1.59 × UC) – 59 = (1.59 × 86.7) – 59 = 78.8%

2.3.2 UNIFORMITY COEFFICIENT (% UC) AND DISTRIBUTION UNIFORMITY (% DU) FOR SEMI PORTABLE SPRINKLER IRRIGATION SYSTEM

The data in Table 2.3 and Eqs. (1) and (4) were used to calculate % UC and % DU for semi portable sprinkler irrigation system (SPSIS).

d_{LQ} (Number of quarter cans ranked) = average of cans 1 to 5 in ascending order = 2.98

M = average absolute deviation of irrigation amounts in cans (1 to 20) = 3.91 = 3.9

D = average absolute, |di-dz| = 1.00

Substituting these values in Eq. (1), we get: UC = 100 (1-[D/M])

UC (%) = 100 × [1 – (1.0/3.9)] = 100 × 0.74 = 74%

Substituting value of UC in in Eq. (4), we get distribution uniformity (DU, %):

DU = (1.59 × UC) – 59 = (1.59 × 74) – 59 = 58.7%

Uniformity coefficients (% UC) in the two experimental fields were 86.7 and 74.0% in permanent and semi portable sprinkler irrigation systems, respectively. Distribution uniformity (% DU) were 78.8 and 58.7% in permanent and semi portable sprinkler irrigation systems, respectively. These results for the two sprinkler irrigation systems types are according to Kunde [15] and Solomon [23, 24].

TABLE 2.3   Field Evaluation of Parameters for Permanent and Semi-Portable Sprinkler Irrigation Systems

CAN No.	Permanent sprinkler irrigation system			Semi portable sprinkler irrigation system		
	Water depth (cm)	di (Ascending ranking)	Absolute |di-dz|	Water depth (cm)	di (Ascending ranking)	Absolute |di-dz|
1	4.5	2.8	1.7	5.1	2.4	0.6
2	4.9	3.3	1.2	3.3	2.8	0.2
3	4.1	3.5	1.0	4.2	3.2	0.2
4	5.3	3.6	0.9	3.8	3.2	0.2
5	4.7	3.8	0.7	4.1	3.3	0.3
6	3.5	4.1	0.4	4.7	3.5	0.5
7	5.6	4.2	0.3	2.8	3.6	0.6
8	3.8	4.3	0.2	2.4	3.7	0.7
9	4.8	4.4	0.1	4.5	3.8	0.8
10	4.4	4.5	0.0	3.6	3.8	0.8
11	4.6	4.6	0.1	3.7	4.1	1.1
12	5.3	4.7	0.2	4.2	4.1	1.1
13	3.3	4.8	0.3	3.2	4.2	1.2
14	2.8	4.8	0.3	5.1	4.2	1.2
15	5.1	4.9	0.4	3.8	4.2	1.2
16	4.2	5.1	0.6	4.2	4.5	1.5
17	4.3	5.3	0.8	3.5	4.7	1.7
18	4.8	5.3	0.8	4.8	4.8	1.8
19	3.6	5.6	1.1	4.1	5.1	2.1
20	5.6	5.6	1.1	3.2	5.1	2.1
Average	4.46	Quarter avg. (cans 1 to 5) =3.4	0.6	3.91	Quarter avg. (cans 1 to 5) =2.98	1.0

2.3.3 EFFECTS OF TREATMENTS ON THE WHEAT GROWTH PARAMETERS

Tables 2.4 and 2.6 show the effects of types of sprinkler irrigation systems, and irrigation treatments on number of spikes per m^2, seed index, peduncle length (cm), plant height (cm), grain and straw yields (Kg per fed.), and water use efficiency (WUEg, Kg of grain per m^3 of irrigation water).

Based on all values of growth parameters, the growing seasons can be ranked in the ascending order: 2011–2012 < 2012–2013. Differences among the values of growth parameters between the two seasons were significant at

TABLE 2.4 Effects of Types of Sprinkler Irrigation Systems and Three Irrigation Depths on Vegetative Growth Parameters of Wheat, During Two Growing Seasons

Types of irrigation systems (II)	Irrigation depth (% of ETo) (III)	No of spikes, per m²	Weight of 1000 grains, g	Peduncle length, cm	Plant height, cm
Growing season I, 2011–2012					
Semi-portable	100	523 a	38.8 a	29.5 a	98.2 a
	75	516 b	36.2 b	27.3 b	95.7 b
	50	508 c	22.5 c	21.7 c	75.5 c
Permanent	100	531 a	40.6 a	30.2 a	99.7 a
	75	520 b	37.3 b	28.6 b	96.6 b
	50	511 c	28.1 c	22.8 c	76.9 c
Means		518	34	27	90
Growing season I, 2012–2013					
Semi-portable	100	526 a	42.5 a	32.9 a	102.6 a
	75	520 b	41.6 ba	30.1 b	100.2 b
	50	518 cb	29.2 c	23.5 c	79.6 c
Permanent	100	537 a	44.6 a	34.6 a	105.8 a
	75	530 b	42.7 b	33.2 b	103.6 b
	50	527cb	33.1 c	25.7 c	84.3 c
Means		526	39	30	96
LSD at P = 0.05		4	2.3	1.5	0.6

LSD = least square difference. Values followed by same letters were not significant at 5%.

5% level. Also, the irrigation systems types can be written in the ascending order: SPSIS < PSIS. The effects of the types of sprinkler irrigation systems on all growth parameters were significant at 5% level, except the seed index.

Decreasing the irrigation water from I_{100} to I_{50} showed positive effects on WUEg and negative ones on all other parameters. Irrigation treatments can be arranged in the ascending order: $I_{100} < I_{75} = I_{50}$; and $IR_{50} < IR_{75} < IR_{100}$ for WUEg and the other parameters, respectively. Differences in most parameters among the irrigation treatments were significant at 5% level. The exceptions were between irrigation treatments $(I_{100}; I_{75})$ for seed index and $(I_{75}; I_{50})$ in the case of WUEg, respectively.

2.3.3.1   Effects of Interactions on the Treatments

Table 2.5 indicates that the interaction, semiportable type × seasons, had significant effects on the biomass and grain yield, and WUEg at 5% level. The maximum and the minimum values were obtained in the interactions, PS × I_{100} × 2013 and SPSIS × I_{50} × 2012, respectively. Taking into consideration the effects of irrigation treatments on soil moisture stress before the next irrigation, these can be arranged in the ascending order: $I_{100} < I_{75} < I_{50}$. The cumulative effect of soil moisture stress was increased with time from germination to maturity.

2.4   CONCLUSIONS

Based on results of this study and data in Tables 2.4–2.6, following observations can be made:

- The increase in soil moisture stress in the root zone has a depressive effects on lower leaves flowering and tillering.
- Decreasing tillering led to lower photosynthesis process and subsequently affected all growth parameters.
- At the time of flowering, root growth may be reduced by soil moisture stress and may even cease and considerable damage can be caused leading to yield loss. Farmers must note that this loss cannot be recovered by providing adequate water supply during the later growth period.

TABLE 2.5 Effects of Types Sprinkler Irrigation Systems and Three Water Amounts on Biomass (straw) and Grain Yield, and Water Use Efficiency (WUEg)

Types of irrigation systems (II)	Irrigation depth (% of ETo) (III)	Water amount (m³/fed.)	Biomass (straw) (kg/fed.)	Grain yield (kg/fed.)	WUEg (Kg/m³)
Growing season I, 2011–2012					
Semi-portable	100	2185.0	4396 a	3767 a	1.7 a
	75	1638.8	4384 b	3521 b	2.1 b
	50	1092.5	4189 c	2207 c	2.0cb
Permanent	100	2185.0	4487 a	3858 a	1.8 a
	75	1638.8	4465 b	3632 b	2.2 b
	50	1092.5	4268 c	2387 c	2.1cb
Means	—		4365	3229	2.0
Growing season I, 2012–2013					
Semi-portable	100	1987.0	4496 a	3975 a	2.0 a
	75	1490.3	4476 b	3841 b	2.6 b
	50	993.5	4286 c	2437 c	2.5cb
Permanent	100	1987.0	4585 a	4088 a	2.1 a
	75	1490.3	4564 b	3953 b	2.7 b
	50	993.5	4364 c	2523 c	2.5cb
Means	—		4462	3470	2.4
LSD at P = 0.05	—		8	12	0.6

LSD = least square difference. Values followed by same letters were not significant at 5%.

- Pollen formation and fertilization can be seriously affected under heavy soil moisture stress.
- During the time of head development and flowering, water shortage will number of spikes/plant, head length and number of grains/head.
- Water deficit during the grain formation caused grains shriveling and grain weight reduction.
- The hot climate and stronger wind were observed during the yield formation during 2011–2012 compared to 2012–2013.
- The residual effect of the manure added in the 1st year was extended to the 2nd year.

TABLE 2.6 Effects of Types Sprinkler Irrigation Systems and Irrigation Treatments on Wheat Vegetative Growth and Yield, During Two Seasons

Treat-ments	No. of spikes per m²	Weight of 1000 grains, (g)	Peduncle length, (cm)	Plant height, (cm)	Biomass (kg/fed.)	Grain Yield, (kg/fed)	WUEg (Kg/m³)
2012	518.2b	33.9d	26.7c	90.4f	4364.8e	3228.6f	1.9b
2013	526.3a	39.0b	30.0b	96.0c	4461.9c	3469.5c	2.4a
SPS	518.5b	35.1c	27.5c	92.0e	4371.2e	3291.3e	2.1b
PS	526.0a	36.5c	29.2b	94.5d	4455.5d	3406.8d	2.2a
I = 100	529.3a	41.6a	31.8a	100.8a	4491.0a	3922.0a	1.9b
I = 75	521.5b	39.5a	29.8b	99.0b	4472.3b	3736.8b	2.4a
I = 50	516.0c	28.2e	23.4d	79.1 g	4276.8f	2388.5 g	2.3a

SPS: semi portable sprinkler irrigation system, PS: permanent sprinkler irrigation system, and I: irrigation amount based on ETo. Values followed by same letter were not significant at P = 0.05.

- The SPSIS system resulted in undesirable mechanical damage of some wheat plants.
- Moving the lateral lines in SPSIS, after irrigation every 4 days in a wet soil, caused soil compaction, poor aeration and root growth impedance.
- In both SIS, due to dense and to some extent high height of wheat plants, we do expect lower water distribution uniformity due to: the difficulty in straight forward lateral lines; and water leakage from the joints.
- WUEg followed the order: SPSIS < PSIS.

2.5 SUMMARY

It is common to use the sprinkler irrigation system on large-scale irrigation intensive crops in the Egyptian desert. In Egypt, the cultivation and production of wheat crop depends basically on sprinkler irrigation system. Two field experiments were conducted, during two successive seasons (2011–2012 and 2012–2013) at two sites El-Emam Malek village and NRC Farm,

Nubaria, Behaira Governorate, to study the effects of two types of sprinkler irrigation system and three irrigation depths on vegetative growth, WUE, and yield of wheat crop (*Triticuma estivum L.* cv. Gemmaiza 9). Uniformity coefficient (% UC) was 86.7 and 74.0% in permanent and semi portable sprinkler irrigation systems, respectively. Distribution uniformity (% DU) was 78.8 and 58.7% in permanent and semi portable sprinkler irrigation systems, respectively. The effects of types sprinkler irrigation on all growth parameters were significant at 5% level, except on seed index. Decreasing the irrigation depth from I_{100} to I_{50} had positive effects on (WUE)g and negative ones on all other growth parameters. Differences in most parameters among the irrigation treatments were significant at 5% level. The better production of wheat was in the second season 2012–2013. The permanent sprinkler system gave the highest production of grain and biomass.

KEYWORDS

- aeration
- distribution uniformity
- drip irrigation
- Egypt
- grain yield
- irrigation water
- root growth
- soil compaction
- sprinkler irrigation
- sprinkler irrigation, permanent
- sprinkler irrigation, semiportable
- uniformity coefficient
- vegetative growth
- water use efficiency
- wheat

REFERENCES

1. Abd El-Kader, A. A., Shaaban, S. M. and Abd El-Fattah, M. S., 2010. Effect of irrigation levels and organic compost on okra plants grown in sandy calcareous soil. *Agriculture and Biology Journal of North America*, 1:225–231.
2. Allen, R. G., Pereira, L. S., Raes, D. and Smith, M., 1998. *Crop evapotranspiration – Guidelines for computing crop water requirements*. FAO Irrigation and drainage paper 56. Food and Agriculture Organization of the United Nations Rome, Italy.
3. Ayars, J. E., Hutmacher, R. B., Vali, S. S. and Schoneman, R. A., 1991. Cotton response to nonuniform and varying depths of irrigation. *Agri. Water Mgmt.*, 19:151–166.
4. Cai, X. L. and Sharma, B. R., 2010. Integrating remote sensing, census and weather data for an assessment of rice yield, water consumption and water productivity in the Indo-Gangetic river basin. *Agri. Water Mgmt.*, 97:309–316.
5. Colaizzi, P. D., Schneider, A. D. and Howell, T. A., 2005. Comparison of SDI, LEPA and spray irrigation performance for grain sorghum. *Am. Soc. of Agri. Eng.*, 47:1477–1492.
6. English, M., 1990. Deficit irrigation, I: An analytical framework. *J. Irri. Drain. Eng.* (ASCE), 116:399–412.
7. FAO, 1979. *Yield response to water*. FAO Irrigation and Drainage Paper No. 33. Rome.
8. Megh R. Goyal, 2012. *Management of Drip/Trickle or Micro Irrigation Management*. Apple Academic Press Inc., Oaksville, ON, Canada.
9. Grassini, P., Thorburn, J., Burr, C. and Cassman, K. G., 2011. High yield irrigated maize in the Western U.S. Corn Belt, I: On-farm yield, yield potential, and impact of agronomic practices. *Field Crops Res.*, 120:142-150.
10. Haise, H. R., Haas, H. J. and Jensen, L.R., 1955. Soil moisture studies of some Great Plains soil, II: Field capacity as related to 1/3-atmosphere percentage and "minimum point" as related to 15- and 26-atmosphere percentages. *Soil Sci. Soc. Amer. Proc.*, 19:20–25.
11. Kijne, J., Barker, R. and Molden, D., 2003. Improving water productivity in agriculture: Editors' Overview. In: *Water Productivity in Agriculture: Limits and Opportunities for Improvement, Comprehensive Assessment of Water Management in Agriculture*, eds. J. W. Kijne, D. Molden, and R. Barker. UK: CABI Publishing in Association with International Water Management Institute.
12. Kirda, C. and Kanber, R., 1999.Water no longer a plentiful resource, should be used sparingly in irrigated agriculture. In: C. Kirda, P. Moutont, C. Hera and D. R. Nielsen, eds. *Crop Yield Response to Deficit Irrigation*. Dordrecht: the Netherlands and Kluwer Academic Publishers.
13. Klute, A. and Dirksen, A., 1986. Water retention: Laboratory methods. In: A. Klute (ed.), *Methods of Soil Analysis, Part 1: Physical and mineralogical methods*. Pages 635–662, ASA and SSSA Publication 9, Madison, WI.
14. Klute, A. and Dirksen, A., 1986. Hydraulic conductivity. In: A. Klute (ed.), *Methods of Soil Analysis, Part 1: Physical and mineralogical methods*. Pages 678–734, ASA and SSSA Publication 9, Madison, WI.

15. Kunde, R. J., 1985. Life cycle costs resulting from various design emission uniformities. Proceedings 3rd International Drip/Trickle Irrigation Congress, November 18–21, Fresno, CA, Volume II:859–866.

16. Malr, 2012. *Study of Important Indicators of Agricultural Statistics*. Ministry of Agriculture and Land Reclamation, Egypt: Winter Crops, 4:(1170).

17. Michael, A.M., 1978. *Irrigation theory and practice*. Vikas Publishing House Pvt. Ltd., New Delhi.

18. Moteos L., Montovani, E. E. and Villalobos, F. J., 1997. Cotton response to nonuniformity of conventional sprinkler irrigation. *Irri Sci.*, 17:47–52

19. Ortega Alvarez, J. F., Tarjuelo Martin-Baito, J. M., De Juan Valero, J. A. and Perez, P. C., 2004. Uniformity distribution and its economic effect on irrigation management in semiarid zone. *J. Irri. Drain. Eng.*, 130:112–120.

20. Phocaides, A., 2000. *Technical hand book on pressurized irrigation techniques*. Food and Agriculture Organization of the United Nations (FAO), Rome.

21. Rebecca, B., 2004. *Soil Survey Laboratory Methods Manual*. Soil Survey Investigations Report No. 42, Natural Resources Conservation Services, USDA.

22. SAS Institute, 2001. *SAS statistics users' guide*. Release 8.2. SAS Institute, Cary, NC.

23. Solomon, K. H., 1983. *Irrigation Uniformity and Yield Theory*. PhD dissertation, Department of Agricultural and Irrigation Engineering, Utah State University, Logan UT, 287 p.

24. Solomon, K. H., 1987. *Sprinkler Irrigation Uniformity*. Extension Bulletin No. 247, Food & Fertilizer Technology Center, Tapei City, Taiwan, Republic of China.

25. Van Halsema, G. E. and Vincent, L., 2012. Efficiency and productivity terms for water management: A matter of contextual relativism versus general absolutism. *Agri. Water Mgmt.*, 108:9–15.

26. Warrick, A. W. and Gardner, W. R., 1983. Crop yield as affected by spatial variation of soil and irrigation. *Water Resour. Res.*, 19:181–186.

27. Wichelns, D., 2002. An economic perspective on the potential gains from improvements in irrigation water management. *Agri. Water Mgmt.*, 52:233–248.

CHAPTER 3

PERFORMANCE OF SPRINKLER IRRIGATED WHEAT – PART II

E. ELDARDIRY, F. HELLAL, H. A. MANSOUR, and M. A. EL HADY

CONTENTS

3.1   INTRODUCTION

Due to the increase in cereal demand for human consumption as a result of population growth, sustainable increases in crop yield are needed to ensure

Modified and printed from *E. Eldardiry, F. Hellal, H. A. Mansour and M. A. El Hady, 2013. Assessment cultivated period and farmyard manure addition on some soil properties, nutrient content and wheat yield under sprinkler irrigation system. Agricultural Sciences, 4(1), 14–22. Open access article at:* http://www.scirp.org/journal/as/.

In this chapter: 1 feddan = 0.42 hectares = 4200 m² = 1.038 acres = 24 kirat. A feddan (Arabic) is a unit of area. It is used in Egypt, Sudan, and Syria. The feddan is not an SI unit and in Classical Arabic, the word means 'a yoke of oxen': implying the area of ground that can be tilled in a certain time. In Egypt the feddan is the only nonmetric unit, which remained in use following the switch to the metric system. A feddan is divided into 24 kirats (175 m²). In Syria, the feddan ranges from 2295 square meters (m²) to 3443 square meters (m²).

food security in Egypt. Consequently, the judicious and scientific management of soil and water resources is essential to meet the increasing demand of cereals. Soil management practices have profound impacts on soil fertility, which is closely linked to land productivity. As soil processes are often slow, it is only through long-term management that they can improve soil characteristics to enhance crop production. The need to increase the grain yield of cereals per unit is of utmost importance in the developing countries [19]. Wheat is considered one of the most important and strategically cereal crop in Egypt, but cultivated area only produces about 30% of the domestic need. One of several technologies to increase wheat production is an appropriate application of organic matters, especially in the newly reclaimed areas [29].

Organic matter is a key component of the soil because of its many functions in agro-ecosystems and it is applied the soil to improve its physical, chemical and biological properties [11]. Fliessbach et al. [9] suggested that application of farm yard manure (FYM) increased the transfer elements between the solid phase and soil solution in addition to higher microbial activity. They also reported that organic soil management improved the soil structure, thus reducing the risk of soil erosion and promoted the development of the soil conditions for plant. The activity of soil microorganisms was higher in the organic farming system, which helped faster nutrient uptake. Dalal et al. [4, 5] found that dry matter yield and N-uptake of winter cereal crops (wheat and barley) showed significant decreasing trends with periods of cultivation in all soils. They added that most of the newly reclaimed areas in the deserts are very poor in the organic matter contents as well as the primitive fertility. Enke et al. [7] indicated that long-term additions of organic manure have the most beneficial effects on grain yield of wheat and maize.

It is customary to consider that soils in arid and semiarid regions have a pH of about 8.5 and this value is strongly affected by continuous cultivation to encourage growing of microorganisms and plant root, period of cultivation, and continuous application of organic manure [6]. Part of the solution to poor soil is continuous addition of FYM, which is related with the assessment of land performance for maximizing crop production.

Water supply is another major constraint to crop production. Efficient use of irrigation water is becoming increasingly important, and alternative water application methods such as sprinkler irrigation, may contribute substantially toward making the best use of water for agriculture and especially for cereal production [2, 12, 22].

This chapter investigates the effects of FYM in sprinkler-irrigated wheat on the changes in soil hydrophysical and chemical properties and performance of wheat production (grain and straw yield, NPK/Protein and carbohydrate contents), during the growing season.

3.2 MATERIALS AND METHODS

During two successive season of 2011/2012, field experiments were established at two sites in El-Emam Malek and El-Shagaah villages, Nubaria, Behaira Governorate of Egypt. The study area is located between latitudes 30°31′44″ and 30°36′44″ N and longitudes 30°20′19′ and 30°26′50″ E (Fig. 3.1) to study effects of different cultivated soil periods (10 and 25 years), FYM as a source of organic matter (OM: continuous application) on performance parameters of wheat under sprinkler irrigation system.

Soil physical, chemical properties were determined [10]. Soil moisture retention at field capacity and wilting point were also determined [18]. Soils at the site are sandy loam. Analysis of farm yard manure resulted in 4.85 dS/m (EC, 1:20), 7.77 of pH (1:20), 11.2% of OM, 5.4% of N, 0.85% of P, 1.12% of K and 1:16.5 as C:N ratio.

The experiment design was randomized complete block with three replications. The area of the experimental plot was 12×14 m^2. FYM was added @ of 10 m^3/fed. The organic manure was thoroughly mixed in the 0–30 cm of the surface soil layer before planting. Recommended doses of NPK were applied. Wheat seeds (*Triticum aestivum* L. *cv.* Gemmaiza 9) @ 100 kg/fed. were broadcasted. At the maturity stage, the plants were harvested and separated into grains and straw. Soil samples at 0–20 cm depth were taken after last harvesting to determine soil hydrophysical and chemical characteristics and soil bulk density (gm/cm^3), [24].

Soil hydraulic conductivity (HC) under saturated conditions was also measured in the laboratory under a constant head technique [17].

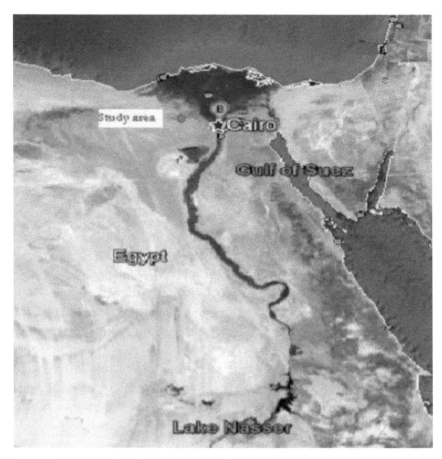

FIGURE 3.1   Location of the study area in the North Nile Delta near Cairo, Egypt.

Soil water retention at field capacity and at wilting point and available water were also measured [18]. Total nitrogen, phosphorus and potassium contents in grain and straw were determined [8]. Carbohydrate content of wheat grains was determined in hot water extract [30]. Dry leaf sample (0.2 g) was put in 10 mL distilled water and the mixture was kept in the boiling water bath for 30 min. For the determination of carbohydrates, 1 mL of diluted sample was mixed with 5 mL of Anthron's reagent and was then kept in the boiling water bath for another 30 min. The absorbance was measured at 620 nm, after cooling against glucose as standard.

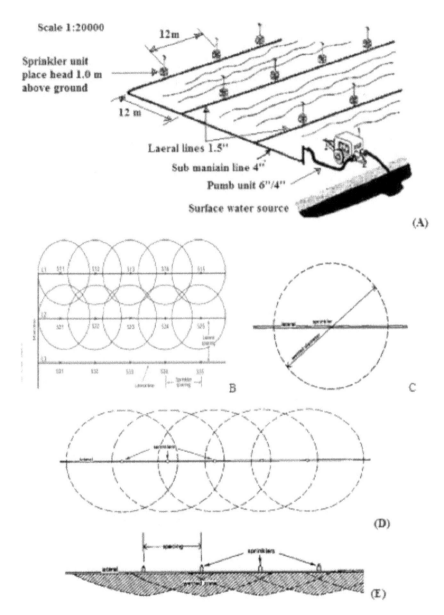

FIGURE 3.2 Layout and operation of fixed sprinkler irrigation system [12].

Irrigation water was applied using sprinkler irrigation system with fixed lateral lines of 1.5 inch in diameter at a lateral spacing of 12 m, supports height was 1.0 m. Mean sprinkler discharge was 1.2 m3/h at an operating pressure of 2 bars. Number of sprinklers on each lateral line were 14 (Fig. 3.2).

Irrigation water amounts were estimated using the reference evapotranspiration and crop coefficients [15]. Calculated amount of water requirements was 550 mm or 2310 m³/fed during the season. Water consumptive use (WCU) was calculated [16] by the following equation:

$$\text{Wcu} = \Sigma^{i=n} (\theta2 - \theta1)/[100\ x\ \text{DB} \times (60/100)\ c\ (4200)] \tag{1}$$

where: Wcu = water consumptive use (m³/fed.), n = number of irrigations, $\theta2$ = soil moisture (%) after irrigation, $\theta1$ = soil moisture (%) before the next irrigation, DB = bulk density of soil (g/cm³), I = from 1 to nth observation.

Field water use efficiency (FWUE) and crop water use efficiency (CWUE) were calculated according to Michael [20]. Data were subjected to analysis of variance in randomized complete block design. Means were separated according to LSD test and correlation coefficients, using the SAS program [26].

3.3   RESULTS AND DISCUSSION

3.3.1   WATER REQUIREMENTS

The Table 3.1 shows that wheat ETc varied from 0.8 (May) to 5.4 (March) mm/month. Wheat water requirement (CWR) is increasing with the growing period and it was maximum amount at the crop development and mid-season stage. The maximum CWR was observed in the month of March (167.4 mm/month) while the minimum was (24.8 mm/month) observed in the month of May. Values of CWR increase in the month of March (28.4 mm/month) as wheat was in maturity stage. It was also found that crop water requirement was less in the maturity stage as compared to the seed filling stage. These findings agree with those reported by Allen et al. [3].

TABLE 3.1 Water Requirements for Wheat Crop at Nubaria Sites, Egypt

Item	Month					
	Dec	**Jan**	**Feb**	**March**	**April**	**May**
Days	31	31	29	31	30	31
ET (mm/day)	2.8	6.3	5.9	4.2	7.4	2.0
Kc	0.4	0.4	0.8	1.3	0.5	0.4
ETc (mm/day)	1.1	2.5	4.7	5.4	3.7	0.8
Growth stage	**Planting estab-lishment**	**Rapid vegetative growth**		**Flower-ing and seed filling**	**Maturity and harvesting**	
IRn (mm/month)	33.1	77.6	136.4	167.5	111.1	24.9
IRg (mm/month)	36.4	85.3	150.0	184.4	122.0	27.4

IRn = net irrigation requirements, IRg = gross irrigation requirements,
ET = evapotranspiration, Kc = crop coefficient, ETc = crop evapotranspiration.

3.3.2 SOIL PROPERTIES AFTER LAST HARVEST

Table 3.2 shows the soil properties after the last harvest of wheat. Data revealed that addition of FYM decreased soil bulk density (BD) by about −1.2 to −3.8% compared to the control treatment (no FYM) under 10 and 25 years of cultivated periods, respectively. Also, BD after 25 years cultivation period decreased by about −4.3% comparing to 10 years period. Soil BD was lower in the long period-cultivated site than short one. However, the results agree with the observations by other researchers [5, 19], who indicate that BD decreases with increase in the period of cultivation. This indicates that as cultivation causes accumulation of OM in the soil, the BD decreases. While Saeed et al. [25] reported that there was no clear effect of short-term cultivation on BD.

Hydraulic conductivity (HC) values were strongly affected by cultivation period and addition of FYM. It can be observed in Table 3.2 that both the cultivated period and FYM addition significantly decreased HC values by about 18.9 and 12.1% in same sequence.

Based on values of soil water content at field capacity (FC), wilting point (WP) and available water (AW), data conclude that cultivation period has

TABLE 3.2 Effects of Cultivated Period and Organic Manure on Soil Properties at the End of the Wheat Growing Season

Cultivated period (A)	FYM addition (B)	HC cm/h	BD gm/cm³	FC %wb	WP	AW	pH	EC dS/m	OM %	K ppm	P	TSN
10 years	–	11.5	1.62	9.3	3.8	5.5	8.22	2.37	0.65	36.4	7.86	52.5
	+	10.7	1.60	10.8	4.3	6.5	8.05	2.52	0.78	42.9	9.16	95.1
	mean	**11.1**	**1.61**	**10.05**	**4.05**	**6.00**	**8.14**	**2.45**	**0.72**	**39.7**	**8.51**	**73.8**
25 years	–	9.9	1.57	11.8	4.7	7.1	8.12	1.98	0.85	38.4	7.92	108.5
	+	8.1	1.51	13.5	5.1	8.4	8.01	2.06	1.21	54.1	9.78	122.5
	mean	**9.0**	**1.54**	**12.7**	**4.9**	**7.8**	**8.07**	**2.02**	**1.03**	**46.3**	**8.85**	**115.5**
LSD 5%	A or B	1.2	0.03	1.4	0.2	0.4	0.05	0.06	0.12	2.3	0.7	6.5
	A x B	0.8	0.02	1.1	0.1	0.2	0.03	0.04	0.09	1.7	0.5	4.8

FYM: farm yard manure, –: noncontinuous, +: continuous, OM: organic matter, FC: field capacity, WP: wilting point, AW: available water, TSN: total soluble nitrogen, EC: Electrical conductivity, HC: Saturated hydraulic conductivity, BD: Bulk density.

more pronounced effect than the addition of FYM. The percentages of the increases of these parameters were 15.1 and 9.3; 19.0 and 25.7; 19.5 and 30.0 for FYM and cultivation period comparing to the control.

Data in Tables 3.2–3.5 show that soil HC is strongly positive correlated with BD, soil pH, soil EC, while is negatively correlated at 1% with soil water constants (FC, WP and AW), FYM content and soil N and P contents. Whereas, soil water constants are positively correlated with soluble N, P and K in soil. This is mainly attributed to the role in modification of BD and hence water retained in soil at FC and WP. Also, improvement in soil structure was associated with improvement in water movement in soil under saturated flow (HC). Also, root plant growth is strongly correlated with both cultivation period and addition of FYM addition. This agrees with the findings of Schumacher et al. [27] who found that cultivation practices can change the soil water content, aeration, and the degree of mixing of crop residues within the soil matrix, thereby affecting activity of soil organisms, which have important functions in soils such as structure improvement, nutrient cycling and organic manure decomposition [16].

Total soluble N, P and K in soil were measured at the end of the growing season of wheat crop. Results indicate that control treatments gave lowest values of these nutrients under 10 years than 25 years cultivated period by about −5.2, −0.8 and −51.6, respectively. Regarding the effect of cultivation period on these nutrients, results pointed out that there is an increase in soil nutrients resulting in 21.2, 4.6; 56.5, 5% for the cultivation periods; and 29.7, 20.0 and 35.2% for FYM addition. These results agree with those reported by Abbas et al. [1].

3.3.3 MACRONUTRIENTS, PROTEIN AND CARBOHYDRATES CONTENTS IN WHEAT GRAIN AND STRAW

Table 3.3 illustrates N, P and K contents in grain and straw. Data conclude that cultivated period had less effect on these nutrient contents in grain than those under FYM. The percentage increases were 5.2, 13.5; 3.8 and 26.5, 21.3; 22.6 with 25 and 10 years periods and addition of FYM compared to control, respectively. We can observe that effects of FYM individually under two studied cultivated periods is more effective under 10 years (28.0, 25.2;

TABLE 3.3 Effects of Cultivated Period and Organic Manure on Chemical Properties of Wheat

Cultivated period (A)	FYM addition (B)	Grain					Straw			
		N	P	K	Protein	Carbohy-drates	N	P	K	Protein
10 years	–	1.68	0.276	1.13	10.29	55.16	0.73	0.220	3.22	4.20
	+	2.15	0.323	1.48	11.68	58.92	0.78	0.271	5.04	4.48
	mean	1.92	0.30	1.31	10.99	57.04	0.76	0.246	4.13	4.34
25 years	–	1.79	0.302	1.26	11.63	64.19	0.62	0.267	3.76	3.57
	+	2.24	0.378	1.45	12.86	67.43	1.01	0.283	4.88	5.78
	mean	2.02	0.34	1.36	12.25	65.81	0.82	0.275	4.32	4.68
LSD 5%	A or B	0.78	0.08	0.14	0.95	2.34	0.11	0.021	0.23	0.95
	A × B	0.53	0.05	0.11	0.65	1.33	0.07	0.015	0.17	0.73

FYM: farm yard manure, –: un-continuous, +: continuous.

15.1%) than the 2nd one (25.1, 25.2; 15.1%) comparing with untreated FYM plots. The N, P and K contents in wheat straw values had unclear trend and the percentage increase of these nutrients were 6.8, 23.23; 56.5% and 62.9, 6.0; 29.8 as a result of addition of FYM under 10 and 25 years cultivated periods.

With respect to the protein and carbohydrates contents in wheat grains as affected by cultivated period and OM addition, data in Table 3.3 indicated that the highest values (12.86 and 67.43%) were obtained under cultivated period 25 years after FYM addition. Also, the lowest ones (10.29 and 55.16%) were attained in cultivated periods 10 years without FYM addition for protein and carbohydrates contents in wheat grains. Effects of both investigated factors, cultivated period and FYM addition, showed no significant difference between increased values relative to their effects (11.5 and 12.0%) for protein content, while there was significant difference between their effects on carbohydrates content (15.4 and 5.9%), respectively.

Regarding to the effects of both investigated factors, FYM showed strong effect on the above mentioned nutrient contents in straw and the increase was 32.6, 13.8 and 27.8% of N, P, K. Cultivated period 25 years did not affected significantly on nutrient status in wheat straw with values of 12.0, 4.6 and 1.7% of N-P-K relative to 10 years cultivated period.

According to protein content in wheat straw, data revealed that addition of FYM had a pronounce effect under 25 years than 10 years cultivated period with values of 61.9 and 6.7% comparing to control, respectively, while FYM alone improved the protein content in straw by about 31.9% comparing to untreated one. Low increase in protein in straw was attained compared to the increase in cultivated period by about 7.8%.

Randall et al. [23] indicated that continuous cultivation for long periods can enhance leaching of nitrate through the soil profile. It is well established that the presence of organic manure on the surface protects the soil from erosion, improves infiltration and release nutrients. Lack of soil disturbance was found to result in stratification of soil organic manure [15]. Also, Tawfik et al. [28] found that significant increase in N uptake by maize and wheat was observed with continuous application of organic manures. Hellal et al. [13] found that farmyard manure application significantly enhanced the yield and N, P and K uptake of wheat. Abd El-Kader et al. [2] found that the organic manure enhanced grain yield and total N uptake of wheat compared with unfertilized plots.

3.3.4 SEED INDEX, GRAIN AND STRAW YIELD AND WATER USE EFFICIENCY

Table 3.4 demonstrates that the overall status of seed index (SI), wheat grain and straw yield, and water use efficiency (WUE) were affected by cultivated period and organic matter. Results showed that addition of FYM increased SI by 25.5 and 47.6% under 10 and 25 years cultivated periods, respectively. However, FYM alone increased SI by 36.6% (relative to untreated one) Cultivated periods had less significant effects on increasing SI (10.4%) when comparing 25 with 10 years cultivated periods. Based on the water consumptive use (WCU) values through growing season of wheat crop, data showed that the lowest and highest values were recorded at 10 years cultivated period with untreated FYM (2146.6 m³/fed/season) and FYM treated plot (2123.2 m³/fed/season) under 25 years. Data also indicated that the percentage of water saving according to increase in cultivated period from 10 to 25 years and addition of FYM were −1.6 and 1.3%, respectively.

For the grain and straw yield of wheat crop, results showed that the highest values were recorded at 10 years cultivated periods + treated FYM (2966.8 kg/fed) and at 25 years cultivated periods treated with FYM (3835.6 kg/fed). Regardless of effects of FYM, cultivated periods increased the grain and straw yields by about 57.6 and 8.3%. Whereas, FYM increased the grain and straw yield by about 39.8 and 58.8% compared to the control, respectively.

Based on WUE values of wheat grain and straw, data showed that cultivated periods had a significant effect on the WUE values of grain with a percentage increase of 59.8% and less significant on straw (9.8%). The opposite was true in case of effects of FYM, where WUE for grain and straw of wheat crop improved by about 41.4 and 60.8% in same sequences. Considering effects of addition of FYM on individually two studied factors, one can notice that effects in 10 years cultivated periods were more pronounced than 25 years, where the improvement under two factors were 75.3% (cultivated periods) and 46.5% (FYM addition). These finding are agreement with those obtained by Abbas et al. [1] and Enke et al. [6].

With respect to effects of soil pH, EC and FYM from side and soil water constants (FC, WP and AW) from the other side, data shows that there is a

TABLE 3.4 Effects of Cultivated Period and Organic Manure on Wheat Yield, Seed Index and Water Use Efficiency (WUE)

Cultivated period (A)	FYM addition (B)	SI	WCU m³/season	Yield (kg/fed)		FWUE (m³ water/kg yield)	
				Grain	Straw	Grain	Straw
10 years	−	37.2	2185.7	1562.4	1854.7	0.71	0.85
	+	46.7	2151.5	2966.8	3198.7	1.38	1.49
	mean	**42.0**	**2168.6**	**2264.6**	**2526.7**	**1.05**	**1.17**
25 years	−	37.4	2146.6	3302.8	2211.7	1.54	1.03
	+	55.2	2123.2	3835.6	3258.7	1.81	1.53
	mean	**46.3**	**2134.9**	**3569.2**	**2735.2**	**1.67**	**1.28**
LSD 5%	A or B	1.3	11.8	78.2	44.6	0.25	0.09
	A × B	0.9	8.7	45.6	37.1	0.14	0.05

FYM: farm yard manure, −: un-continuous, +: continuous, SI: seed index (g/1000 grain), WCU: water consumptive use, m³/fed/season), FWUE (m³/kg grain yield).

TABLE 3.5   Correlations Among Some Soil Properties

	HC	BD	FC	WP	AW
BD	0.999**				
FC	-0.991**	-0.983**			
WP	-0.978**	-0.968**	0.996**		
AW	-0.994**	-0.988**	0.999**	0.991**	
pH	0.790**	0.765**	-0.830**	-0.812**	-0.837**
EC	0.718**	0.736**	-0.687*	-0.707*	-0.675*
OM	-0.991**	-0.991**	0.971**	0.946**	0.980*
K, ppm	-0.331	-0.301	0.436	0.516	0.398
P, ppm	-0.664**	-0.652*	0.657*	0.601	0.681*
TSN, ppm	-0.919**	-0.899**	0.963**	0.976**	0.955**

*: significant at 5%, **: significant at 1%.

significant correlation in case of pH and EC, which it is positively correlated in FYM.

Simple correlation coefficients were estimated between WCU and wheat grain/straw yields, N/P/K/protein/carbohydrate contents in wheat grains; and N/P/K/protein contents in wheat straw Table 3.5. The negative correlations were observed among all these variables.

Highly significant correlation coefficients at 1% level were attained with grain yield (-0.991**), P content in grain (-0.920**), protein and carbohydrates content in grain (-0.993** and -0.954**) and P content in straw (-0.964**), while at significant level of 5% negative correlation coefficients were observed with straw (-0.734*), grain content from N and K (-0.762* and -0.736*). Also WCU was not significant correlated with N, K and protein content in wheat straw.

3.4   CONCLUSIONS

There is a strong need to use appropriate assessment techniques to determine the long-term effects of cultivation periods and continuous application of FYM on soil properties. They may have greatly increase wheat productivity compared to untreated ones, resulting in improved soil properties. Also, FYM improves root growth and hence create stable aggregate, which is reflected in improvement of water movement and retention in soil. FYM decreases soil pH as a result of decomposition process and activity of microorganisms, which affect directly in improving root growth.

3.5 SUMMARY

This study examined changes in hydrophysical and chemical properties of soils; and effects on wheat yield (grain and straw yield, N, P, K, Protein and carbohydrates contents) under two cultivated period of 10 and 25 year and addition of FYM in sprinkler irrigated wheat on a newly reclaimed soils, Nubaria, Beheira Governorate, Egypt.

It was observed that cultivation period has more pronounced effect than addition of FYM on soil water content constants (field capacity, wilting point and available water). Hydraulic conductivity values were strongly affected by cultivation period and addition of FYM. Wheat straw content from protein had a superior effect under 25 than 10 years cultivated periods with values of 61.9 and 6.7% compared to control, respectively. FYM alone-improved protein content in straw by about 31.9% compared to untreated one. Slight increase in straw protein content was attained in cultivated period by about 7.8%.

FYM individually, under two studied cultivated periods, is more effective under 10 years (28.0, 25.2; 15.1%) than under 10 years (25.1, 25.2; 15.1%) comparing to untreated FYM plots. While N, P and K content in wheat straw had unclear trend and the increases were 6.8, 23.23; 56.5% and 62.9, 6.0; 29.8 as a result of FYM addition under 10 and 25 years cultivated periods, respectively. The highest values of protein and carbohydrates content in wheat grains as affected by studied factors were 12.86 and 67.43% under cultivated period 25 years.

Cultivated periods had a highly significant effect on the field water use efficiency values of grain than the effect of FYM. The highest values of grain and straw yield were recorded at 10 years cultivated periods + treated FYM (2966.8 kg/fed) and 25 years cultivated periods treated with FYM (3835.6 kg/fed). Cultivated periods increased grain and straw yield of wheat crop by about 57.6 and 8.3%. Whereas, FYM increased grain and straw yield by about 39.8 and 58.8% compared to the control, respectively.

KEYWORDS

- aeration
- carbohydrate
- chemical properties
- distribution uniformity
- drip irrigation
- Egypt
- farm yard manure
- field water use efficiency
- grain yield
- hydrophysical properties
- irrigation water
- nutrient content
- protein
- root growth
- sandy soil
- soil compaction
- sprinkler irrigation
- sprinkler irrigation, permanent
- straw
- uniformity
- uniformity coefficient
- vegetative growth
- water use efficiency
- wheat
- wheat yield

REFERENCES

1. Abbas, G., A. Hussain, A. Ahmad and S. A. Wajid, 2005. Water use efficiency of maize as affected by irrigation schedules and nitrogen rates. *J. of Agric. and Soc. Sci.*, 339–342.

2. Abd El-Kader, A.A., S.M. Shaaban and M.S. Abd El-Fattah, 2010. Effect of irrigation levels and organic compost on okra plants (*Abelmoschus esculentus* L.) grown in sandy calcareous soil. *Agric. Biol. J. North Am.*, 1(3):225–231.

3. Allen, R. G., L. S. Pereira, D. Raes and M. Smith, 1998. *Crop evapotranspiration – Guidelines for computing crop water requirements.* FAO Irrigation and Drainage Paper 56. FAO – Food and Agriculture Organization of the United Nations, Rome.

4. Dalal, R.C. and R.S. Mayer, 1986. Long-term trends in fertility of soils under continuous cultivation and cereal cropping in South Western Queensland, I: Overall changes in soil properties and trends in winter cereal yields. *Aust. J. Soil Res.*, 24:265–279.

5. Dalal, R. C., 1982. Organic matter content in relation to the period of cultivation and crop yields in some subtropical soils. 12th International Congress. *Soil Sci.*, 6:59.

6. EL-Gindy, A. M. and A. A. Abdel-Aziz, 2003. Maximizing water use efficiency of maize crop in sandy soils. *Arab. Univ. J. Agric. Sci.*, 11:439–452.

7. Enke, Li., Y. Changrong, M., Xurong, H., Wenqing, S.H. Bing, D. Linping, L. Qin, L. Shuang and F. Tinglu, 2010. Longterm effect of chemical fertilizer, straw, and manure on soil chemical and biological properties in north-west China. *Geoderma,* 158:173–180.

8. Faithfull, N.T., 2002. *Methods in agricultural chemical analysis.* A practical handbook. CABI Publishing. Pages 84–95.

9. Fliessbach, A., H., K. Roland, R. Daneil, F. Robert and E. Frank, 2000. Soil organic matter quality and soil aggregates stability in organic and conventional soil. (http://orgrints.org/00002911/).

10. Franzleubbers, A.J., 2002. Soil organic manure stratification ratio as an indicator of soil quality. *Soil Tillage Res.*, 66:95–106.

11. Franzluebbers, A. J. and J. A. Stuedemann, 2008. Early response of soil organic fractions to tillage and integrated crop–livestock production. *SSSAJ*, 72(3):613–625.

12. Megh R. Goyal, 2012. *Management of Drip/Trickle or Micro Irrigation Management.* Apple Academic Press Inc., Oaksville, ON, Canada.

13. Hellal, F.A., M. Abd El-Hady and A.A. M. Ragab, 2009. Influence of organic amendments on nutrient availability and uptake by faba bean plants fertilized by rock phosphate and feldspar. *American-Eurasian J. Agric. & Environ. Sci.*, 6(3):271–279.

14. Israelsen, O.W. and V.E. Hansen, 1962. Flow of water into and through soils. In: *Irrigation principles and practices.* John Willey and Sons, Inc. New York, USA.

15. Kay, B.D., and Vanden Bygaart, A.J., 2002. Conservation tillage and depth stratification of porosity and soil organic matter. *Soil Tillage Res.*, 66:107–118.

16. Kladivko, E.J., 2001. Tillage systems and soil ecology. *Soil Tillage Res.*, 61:61–76.

17. Klute, A and Dirksen, A. 1986. Hydraulic conductivity. In: A. Klute (ed.), *Methods of Soil Analysi, Part1, Physical and mineralogical methods.* Pages 678–734, Monograph 9 by ASA and SSSA, Madison, WI.

18. Klute, A. 1986. Water retention: Laboratory methods. In: A. Klute (ed.), *Methods of Soil Analysi, Part1, Physical and mineralogical methods.* Pages 635–662, Monograph 9 by ASA and SSSA, Madison, WI.

19. Lal, R., 2007. Anthropogenic influences on world soils and implications to global food security. *Adv. Agron.,* 93:69–93.

20. Michael, A.M., 1978. *Irrigation theory and practice.* Vikas publishing house Pvt. Ltd., New Delhi.

21. Mubarak, A.R., and Rosenani, A.B., 2003. Soil Organic manure fractions in the humid tropics as influenced by application of crop residues. *Commun. Soil Sci. Plant Anal.,* 34:933–943.

22. Phocaides, A., 2000. *Technical Hand Book on Pressurized Irrigation Techniques.* Food And Agriculture Organization Of The United Nations, Rome, Italy.

23. Randall, G.W., and Iragavarapa, T.K., 1995. Impact of long-term tillage systems for continuous corn on nitrate leaching to tile drainage. *J. Environ. Qual.,* 24:360–366.

24. Rebecca, B., 2004. *Soil Survey Laboratory Methods Manual.* Soil Syrvey Investigations Report No. 42 by USDA Natural Resources Conservation Services.

25. Saeed, A.B., and Eissa, H.Y., 2002. Influence of tillage on some properties of heavy cracking clay soils and sorghum yield in the mechanized rain-fed agriculture. *UK J. Agric. Sci.,* 10:267–276.

26. SAS Institute, 2001. *SAS statistics users' guide.* Release 8.2. SAS Institute, Cary, NC, USA.

27. Schumacher, T.E., Lindstrom, M.J.L., Eynard, A., and Malo, D.D., 2000. Tillage system effects on soil structure in the Upper Missouri river basin. In: Proceedings of the 15th Conference of ISTRO, Fort Worth, Dallas, USA, 2–7 July.

28. Tawfik, M.M. and A.M. Gomaa, 2005. Effect of organic and biofertilizer on growth and yield of wheat plants. *Journal Agric. Resource,* 2(2):711–725.

29. Yassen, A.A., S.M. Khaled and M.Z. Sahar, 2010. Response of wheat to different rates and ratios of organic residues on yield and chemical composition under two types of soil. *Journal of American Science,* 6(12):858–864.

30. Yemm, E.W. and A.J. Willis, 1954. The estimation of carbohydrates in plant extracts by anthrone. *Biochemistry,* 57:508–514.

PART II

CLOSED CIRCUIT TRICKLE IRRIGATION DESIGN

CHAPTER 4

DESIGN CONSIDERATIONS FOR CLOSED CIRCUIT DESIGN OF DRIP IRRIGATION SYSTEM

HANI A. MANSOUR

CONTENTS

In this chapter: 1 feddan = 0.42 hectares = 60 × 70 meter = 4200 m² = 1.038 acres = 24 kirat. A feddan (Arabic) is a unit of area. It is used in Egypt, Sudan, and Syria. The feddan is not an SI unit and in Classical Arabic, the word means 'a yoke of oxen': implying the area of ground that can be tilled in a certain time. In Egypt the feddan is the only nonmetric unit, which remained in use following the switch to the metric system. A feddan is divided into 24 kirats (175 m²). In Syria, the feddan ranges from 2295 square meters (m²) to 3443 square meters (m²).

In this chapter: One L.E. = 0.14 US$. The Egyptian pound (Arabic: ʿGeneh Masri-EGP) is the currency of Egypt. It is divided into 100 piastres, or (Arabic: 100 kersh), or 1,000 Millimes (Arabic:ʿMillime). The ISO 4217 code is EGP. Locally, the abbreviation LE or L.E., which stands for (Egyptian pound) is frequently used. E£ and £E are rarely used. The name Geneh is derived from the Guinea coin, which had almost the same value of 100 piastres at the end of the nineteenth century.

4.1   INTRODUCTION

Nowadays, shifting towards using more modified irrigation methods for both saving energy and water is a must. Hence increasing water and energy use efficiency by decreasing their losses in the traditional irrigation systems has been a challenge for irrigation designers.

About 75% of the global freshwater is used for agricultural irrigation. Most of the water is applied by conventional surface irrigation methods. According to US Census Bureau 2002, out of total irrigated land of 52,583,431 acres in the US, only 2,988,101 acres of land were irrigated by trickle/trickle irrigation in 2003 (about 5.68%). If the acreage under trickle irrigation can be increased, the most valuable and limited water resources can be saved substantially. In addition to substantial water saving, water can be applied where it is most needed in a controlled manner according to the requirements of crops in drip irrigation [23].

Trickle irrigation has advantages over conventional furrow irrigation as an efficient means of applying water, especially where water is limited. Vegetables with shallow root systems and some crops like corn (*Zea mays* L.) respond well to trickle irrigation with increased yield and substantially higher fruit or fiber quality with smaller water applications, justifying the use of trickle irrigation [16]. However, high initial investment cost of trickle irrigation systems needs to be offset by increasing production to justify investment over furrow irrigation systems. The main components of a trickle irrigation system are: trickle polyethylene (PE) tubes with emitters equally spaced along the lateral length, pump, filtration system, main lines, manifold, pressure regulators, air release valve, fertigation equipment. A pump is needed to provide the necessary pressure for water emission.

Distributed uniformity of water and nutrients along the laterals in traditional trickle irrigation systems are negatively affected with large reduction in pressure at the ends of laterals. Accordingly, plant growth and yield follow the same trend. This results in reduction in water, energy, nutrients and water use efficiencies. In addition, Egypt is facing problem of fast growing population, limited water resources, and dry hot climate.

Recently, the trickle irrigation lateral lines are assembled of plastic tubes and have become increasingly common in irrigated areas in Egypt.

They make about 80% of all tubes installed to setup the trickle irrigation lateral lines. New materials, technologies to manufacture tubes and assembly techniques are being developed. Trickle irrigation lateral lines are installed using socket polyethylene (PE) connectors that are manufactured using the continuous extrusion method. The inner surface of such tubes is formed using compressed air to give hydraulically smooth surface. The surface roughness, of previously manufactured tubes using the extrude method, was higher and depended on the manufacturing conditions. Therefore, numerous empirical formulas were developed to calculate frictional losses.

The assessments of losses in joints were based on inaccurate assumptions. The flow through trickle irrigation lateral lines is not free-surface type. The layer of air provides additional resistance that depends on the degree of filing of the tube. The analysis of plastic tubes has been performed only for smooth tubes with no joints. Adjusted flow formulas for pipes have been used for trickle irrigation lateral lines. Hydraulic calculations and such formulas suit well, when the values of Reynolds number (Re) are high. When the filling of trickle irrigation lateral lines is low, the Re values are small.

This chapter is an abbreviated version of PhD dissertation by the author [54]. The chapter discusses research studies on the effects of three trickle irrigation closed circuits and three lateral line lengths on:

- solution to the problem of pressure reduction at the end of lateral lines;
- comparison between two types of trickle irrigation circuits with traditional trickle system as a control;
- some hydraulic parameters: Pressure head, friction loss, flow velocity and velocity head (or dynamic head);
- variations in discharge, uniformity coefficient, and coefficient of variation of an emitter;
- corn growth and productivity, water and fertilizers use efficiencies under field conditions; and
- cost analysis of corn production, economic net income and physical net income due to modified trickle irrigation systems.

4.2   REVIEW OF LITERATURE

4.2.1   *TRADITIONAL TRICKLE IRRIGATION SYSTEM*

In Egypt, the first trickle irrigation system was installed and tested in 1975. However, it was operated at very low-pressure head of 40 cm [28]. The trickle irrigation is described using low discharge from small diameter orifices that are either connected to or a part of distribution lateral lines placed on the ground surface or immediately below the soil surface [76]. Trickle irrigation is defined as a slow application of water on the surface or beneath the soil by systems, namely: Surface trickle, subsurface trickle, bubbler spray, mechanical-move, or pulse. Water is applied in discrete or continuous drops, tiny streams, or miniature spray through emitters or applicators placed along the lateral line near the plant [61]. Larry [51] described the trickle irrigation system as a frequent slow application of water on the land surface or subsurface soil in the root-zone of a crop. He also stated that trickle irrigation encompasses several methods of irrigation, including trickle, surface, spray and bubbler irrigation system.

Several problems have been encountered: In the mechanics of applying water with emitters for some soil types; water quality and environmental conditions. Some of the more important possible disadvantages of trickle irrigation compared to other irrigation methods include: (i) emitter clogging, (ii) damage by rodents, insects or other animals, (iii) salt accumulation at the periphery of wetted zone, (iv) inadequate soil, water movement and plant-root development, and (v) high initial cost of installation and technical limitations [44]. Emitter clogging can cause poor uniformity of water application [47]. A special equipment's and methods for controlling clogging, as well as size of pipes, emitter type, valve type, etc., often increases the cost per unit crop area compared to solid-set sprinkler system [44].

4.2.2   *HEAD LOSSES FOR LATERAL LINES FOR TRICKLE IRRIGATION SYSTEM*

The local head loss is mainly due to friction losses in PE pipes and changes in water temperature in the laterals. Friction loss due to the velocity of water can be determined using Darcy-Weisbach equation. Although a single emitter generally produces a small local head loss, yet the total local head losses can become a significant fraction of the total energy loss due to the high number of emitters along a lateral [68]. The emitter flow variation may be caused by variation in injection pressure, heat instability during the manufacturing process, and a heterogeneous mixture of materials used in the production [48].

The effects of clogging in emitters and laterals on the water flow through all laterals have been modeled [11]. Researchers have found that simulated discharge from laterals was decreased due to clogging while the water flow through laterals was increased under no clogging. In addition to decreases in discharge for clogged emitters, the model showed an increase of pressure at the manifold inlet. Due to the increased inlet pressure, a lower discharge rate of the pump was observed. Reductions in the emitter flow ranging from 7 to 23% has been reported [12]. Scouring velocity was reduced from 0.6 m/sec to 0.3 m/sec due to clogging. Lateral lines also developed some slime build-up, as reflected by the reduction in scouring velocities.

Design charts based on spatially varied flow have been developed for a lateral with a longitudinal slots. The solution did not take into consideration the presence of a laminar flow in a considerable length of the downstream part of the lateral [78, 86].

Hathot et al. [42] provided a solution based on uniform emitter discharge, but took into account the change in velocity head and the variation in Reynolds' number. They used the Darcy-Weisbach friction equation to estimate friction losses. Hathot et al. [43] considered individual emitters with variable outflow and presented a step-by-step computer program for designing either the diameter or the lateral length. They considered the pressure head losses due to the emitter's protrusion. These head losses occur when the emitter barb protrusion obstructs the water flow. Three sizes of emitter barbs were specified: small, medium and large – with an area equal or less than 20 mm^2 for small, 21–31 mm^2 for medium, and equal to or more than 32 mm^2 for large emitter barb [79].

4.2.3 RELATIONSHIP BETWEEN EMITTER DISCHARGE RATE AND PRESSURE HEAD

Smajstrla et al. [68] stated that a basic component of emitter characteristics is the discharge rate (Q) vs. pressure head (H) relationship. The development of a Q versus H curve for emitter plays an important role in the emitter type selection and system design. They developed following power equation between the Q and the H:

$$Q = CHx \qquad (1)$$

where: Q is the emitter discharge rate (lph), C is the emitter constant, H is the working pressure head (m), and x is the emitter discharge exponent. Exponent x is an indication of the flow type and emitter type. It is an indirect measure of the sensitivity of discharge rate to the change in pressure. The value of x typically ranges between 0.0 to 1.0. A low value indicates a low sensitivity and a high value indicates a high sensitivity. They also indicated that the major sources of emitter discharge rate variations are emitter design, the material used to manufacture the trickle tubing, and precision. Kirnak et al. [48] investigated hydraulic characteristics of five commercial trickle irrigation laterals and found that these characteristics varied widely as a function of emitter design. Normally, a pump is used to develop the necessary operating pressure for the emission of water and also to protect the trickle lines from clogging.

4.2.4 TRICKLE IRRIGATION HYDRAULIC AND UNIFORMITY COEFFICIENTS

The major factors affecting trickle irrigation uniformity are [57]: (1) Manufacturing variations in emitters and pressure regulators; (2) Pressure variations due to elevation changes, (3) Friction head losses throughout the pipe network, (4) Emitter sensitivity to pressure and irrigation water temperature changes, (4) Emitter type and (6) Emitter clogging. Similarly, according to The manufacturer's coefficient of emitter variation (CVm) values below 10% are considered acceptable, and values >20% are unacceptable [6]. The emitter discharge variation rate (qvar) should be evaluated as a design criterion in trickle irrigation systems; qvar < 10% may be regarded as good and qvar >20% as unacceptable [17, 83]. Table 4.1 shows that acceptability depends on the range of statistical uniformity.

The acceptability of micro irrigation systems has also been classified according to the statistical parameters (Uqs and EU) namely: EU = 94%-

TABLE 4.1   Degree of Acceptability Based on Statistical Uniformity [7, 39, 40]

Degree of Acceptability	Statistical Uniformity, Us (%)
Excellent	100–95
Good	90–85
Fair	80–75
Poor	70–65
Unacceptable	< 60

100% and Uqs = 95%-100% are excellent; and EU <50% and Uqs <60% are unacceptable [5, 39].

Ortega et al. [64] calculated emission uniformity (EU), pressure variation coefficient (VCp), and flow variation coefficient per plant (VCq) for localized irrigation systems and reported that these values were 84.3% of the EU, 0.12 of VCp, and 0.19 of VCq. They classified the systems, unacceptable for VCq >0.4 and excellent for VCq <0.1.

In addition to pressure variation along drip irrigation tapes, variation in emitter structure or emitter geometry has been known to cause poor uniformity of emitter discharge [3, 48, 83].

4.2.5   DESIGN OF LATERALS AND PREDICTION OF PRESSURE HEAD

The inlet pressure is one of the most important factors in trickle irrigation design. If the inlet pressure head becomes greater than the required pressure head, it may cause backflow. And if the inlet pressure head becomes lower than the total required pressure head, it may create negative pressure at the lateral, which will affect the distribution uniformity. Consequently, to avoid both problems, the inlet pressure head must be determined precisely to balance the energy gain due to inlet flow and the total required pressure head within the lateral.

Many researchers [23, 41, 43, 64, 85] have attempted a mathematical approach to calculate the inlet pressure head. In any irrigation system, the energy required to operate the system depends on the required head and the system discharge. Gerrish et al. [37] indicated that the relationship between the flow rate and the pressure head is nonlinear in the transition and the turbulent flow types. Also, they proposed a method to incorporate pipe components into the hydraulic network analysis by adding their contribution to the nodal equations instead of treating them as separate items.

Von Bernuth [77] used the following Darcy-Wiesbach equation to calculate the friction head losses for a continuous flow through plastic pipe.

$$h_{loss} = [8f_sQl] \div [\pi g D^2] = f_s \; x \; l/D \; x \; [V^2/2 \; g] \qquad (2)$$

where: h_{loss} = head loss (m); f_s = Coefficient of friction (m/100 m); Q = Flow through the pipe (l·h⁻¹); l = Pipe length (m); g = acceleration due to gravity (= 9.81 m/sec⁻²); and D = Pipe inside diameter (mm). Hathoot et al. and others [6, 7, 8, 16, 42, 43, 77] used the Darcy-Wiesbach equation to calculate the value of friction coefficient, f_s, based on laminar, transient or turbulent flow.

The head loss due to elbows, tees, fittings, and valves can significantly affect the pressure in an irrigation network [81]. A computer model has been developed to optimize the irrigation system design for small areas in South Dakota, USA [62]. The model took into consideration crop type, soil type, irrigation interval, system layout, and pressure requirements of the emitter. Some of the parameters needed for the system design were calculated using the generalized equation for predicting parameters: the wetting diameter, the shortest irrigation interval, etc.

4.2.6   THE CORN VEGETATIVE GROWTH AND CORN YIELD

Corn (*Zea Mays L.*) is cultivated in areas lying between 58°N and 40°S latitudes and at an altitude up to 3,800 m above the mean sea level. The corn crop is irrigated worldwide. USA is the main maize producing country [30, 60].

Egypt has plans to use its limited water resources efficiently and overcome the gap between supply and demand. In the old lands of the Nile Valley and Delta, most farmers still use primitive methods of irrigation, fertilization, and weed and pest control practices. The application of fertilizers is usually by manual labor with low efficiency, resulting in high cost of production and environmental pollution problems [2]. Abou Kheira [2] stated that Corn (*Zea Mays L.*) is one of the most important cereals, both for peoples and animals consumption, in Egypt and is grown for both grain and forage. The questions often arise, "What is the minimum irrigation capacity for irrigated corn? And what is the suitable irrigation system for irrigating corn?" These are very hard questions to answer because they greatly depend on the climatic/crop/soil/local/ and economic factors, and yield goal necessary for profitability.

In Egypt, he irrigation water requirements of maize ranges from 500 to 800 m^3 per acre to achieve maximum production with a high yielding variety in clay loam soil [26]. On the coarse texture soils, maize production was increased with a combination of deep tillage and the incorporation of hay as mulch, and increase in irrigation depth [38].

Other research scientists [25, 32, 33] have found that irrigation of maize is almost important, from the appearance of the first silk strands until the milky stage in the maturation of the kernels on the cob. Once the milky stage has occurred, the appearance of black layer development on 50% of the maize kernels is a sign that the crop has fully ripened. These aforementioned criteria were used in the experimental plot for the total irrigation process in this chapter.

Most research projects in Egypt have studied the effects of irrigation on corn yield using sprinkler irrigation or furrow irrigation. In contrast, only a few studies have been made in maize cultivation under trickle irrigation [25, 33, 34]. These few studies used the evaporation to pan method to calculate the amount of water needed for irrigation. This method was used in England, in 2001, for irrigation scheduling up to 45% of the irrigated areas of the country in outdoor cultivation [80]. Trickle irrigation also permits use of many available tools for soil moisture measurement [19, 31]. Electronic and electro hydraulic irrigation controllers give the possibility of complete automation of drip irrigation networks [21, 31, 39, 40].

4.2.7 WATER USE AND FERTILIZER USE EFFICIENCIES

Water use efficiency (WUE) of corn depends on the physiological characteristics of maize, genotype, soil characteristics such as soil water holding capacity, meteorological conditions and agronomic practices. To improve WUE, integrative measures should aim to optimize cultivar selection and agronomic practices. The most important management interaction in many drought-stressed corn environments is between soil fertility management and water supply. In areas subject to drought stress, many farmers are reluctant to adopt measures to reduce risks of economic loss by applying fertilizer, and strengthening the link between drought and low soil fertility [8]. Ogola et al. [63] reported that the WUE of corn was increased by application of nitrogen. They added that corn plants are especially sensitive to water stress because their root system is relatively sparse.

Laboski et al. [50] found that corn yield responded to amount of water applied by trickle irrigation. It is therefore essential to use best trickle irrigation management practices. Increasing the plant population density usually increased corn grain yield until an optimum number of plants per unit area was reached by Ref. [45]. Fulton [35] also reported that higher corn plant densities produce higher grain yields. Plant densities of 90,000 corn plants/ha are common in many regions of the world [58].

The nutrient use efficiency (NUE) of plant depends on fertilizer application rate, method, time, type of fertilizer, crop and soil factors. Proper method and time of fertilizer application is inevitable to reduce the loss of plant nutrients and is important for an effective fertility programs. Nitrogen fertilizers should be applied in split doses for the long season crops. Similarly nitrogen should not be applied to sandy soil in a single dose, as there are more chances for nitrate leaching [13]. Phosphate fertilizers applications are also of great concern when applied to soil. They are often fixed or rendered unavailable to plants, even under the most ideal field conditions. In order to prevent rapid reaction of phosphate fertilizer with the soil, the materials are commonly placed in localized bands. To minimize the contact with soil, pelleted or aggregated phosphate fertilizers are also recommended by Brady [15]. He also reported that much of the phosphate is used early stages of plant in row crops. Similarly data collected on the yield of maize showed that application of all phosphorus at sowing was better than its late application. Memon [55] concluded that phosphorus uptake by roots depends upon the phosphorus uptake properties of roots and the phosphorus supplying properties of soil. He also added that maximizing the uniformity of water application is one of the easiest way to save water, at the farm level. It is too frequently forgotten by the farmers. The evaluation of the emission uniformity of the trickle system should be done periodically.

In comparison studies among different irrigation systems [53], it was observed that the increases in both WUE and water utilization efficiency of grape vines in the 2nd season compared to the 1st season were the maximum under drip irrigation system (42, 43%, respectively), followed by the low head bubbler irrigation system (40.7, 37%), while the minimum increases in WUE and water utilization efficiency were (30.6, 32%, respectively) under gated pipe irrigation system. The increases in fertilizers use efficiency (FUE) of N/ P_2O_5/K_2O in the 2nd season compared to the 1st season of grape vines were (24%, 23%, 28%), (22%, 21%, 27%) and (9%, 8%, 14%) under drip irrigation system, low head bubbler irrigation system and gated pipe irrigation system, respectively.

4.2.8 ECONOMIC ANALYSIS FOR TRICKLE IRRIGATED MAIZE

Trickle irrigation offers many unique features of agricultural technologies and economic development [61]. Many authors have studied the effects of irrigation method, irrigation levels, fertilizer treatment and plant species on the net income [20, 22, 24, 27, 29, 56, 66, 71–75, 88 54]. The net income had been over estimated in some of the previous studies, which attributed to missing one or more of the fixed costs, that is, interest on the capital costs, land rent, and subsidized free water to the farmers.

The maximum and the minimum net profit from grape crop were 3335 and 1414 LE/fed. Under trickle irrigation and gated pipe irrigation systems, respectively [53, 74]. El-Shawadfy [29] indicated that depending on irrigation method, irrigation level and bean varieties, the maximum and minimum net income were 5751 and 2045 LE/fed., respectively. The maximum and minimum net income from garlic were 4521 and 709 LE/fed., respectively, depending on the irrigation treatment, phosphorous treatment and type of fertilizer injector [29, 66].

The production efficiency of dry beans was in the range of 1.22 to 2.14 kg of seeds per m^3 of irrigation water [73]. Tayel et al. [73] mentioned that the maximum and the minimum water price varied from 11.6 to 13.0 and from 2.5 to 3.5 LE per cubic meter of irrigation water. They added that this price of irrigation under trickle irrigation was affected by irrigation regime, phosphorous level and faba bean (Vicia Faba) varieties. In western Kansas – USA, surface trickle irrigation system gave lower returns than in-canopy center pivot sprinkler systems for corn production. Initial investment, system longevity, and corn yield affected economic returns rather than pumping costs and application efficiencies [66]. Good irrigation managements, irrigation scheduling decisions and the appropriate evaluation of the economic impacts at farm level are the main constraints of the adoption of deficit irrigation strategies [24].

El-Amami [27] stated that the primary determinant of the cost of the irrigation system is the source of power or energy, while the revenue of capital investment was based on: scope of targets to be achieved, field topography, the availability of water sources, type of crop and soil, the area of farm number to be irrigated, and agricultural equipment requirements.

4.3 MATERIALS AND METHODS

4.3.1 EXPERIMENTAL SITE

The laboratory tests were conducted at Irrigation Devices and Equipment Tests Laboratory, Agricultural Engineering Research Institute, Agriculture Research Center, Giza, Egypt. The field experiment was conducted at the Experimental Farm of Faculty of Agriculture, Southern Illinois University of Carbondale (SIUC, latitude $37^{0.}73''N$ and $89^{0.}16''W$ and elevation of 118 m above mean sea level), Illinois, USA.

Field experiments were conducted in the cornfield during the growing season 2009 and 2010. Soil texture at the experimental site was clay loam [84]. Tables 4.2 and 4.3 indicate physical and chemical characteristics of soil [36, 49, 84]. Table 4.4 indicates chemical analysis of irrigation water analysis.

4.3.2 EXPERIMENTAL DESIGN AND IRRIGATION SYSTEM

A split-plot randomized complete block design with three replications was used for the laboratory and field experiments. Following treatments were considered (Figs. 4.1–4.3):

TABLE 4.2 Soil Physical Properties at Field Site for Carbondale, Illinois, USA

Soil depth cm	Particle size distribution, %				Texture class	F.C. %	W.P. %	AW %
	C. Sand	F. Sand	Silt	Clay				
0–15	3.4	29.6	39.5	27.5	C.L	32.35	17.81	14.54
15–30	3.6	29.7	39.3	27.4	C.L	33.51	18.53	14.98
30–45	3.5	28.5	38.8	28.2	C.L	32.52	17.96	14.56
45–60	3.8	28.7	39.6	27.9	C.L	32.28	18.61	13.67

* Particle Size Distribution after [84] and moisture retention after [36].
Leyend: C.L.: Clay Loam, F.C.: Field Capacity ($\ominus$w %),
W.P.: Permanent wilting point ($\ominus$w %),
AW: Available water ($\ominus$w %).

TABLE 4.3 Soil Chemical Properties at Field Site for Carbondale, Illinois, USA*

Soil depth cm	pH 1:2.5	EC dS/m	Soluble Cations, meq/L				Soluble anions, meq/L			
			Ca^{++}	Mg^{++}	Na^+	K^+	CO_3	HCO_3	SO_4	Cl^-
0–15	7.3	0.35	1.50	0.39	1.52	0.12	0.00	0.31	1.52	1.67
15–30	7.2	0.36	1.51	0.44	1.48	0.14	0.00	0.41	1.56	1.63
30–45	7.3	0.34	1.46	0.41	1.40	0.13	0.00	0.39	1.41	1.63
45–60	7.4	0.73	2.67	1.46	3.04	0.12	0.00	0.67	2.86	3.82

*Chemical properties after [49].

TABLE 4.4 Some Chemical Properties of Irrigation Water

pH	EC dS/m	Soluble cations, meq/L				Soluble anions, meq/l				SAR
		Ca^{++}	Mg^{++}	Na^+	K^+	CO_3	HCO_3	SO_4	Cl_-	
7.3	0.37	0.76	0.24	2.60	0.13	0.00	0.90	0.32	2.51	1.14

1. Three lengths of lateral line: 40, 60, and 80 m.
2. Three drip irrigation closed circuits (DIC):
   • closed circuits with one manifold for lateral lines (CM1DIS);
   • closed circuits with two manifolds for lateral lines (CM2DIS); and
   • traditional drip irrigation system (TDIS) as a control.

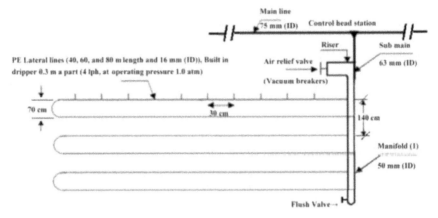

FIGURE 4.1 Layout of drip irrigation closed circuits with one manifold (CM1DIS) for lateral lines.

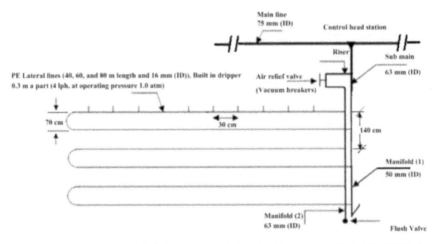

FIGURE 4.2 Layout of drip irrigation closed circuit with two manifolds (CM2DIS) for lateral lines.

Figure 4.4 indicates the direction of water flow inside the manifold and lateral lines in three DIC tests. Details of the pressure and water supply control have been described by Safi et al. [65]. Tests were conducted to come up with recommendations to resolve the problem of lack of pressure head at the end of lateral line in the TDIS.

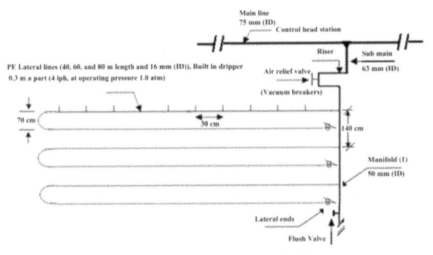

FIGURE 4.3   Layout of traditional drip irrigation system (TDIS).

4.3.3   COMPONENTS OF CLOSED CIRCUIT DRIP IRRIGATION SYSTEM

The components of closed circuits drip irrigation system included supply lines, control valves, supply and return manifolds, lateral drip lines, emitters, check valve, and air relief valve/vacuum breaker. Drip irrigation layout included the following components as shown in Figs. 4.1–4.3:

1. Control head: It was located at the water source. It consisted of a centrifugal pump 3"/3" driven by an electric engine (pump discharge of 80 m³/h and 40 m lift), sand media filter 48" (in pair), screen filter 2" (120 mesh), backflow prevention device, pressure regulator, pressure gages, flow meter, control valves and chemical injection device.

2. Main line: PVC pipes of 75 mm ID to convey the water from the source to the main control location in the field.

3. Submain lines: PVC pipes of 75 mm ID were connected to the main line through a 2" ball valve and pressure gage.

4. Manifold lines: PVC pipes of 50 mm ID were connected to the submain line through a gate valve of 1.5."

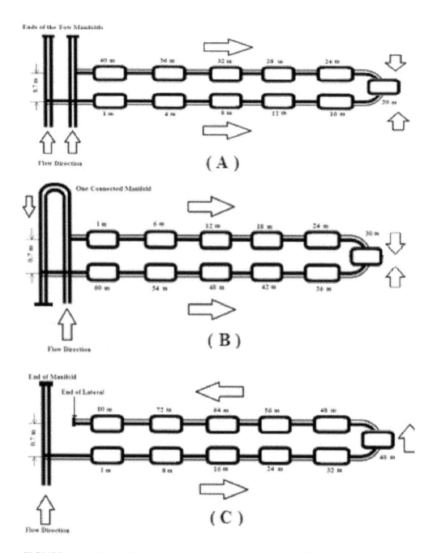

FIGURE 4.4 Water flow direction in lateral lines in different closed circuits lateral lengths (A and B), and traditional trickle system (C).

5. Lateral lines: PE tubes of 16 mm ID were connected to the manifold lines.
6. Emitters: GR in-line emitters (built in PE tubes 16 mm) of 4 lph discharge at an operating pressure of one atmosphere (Figs. 4.5 and 4.6). Emitter spacing down the row was 30 cm. Following equation

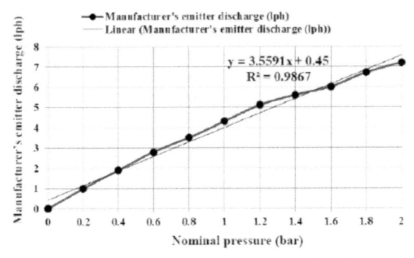

FIGURE 4.5   Discharge versus nominal pressure from the manufacturer's measurements.

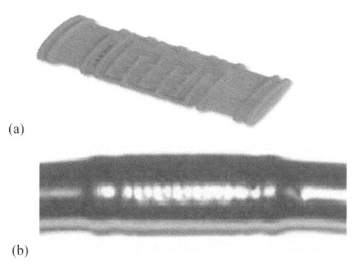

(a)

(b)

FIGURE 4.6   Built-in emitter: (a) the part, which installed inside lateral line; (b) built-in emitter of lateral line tube (inline drip line).

($R^2 = 0.987$) described the relationship between emitter discharge (Y) and nominal pressure head (X), based on the tests:

$$y = 3.5591(x) + 0.45 \qquad (3)$$

4.3.4 HEAD LOSS IN A PIPE

The flow rate through the pipe depends on pipe surface roughness and air layer resistance. The change in hydraulic friction coefficients depends on the Re number. Hydraulic losses for plastic pipes may be calculated as losses for hydraulically smooth pipes, and multiplied by correction coefficient that assesses losses due to pipe joints and air resistance. Coefficient of friction loss was given by Refs. [59, 67]. The head loss due to friction is calculated by Hazen-Williams equation:

$$\Delta H = JL/100 = 1.21 \times 10^{10} [Q/C]^{1.852}[LD]^{-4.87} \qquad (4)$$

where: ΔH = head loss due to friction (m); J = coefficient of head loss (m/100 m) or %; Q = flow rate is (m³/h); L = pipe length (m); D = inner diameter (ID Ø) of a pipe (mm); and C = Hazen-Williams coefficient of smoothness (the roughness) of the internal surface pipe = the range for a commercial pipe, 80–150. For polyethylene tubes with ID Ø <40 mm, C = 150 [59, 67]. Reynolds' number is defined below:

$$Re = vD /\mu \qquad (5)$$

where: Re = Reynolds' number; v = fluid velocity, m/sec; D = inner diameter Ø of lateral, m; and μ= kinematic viscosity of water = absolute viscosity/density = 1×10^{-6} m²/sec, at 20 °C. Based on values of Re, flow is laminar for Re $\leq$ 3000, flow is turbulent for Re > 3000. For a laminar flow: friction factor, f is 64/Re. For a turbulent flow: friction factor, f = $0.32[Re^{-0.25}]$. Velocity (v, m/s) is calculated from the continuity equation:

$$v = Q/A \qquad (6)$$

where: Q = lateral flow rate (m³/sec) = average flow rate per emitter × number of emitters; and A= $\pi D^2/4$ = cross sectional area of lateral (m²). The calculated emission flow rates were then compared with the measured values. Pressure head was measured by the needle pressure gage. Friction head losses and velocities were calculated with Hazen-Williams and continuity equations [Eqs. (4) and (6)].

4.3.5   IRRIGATION APPLICATION UNIFORMITY

The water application uniformity was calculated using discharge and pressure measurement data (Eqs. (7)–(9)). The following equations [17, 61] were used to compute statistical parameters and analyze uniformity of the subsurface trickle irrigation system. The simple method is still widely used.

$$q_{var} = \frac{q_{max} - q_{min}}{q_{max}} \tag{7}$$

$$CV = \frac{S}{q} \tag{8}$$

$$UC = \frac{\frac{1}{n} \sum_{i=1}^{n} |q_i - \bar{q}|}{\bar{q}} \tag{9}$$

where: q_{max} and q_{min} are maximum and minimum emitter discharge rates, respectively; CV = coefficient of variation; S = standard deviation; q = discharge rate; UC = statistical uniformity coefficient (%); q_{var} = variation in discharge; n is the number of emitters; = mean deviation of discharge; Δq = manufacturing coefficient of variation of discharge; q_i = discharge for i-th emitter

ASAE [5, 7] defined statistical uniformity as follows:

$$UC = 1 - \frac{\bar{q}}{q} \tag{10}$$

The coefficient of variation in Eq. (10) is referred to the depth of water applied. This statistical uniformity coefficient describes the uniformity of water distribution assuming a normal distribution of flow rates from the emitters.

4.3.6   USING COMPUTER PROGRAM FOR HYDRAULIC CALCULATIONS

HydroCalc irrigation system planning software is designed to help the designer to define the parameters of an irrigation system [40]. The user can run the program with any suitable parameters, review the output, and change input data in order to match it to the appropriate irrigation system set up (Figs. 4.7–4.9). Some parameters may be selected from a systematic list, whereas others are entered by the user according to the actual needs so that these do not conflict with the limitations of the program. The software package includes: main window, five calculation programs, one language setting window and a database that can be modified and updated by the user. HydroCalc includes several subprograms that are listed below:

- The Emitters program calculates the cumulative pressure loss, the average flow rate, the water flow velocity, etc. for the selected emitter. It can be changed to suit the desired irrigation system parameters.

FIGURE 4.7   HydroCalc irrigation planning [40].

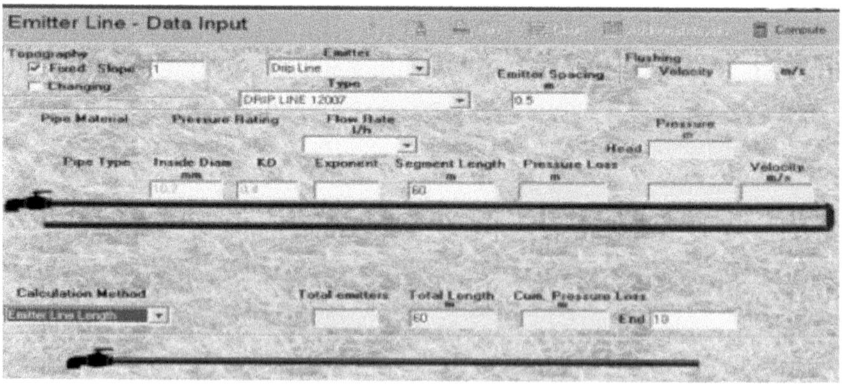

FIGURE 4.8   HydroCalc work sheet before the computation procedure.

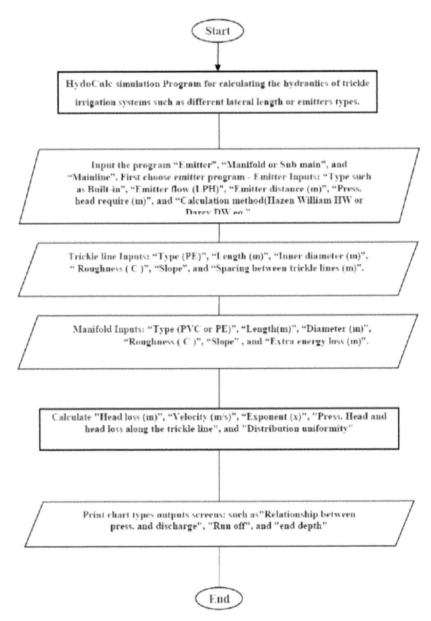

FIGURE 4.9 Flow chart showing the components of HydroCalc simulation program for planning, design, and the hydraulic analysis of trickle irrigation system.

- The SubMain program calculates the cumulative pressure loss and the water flow velocity in the submain distributing water pipe (single or telescopic). It can be changed to suit the required irrigation system parameters.
- The Main pipe program calculates the cumulative pressure loss and the water flow velocity in the main conducting water pipe (single or telescopic). It can be changed to suit the required irrigation system parameters.
- The Shape Wizard program helps transfer the required system parameters (inlet lateral flow rate, minimum head pressure) from the emitters program to the submain program.
- The Valve's program calculates the valve friction loss, according to the given parameters.
- The Shifts program calculates the irrigation rate and number of shifts needed, according to the given parameters.

The Emitters program is the first application, which can be used in the HydroCalc software program. There are four basic types of emitters, which can be used: trickle line, on line, sprinklers and microsprinklers. According to the previous selection, the user can opt for a specific emitter, which can be a pressure, compensated or a nonpressure compensated. Each emitter has its own set of nominal flow rate values available. After the previous mentioned fields are completed, the program automatically fills the following fields: "Inside Diameter," "ID" and "Exponent," values which cannot be changed unless the change will be made in the database. The segment length is next field in which the user must introduce a value. The end pressure represents the actual value of pressure at the farthest emitter.

The computation results also show the maximum lateral length for the given design conditions. "Flow Rate Variation" represents the third computation field, which can be executed to achieve the required flow variation (%) and will generate the maximum lateral length under these conditions. The common values for this field are between 10–15%. The last computation field is "Emission Uniformity (%)" which is similar to "flow rate variation," and is executed to achieve the maximum lateral length. The common value for this field is any value above 85%.

4.3.7   IRRIGATION SCHEDULING

Irrigation interval (I, days) is calculated with the following equation:

$$I = d/ETc \qquad (11)$$

where: I = irrigation interval (days); d = net irrigation depth applied in each irrigation (mm); and ETc = crop evapotranspiration (mm/day). Net irrigation depth applied in each irrigation is calculated as follows:

$$d = AMD \times ASW \times Rd \times p \qquad (12)$$

$$AW(v/v\ \%) = ASW(w/w\ \%) \times B.D \qquad (13)$$

where: AMD = allowable soil moisture depletion (%); ASW = available soil water (mm of water per m depth); Rd = effective root zone depth (m); d = irrigation depth (m), p = percentage of soil area wetted (%).; AW = available water on a volume basis (v/v, %); and B.D. = Soil bulk density (g cm^3). In this chapter, the irrigation interval was 4 days depending on the gross irrigation water requirements (IWRg), which calculated by Class A pan evaporation (Ep) for both closed circuit trickle irrigation and traditional trickle irrigation systems.

4.3.8   MEASUREMENTS OF SEASONAL EVAPOTRANSPIRATION FOR CORN (ETC)

The ETc was computed using Class A Pan evaporation method for estimating (ETo) on a daily basis. Climatic data were taken from the nearest meteorological station as shown in Table 4.6. The modified pan evaporation equation is as follows:

$$ETo = Kp \times Ep \qquad (14)$$

$$ETc = ET \times Kc \qquad (15)$$

$$Kr = GC + [½ (1 - GC)] \qquad (16)$$

where: ETo = reference evapotranspiration (mm/day); KP = pans coefficient of 0.76 for Class A pan placed in short green cropped and medium wind area; GC = ground cover percentage; Ep = daily pan evaporation (mm/day) = seasonal average value of 7.5 mm/day [4].

The reference evapotranspiration (ETo) is multiplied by a crop coefficient Kc for a particular growth stage to determine crop consumptive use at that particular stage of corn growth. The reduction factor (Kr) was calculated using Eq. (16). Basra [11] stated that the reduction factor of soil wetted (Ks) were taken from Table 4.5, according to effective spacing between laterals (m), emission-point spacing, discharge, and textured soils. The irrigation efficiency (Ea) was calculated as below [11]:

$$Ea = Ks \times Eu \tag{17}$$

where: Ea = irrigation efficiency; Eu = emission uniformity (%); and Ks = reduction factor of soil wetted.

Bazaraa [11] also stated that the gross irrigation water requirements IWRg (mm depth) can be calculated with Eq. (18):

$$IWRg = IWRn \times Ea + Lr \tag{18}$$

where: IWRg = the gross irrigation water requirements; IWRn = the net irrigation water requirements; and Lr = the extra amount of water needed for leaching (Table 4.6).

TABLE 4.5 Percentage of Soil Wetted by Various Discharge Rates and Spacing for a Single Row of Uniformly Spaced Distributors in a Straight Line Applying 40 mm of Water Per Cycle Over the Wetted Area

Spacing of Laterals, m	Emission point discharge					
	2 lph			4 lph		
	Recommended spacing of emission points along the lateral for coarse (C), medium (M), fine (F) textured soils					
	C	M	F	C	M	F
	0.3	0.7	1.0	0.6	1.0	1.3
	Percentage of wetted soil					
0.8	50	100	100	100	100	100

TABLE 4.6   The Water Requirements of Corn Grown on Carbondale Site, IL-USA, 2010

Variable	Month					
	Apr	May	Jun	Jul	Aug	Sep
Epan (mm/ day)	6.34	6.92	7.97	9.59	9.32	7.17
Kp	0.76					
Kc	1.05	1.08	1.15	1.17	1.22	1.25
Kr	0.45	0.90	0.95	1.00	1.00	1.00
ETo (mm/day)	4.82	5.26	6.06	7.29	7.08	5.45
ETc (mm/day)	2.28	5.12	6.62	8.53	8.64	6.82
Ks			100% (1.00)			
Eu			90% (1.11)			
Lr			10%			

Growth stage	Planting (Establishment)	Vegetative	Flowering		Ribbing and yield harvesting	
Duration of growth stage	9 April–30 April	1 May–12 Jun	13 Jun–28 Jul		29 Jul–15 Sep	
Number of days (Irri season)	22	43	46		38	
IRg (mm/month)	49.3	158.8	198.6	264.5	268.2	27.3
IRn (mm/month)	40.7	131.1	164.2	218.6	221.7	22.6

IRg = gross irrigation water; IRn = net irrigation water.

4.3.9   CORN CROP ESTABLISHMENT

Transgenic Corn (*Zea mays* L., GDH-LL3–272xB73, genotype) was planted at SIUC farm on April 9. The distance between rows was 0.7 m and 0.25 m between plants down the row. Each row was irrigated by a single straight lateral line in the closed circuits and traditional trickle irrigation plots. Figure 4.10 shows that the total experimental area was 4536 m². For each of the tested trickle irrigation circuits, plot area of Lateral

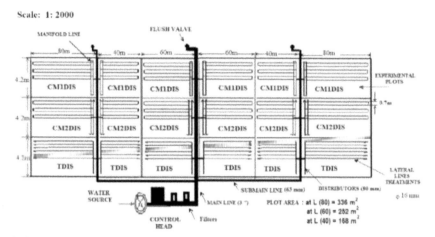

FIGURE 4.10 Layout of the field experimental plots: using three DIC (CM2DIS, CM1DIS and TDIS); and three LLL (LLL1 = 40 m; LLL2 = 60 m and LLL3 = 80 m).

lines lengths was 168, 252 and 336 m² with LLL1 = 40 m, LLL2 = 60 m and LLL3 = 80 m, respectively. Plants density was 24,000 plants per fed according to (ISU), North-east Research and Demonstration Farm. Irrigation of corn was terminated 11 days before the harvest. Corn was harvested on September 15.

Fertilization program was based on the recommended doses throughout the growing season (2009–2010) for drought tolerance, corn crop under the investigated irrigation systems using fertigation technique. These amounts of fertilizers NPK (20–20–10) were 60.48 kg/fed of (20% N) and 71.4 kg/fed of 20% K_2O. The dosage of P_2O_5 was 68.52 kg/fed of 10% P_2O_5. In all plots, weed and pest control measures were, according to recommendations for corn crop in Illinois state, USA.

4.3.10 PLANT MEASUREMENTS AND WATER USE EFFICIENCY

4.3.10.1 Plant Measurements

Plant measurements included: plant height (cm), leaf length (cm) by ruler, leaf area (cm²) by planimeter, number of leaves/plant, total grain weight (kg/fed), and Stover yield (kg/fed) by digital balance with four decimal numbers. All measurements and observations were started 21 days after planting, and were terminated on the harvest date. All plant samples were dried at 65 °C until constant weight was achieved. Grain yield was determined by hand harvesting the 8 m sections of three adjacent center rows in each plot on 2010 and was adjusted to 15.5% water content. In all treatment plots, the grain yields of individual rows were determined in order to evaluate the yield production uniformity among the rows.

4.3.10.2   Water Use Efficiency (WUE)

WUE is an indicator of the effectiveness of using irrigation water [46]. The WUE is defined as a ratio of crop yield (kg/ha) to the total amount of irrigation during the growing season (m³/ha). WUE of corn was calculated as follows:

$$\text{WUE of grain yield (kg/m}^3) = [\text{Total yield (kg/fed.)}]/ [\text{Total amount of IW (m}^3/\text{fed.)}] \quad (19)$$

4.3.10.3   Fertilizers Use Efficiency

Fertilizers use efficiencies (FUE: NUE, PUE, and KUE) are an indicator of effectiveness of fertilizer use. Fertilizers use efficiencies of corn was calculated as follows, according to Barber [10].

$$\text{FUE of corn (kg/kg)} = [\text{Total grain yield (kg/fed.)}]/ [\text{fertilizer type applied (kg/fed.)}] \quad (20)$$

4.3.11   ESTIMATION OF FEASIBILITY COST

4.3.11.1   Total Production Costs

Total production costs of corn yield included costs due to irrigation, fertigation, weed control, and pest control.

About Kheira [2] stated that the capital cost of trickle irrigation system was 5161 LE/fed. According to the market price of 2008 for equipment and installation. The annual cost (fixed and operating costs) of different DIC systems for corn yield and Stover yield were computed according to Abou Kheira [2]. The annual fixed cost of the irrigation system was calculated by Eq. (21):

$$F.C = D + I + T \qquad (21)$$

$$D = (I.C. - Sv) / E \qquad (22)$$

$$I = (I.C. + Sv) \times [I.R./2] \qquad (23)$$

$$O.C. = L.C + E.C + (R\&M) \qquad (24)$$

$$L.C = T \times N \times P \qquad (25)$$

$$E.C = Bp \times T \times Pr \qquad (26)$$

$$\text{Total annual irrigation cost} = F.C. + O.C. \qquad (27)$$

where: F.C. = annual fixed cost (LE/year); D = depreciation rate (LE/year) = 2.678% of initial cost; I = interest (LE/year) = 4% initial cost; T = taxes and overhead costs (LE/year); I.C. = initial cost of the irrigation system (LE); Sv = salvage value after depreciation (LE); E = expectancy life (year); I.R. = interest rate per year = 4% of initial cost; O.C. = annual operating costs (LE/year/feddan); L.C = labor costs (LE/year/fed); E.C = energy costs (LE/year/fed); R&M = repair and maintenance costs (LE/year/fed); L.C = annual labor cost (LE/year), T = annual irrigation time (hours/year), N = number of farmers per feddan; P = labor rate (LE/hour); E.C. = energy costs (LE/year); Bp = the brake power (kW/h); T = annual operating time in hours; and Pr = cost of electrical power (LE/kW-h).

Depreciation can be calculated with Eq. (22). The current interest is calculated with Eq. (23). Taxes and overhead costs were taken as 1.5–2.0% from the initial costs. Operating costs were calculated with Eq. (24). Labor to operate the system and to check the system components depends on irrigation operating time, which will change for each irrigation system according to irrigation water application rate. Labor cost was estimated with Eq. (25). Abdel-Aziz [1] stated that energy costs can be calculated by Eq. (26). Repair and maintenance costs were taken as 3% of the initial cost of trickle irrigation system. Total annual irrigation cost was estimated with Eq. (27).

4.3.12   STATISTICAL ANALYSIS

MSTATC program (Michigan State University) was used to carry out statistical analysis. Treatments mean were compared using the technique of analysis of variance (ANOVA) and the least significant difference (L.S.D.) between systems at 1%, [69].

4.4   RESULTS AND DISCUSSION

As electricity and heat energy, water flows within irrigation lines from points of higher energy to the ones of lower energy. It is well known that energy within the closed systems is constant, but can change from one form to another. Irrigation energy components within the irrigation laterals are: pressure head (hydraulic head), velocity head, friction head, gravity head and thermal head.

4.4.1   EFFECTS OF TRICKLE IRRIGATION CLOSED CIRCUITS (DIC) AND LATERAL LINE LENGTHS (LLL) ON PRESSURE HEAD AND SOME HYDRAULIC CHARACTERISTICS (FOR AN OPERATING PRESSURE = 1 ATM AND LAND SLOPE = 0%)

4.4.1.1   Pressure Head

Table 4.7 and Fig. 4.11 show the effects of the closed circuit designs and different lengths of lateral line on the pressure head (PH). It can be noticed in Fig. 4.11 that PH decreased along the LLL upto 5.1 and 6.3 indicating 18.5% variation between the highest and lowest pressure head under CM2DIS, CM1DIS and TDIS. It increased again to reach nearly its inlet

TABLE 4.7　Effects of Trickle Irrigation Closed Circuits (DIC) and Lateral Line Lengths (LLL) on Some Hydraulic Parameters of Lateral Lines at Operating Pressure = 1 atm and Land Slope = 0%

DIC	LLL	Pressure head, PH (m)	Friction loss, h_f (m)	Flow velocity, FV (m/sec)	Velocity head, VH (m)
CM2DIS	40	9.50 a	0.50 i	0.786 f	0.030 fg
	60	8.70 dc	1.30 f	1.033 c	0.054 c
	80	8.30 fe	1.70 d	1.376 a	0.096 a
CM1DIS	40	9.23 b	0.80 h	0.751 g	0.029 g
	60	8.33e	1.70 e	0.975 d	0.048 d
	80	7.50 h	2.50 b	1.332 b	0.090 b
TDIS	40	8.86 c	1.14 g	0.593 i	0.018 i
	60	7.99 g	2.21 c	0.722 h	0.027 h
	80	6.05 i	4.00 a	0.801 e	0.033 e
LSD at P= 0.01 X		**0.05**	**0.02**	**0.023**	**0.005**
Means	CM2DIS	8.83 a	1.17 c	1.065 a	0.060 a
	CM1DIS	8.35 b	1.67 b	1.019 b	0.056 ba
	TDIS	7.63 c	2.45 a	0.705 c	0.026 c
	LSD 0.01	0.12	0.06	0.041	0.007
Means	40	9.20 a	0.81 c	0.710 c	0.026 c
	60	8.34 b	1.74 b	0.910 b	0.043 b
	80	7.28 c	2.73 a	1.170 a	0.073 a
LSD 0.01		**0.13**	**0.07**	**0.022**	**0.003**

DIC: trickle Irrigation circuits, LLL: lateral line length, CM2DIS: closed circuit with two manifolds separately, CM1DIS: closed circuit with one manifold, TDIS: traditional trickle irrigation system. Values followed by the same letter are not significant at P = 0.01.

pressure head in both CM2DIS and CM1DIS, respectively. On the other hand, PH decreased continuously with distance from the lateral line inlet in TDIS. This may be due to the existence of two inlets in both CM2DIS and CM1DIS, which caused lowest drop along the LLL by about 5.1 to 6.3% between the lowest and highest values of pressure head values.

According to the Hazen-Williams equation, there is a direct relation between LLL and friction loss. Differences in PH between CM2DIS and

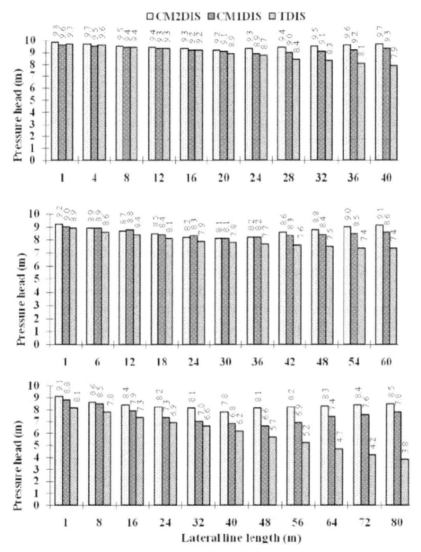

FIGURE 4.11 Effects of three irrigation circuit designs on pressure head along three lateral line lengths at operating pressure = 1.0 atm and slope = 0%. Leyend: CM2DIS = Closed circuit with two manifolds separately, CM1DIS = Closed circuit with one manifold, TDIS = Traditional trickle irrigation system, PH = Pressure head.

CM1DIS may be explained on the basis that lateral lines are supplied with water from two manifolds and one manifold, respectively. In other words, the inlet pressure was higher in CM2DIS compared to CM1DIS, due to doubling the cross sectional area of the manifold and that these are connected in parallel in CM2DIS. Whereas in CM1DIS, manifold is connected in series; and both manifold lengths (L, m) and resistance increased (*see* Figs. 4.1–4.4). Regardless of LLL, and according to the PH values, DIC can be ranked in the following ascending order: TDIS < CM1DIS < CM2DIS. Difference in PH between any two DIC values was significant at the 1% level. Concerning the depressive effects of LLL on PH, LLLs can be ranked in the following ascending order: LLL1 < LLL2 < LLL3. Differences in PH values between LLL1 from one side and both LLL2 and LLL3 from the other side were significant at the 1% level. This is due to the direct relation between friction and both lateral line discharge and its length. The effect of DIC × LLL on PH was significant at the 1% level except between the two interactions: CM2DIS × LLL3 and CM1DIS × LLL2. The highest (9.5 m) and the lowest (6.05 m) values of PH were achieved in the interactions: CM2DIS × LLL1 and TDIS × LLL3, respectively. It can be concluded that the allowable pressure drop between the maximum and minimum pressures along the lateral lines must be ≤1.1 meters under turbulent flow conditions. This is necessary for drip irrigation system to be economical and for acceptable water and fertilizer distribution along the lateral.

Data in Table 4.5 and Figs. 4.11 indicate that all LLL of 16 mm ID under TDIS and 80 m length under CM2DIS and CM1DIS are not recommended to avoid the high cost and the lower uniformity of water and fertilizer distributions along the LLL. Therefore, for 16 mm ID and 80 m length laterals, either LLL should be shortened or the inside diameter of LLL should be increased. As the flow rate in lateral line decreases with respect to its length due to emitter discharges from the lateral lines, the energy gradient line will not be a straight line but a curve of exponential type as shown in Figs. 4.12–4.14. This is in agreement with Wu and others [11, 82]. Wu et al. [82] mentioned that only the total friction drop ratio (ΔH/H) affected the shape of the energy gradient lines. From these figures, it is clear that all factors affecting the ratio (ΔH/H) including DIC and LLL in this study also affected the shape of the energy gradient lines.

According to Table 4.8, pressure head variations gave acceptable results in all cases, except interactions between CM1DIS × LLL3 and TDIS × LLL3.

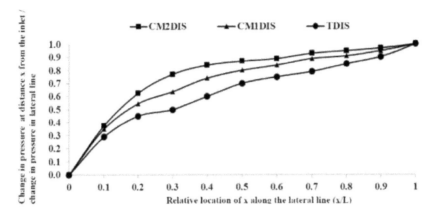

FIGURE 4.12 Dimensionless curve showing the friction drop patterns in trickle lateral line under three irrigation circuits for lateral line length = 40 m, operating pressure = 1.0 atm and slope = 0%. Leyend: DIC = trickle irrigation circuit, LLL = lateral line length, CM2DIS = closed circuit with two manifolds separately, CM1DIS = closed circuit with one manifold, TDIS = traditional trickle irrigation system, PH = pressure head.

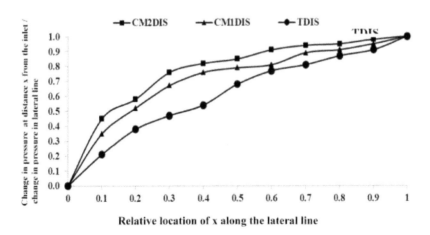

FIGURE 4.13 Dimensionless curve showing the friction drop patterns in trickle lateral line under three irrigation circuits at lateral line length = 60 m, operating pressure = 1.0 atm and slope = 0%. Leyend: DIC = trickle irrigation circuit, LLL = lateral line length, CM2DIS = closed circuit with two manifolds separately, CM1DIS = closed circuit with one manifold, TDIS = traditional trickle irrigation system, PH = pressure head.

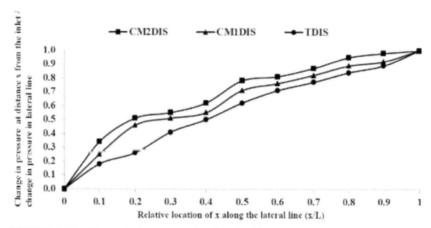

FIGURE 4.14 Dimensionless curve showing the friction drop patterns in trickle lateral line under three irrigation circuits at lateral line length = 80 m, operating pressure = 1.0 atm and slope = 0%. Leyend: DIC = trickle irrigation circuit, LLL = lateral line length, CM2DIS = closed circuit with two manifolds separately, CM1DIS = closed circuit with one manifold, TDIS = traditional trickle irrigation system, PH = pressure head.

TABLE 4.8   Effects of Trickle Irrigation Closed Circuits (DIC) and Lateral Line Lengths (LLL) on the Pressure Head (H) Variation

DIC	LLL	Hmax (m)	Hmin (m)	Hvar (%)	Acceptability by ASAE Standard 2003 [6]
	40	9.8	9.2	6.122	+++
CM2DIS	60	9.2	8.1	11.957	+++
	80	9.1	7.8	14.286	+++
	40	9.6	8.9	7.292	+++
CM1DIS	60	9.0	8.1	10.000	+++
	80	8.8	6.6	25.000	++
	40	9.7	7.9	18.557	+++
TDIS	60	8.9	7.4	16.854	+++
	80	8.1	3.8	53.086	++

DIC: Trickle irrigation circuit, LLL: Lateral line lengths, CM2DIS: Closed circuit with two manifolds separately, CM1DIS: Closed circuit with one manifold, TDIS: Traditional trickle irrigation system, Hmax: The highest-pressure head, H min: The lowest pressure head, Hvar: Pressure head variation.

+++: acceptable and ++: unacceptable.

4.4.1.2 FRICTION LOSS

Data in Table 4.7 is plotted in Figs. 4.12–4.16. Figures 4.15 and 4.16 indicate that the friction loss (FL) followed an opposite trend compared to PH (*see* section 4.4.1.1). Friction loss increased with distance from upstream of lateral inlet reaching its maximum at 50% to 60% of lateral length, then it decreased again up to the downstream end of lateral line in the case of CM2DIS and CM1DIS. In other words, the minimum values of friction loss existed at both the inlets and the ends of the lateral lines. This may be due to direct relation between friction loss from one side and its length and discharge from the other side. According to the friction loss values, DIC can be ranked in the following descending order: TDIS > CM1DIS > CM2DIS. D ifferences in friction loss between any two DIC systems were significant at 1% level. The ascending order of LLL1 < LLL2 < LLL3 illustrated the mean effect of LLL on friction loss. Differences in friction loss among LLL treatments were significant at the 1% level. The effects of the DIC × LLL on friction loss were significant at the 1% level.

The maximum and minimum values of friction loss were obtained in the interactions: TDIS × LLL3 and CM2DIS × LLL1, respectively. As the flow rate in lateral line decreases (with respect to its length due to dripper discharge from the lateral lines), the energy gradient line was of

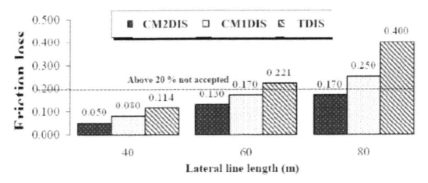

FIGURE 4.15 Effects of three closed circuits and lateral line lengths on friction loss. Leyend: DIC = trickle irrigation circuit, LLL = lateral line length, CM2DIS = closed

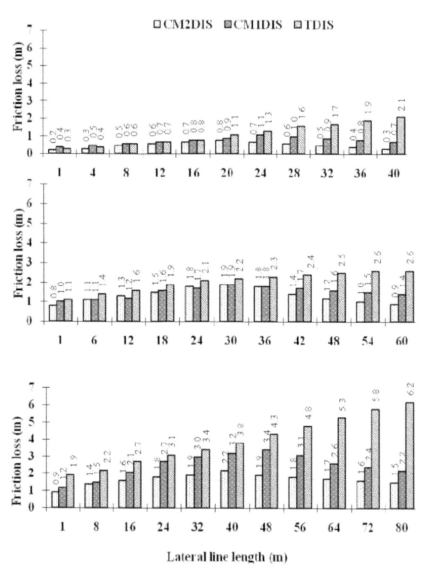

circuit with two manifolds separately, CM1DIS = closed circuit with one manifold, TDIS = traditional trickle irrigation system.

FIGURE 4.16 Effects of three irrigation closed circuit designs on friction loss along three lateral line lengths at operating pressure 1.0 atm and slope = 0%. Leyend: DIC = trickle irrigation circuit, LLL = lateral line length, CM2DIS = closed circuit with two manifolds separately, CM1DIS = closed circuit with one manifold, TDIS = traditional trickle irrigation system.

exponential type (Figs. 4.12–4.14). This is in agreement with Bazaraa [11] and Wu et al. [82, 83]. Wu et al. [82, 83] mentioned that only the total friction loss ratio ($\Delta H/H$) affected the shape of the energy gradient line. It is clear from Figs. 4.12–4.14 that all factors (including DIC and LLL), affecting the ratio, $\Delta H/H$, also affected the shape of the energy gradient lines.

4.4.1.3   FLOW VELOCITY (FV)

Data in Table 4.7 is plotted in Fig. 4.17 and it indicates the effects of DIC and LLL on flow velocity. The reader can conclude that the change in FV took the same trend as of pressure head (H), whereas, it was opposite to that for the friction loss. This may be due to the effect of DIC on H and friction loss. Also, increasing LLL increased its discharge and decreased the amount of water flowing through the lateral lines. The constant cross section area of laterals can be other reason.

According to the FV values, the DIC can be put in the following ascending order: TDIS < CM1DIS < CM2DIS. Differences in FV between any two DIC systems was significant at the 1% level. FV values varied from 0.722 to 1.376 m/sec (i.e., FV < 5 ft/sec) and this is necessary to avoid the effect of water hammer in the main and submain lines. However, in lateral line, it can cause silt and clay precipitation problems.

From the effects of LLL on FV, it is obvious that the FV through LLL3 exceeds that of LLL1, while that of LLL2 occupied intermediate position. Differences in FV among LLL treatments were significant at the 1% level. The effects of the DIC × LLL on FV were significant at 1% level. The maximum and minimum flow velocities were achieved for the interactions: CM2DIS × LLL3 and TDIS × LLL2, respectively.

4.4.1.4   Velocity Head or Dynamic Head

The velocity head (VH or dynamic head) is calculated from the equation: $VH = (v)^2/2\ g$ and $\equiv (m^2\ sec^{-2})/2(m\ sec^{-2}) \equiv m$, units of energy. Therefore, VH took the same trend as of flow velocity (FV). Based on data in Table 4.7 and Fig. 4.18 for velocity head values, DIC can be stated in the ascending

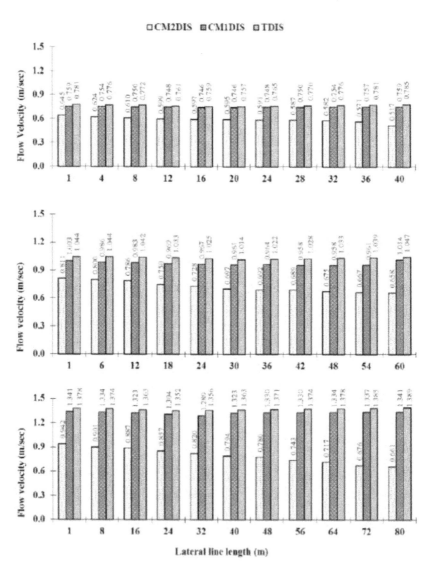

FIGURE 4.17 Effects of three irrigation closed circuit designs on flow velocity (FV) along three lateral line lengths at operating pressure 1.0 atm and slope = 0%. Leyend: DIC = trickle irrigation circuit, LLL = lateral line length, CM2DIS = closed circuit with two manifolds separately, CM1DIS = closed circuit with one manifold, TDIS = traditional trickle irrigation system.

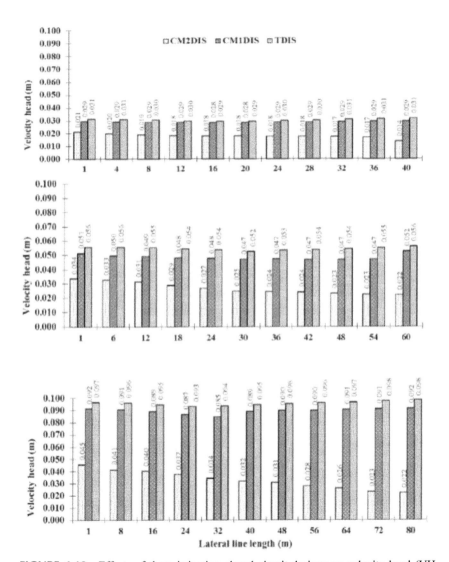

FIGURE 4.18 Effects of three irrigation closed circuit designs on velocity head (VH, dynamic head) along three lateral line lengths at operating pressure 1.0 atm and slope = 0%. Leyend: DIC = trickle irrigation circuit, LLL = lateral line length, CM2DIS = closed circuit with two manifolds separately, CM1DIS = closed circuit with one manifold, TDIS = traditional trickle irrigation system.

order: TDIS < CM1DIS < CM2DIS. Differences in VH among DIC were significant at the 1% level except that between CM2DIS and CM1DIS.

Based on the effects of LLL on velocity head, LLL can be written in the ascending order: LLL1 < LLL2 < LLL3. Differences in VH among LLL treatments were significant at the 1% level.

The effects of the DIC × LLL on velocity head were significant at the 1% level except CM2DIS × LLL2, CM1DIS × LLL1and TDIS × LLL3. The maximum and minimum values of VH were found in the following interactions: CM2DIS × LLL3 and TDIS × LLL1, respectively.

4.4.1.5   Variations in Emitter Discharge

The data in Tables 4.9 and 4.10 are plotted in Fig. 4.19. These data show the effects of DIC and LLL on emitter discharge variation (qvar) and emitter discharge (q_d). According to emitter discharge variation values [6], all cases except TDIS with LLL3 were acceptable. According to emitter discharge values, DIC can be stated in the ascending order: TDIS < CM1DIS < CM2DIS.

TABLE 4.9   Effects of DIC Design and LLL on Emitter qvar Percentage

DIC	LLL	qmax (lph)	qmin (lph)	qvar (%)	Acceptability, ASAE 2003 [6]
CM2DIS	40	4.23	4.1	3.07	+++
	60	3.77	3.65	3.18	+++
	80	3.76	3.66	2.66	+++
CM1DIS	40	4.11	4.04	1.70	+++
	60	3.65	3.45	5.48	+++
	80	3.63	3.49	3.86	+++
TDIS	40	3.49	2.8	19.77	++
	60	2.92	2.37	18.84	++
	80	2.55	1.79	29.80	+

DIC: trickle irrigation circuits, LLL: lateral line lengths, CM2DIS: closed circuit with two manifolds separately, CM1DIS: closed circuit with one manifold, TDIS: traditional trickle irrigation system, qmax: the highest discharge, qmin: the lowest discharge, qvar: emitter discharge variation. +++: excellent, ++: acceptable and +: unacceptable.

TABLE 4.10   Effects of DIC and LLL on Lateral Discharge and Uniformity of Emitters at Operating Pressure = 1 atm and slope = 0%

DIC	LLL	Emitter discharge (lph)	Lateral discharge (lph)	Uniformity coefficient, UC, %	Coefficient of variation (CV)	CV accept-ability ASAE 1996
CM2DIS	40	4.18 a	555.9 fe	97.74 a	0.081 g	+++
	60	3.72 c	744.0 c	95.14 cb	0.063 ig	+++
	80	3.71 dc	990.0 a	92.03 d	0.122 fe	++
CM1DIS	40	4.07 ba	541.0 g	95.73 b	0.071 hg	+++
	60	3.51 fe	702.0 dc	89.45 ef	0.162 ec	++
	80	3.59 e	958.0 ba	83.25 h	0.231 b	++
TDIS	40	3.21 g	426.0 i	88.27 f	0.183 de	++
	60	2.60 h	520.0 h	84.73 g	0.221 cb	++
	80	2.16 i	576.7 e	80.53 i	0.280 a	+
LSD 0.01 x		**0.18**	**80.33**	**1.18**	**0.042**	
Means	CM2DIS	3.87 a	762.35 a	94.97 a	0.089 c	+++
	CM1DI	3.72 ba	732.71 ba	89.47 b	0.155 b	++
	TDIS	2.66 c	507.22 c	84.51 cb	0.228 a	++
	LSD 0.01	**0.44**	**205.75**	**5.19**	**0.027**	
Means	40	3.82 a	507.78 cb	93.91 a	0.112 c	+++
	60	3.28 ba	655.64 b	89.77 ba	0.149 b	++
	80	3.15 cba	838.87 a	85.27 cb	0.211 a	++
	LSD 0.01	**0.77**	**177.05**	**6.91**	**0.028**	

DIC: trickle irrigation circuits, LLL: lateral line lengths, CM2DIS: closed circuit with two manifolds separately, CM1DIS: closed circuit with one manifold, TDIS: traditional trickle irrigation system.

+++ = Excellent, ++ = Good, + = Fair, and LSD 0.01: significantly different at P = 1%.

Differences in q_d between any two DIC were significant at the 1% level except that between CM2DIS and CM1DIS. This may be due to the effect of DIC on pressure head and friction loss. The data in Table 10 revealed no significant differences at 1% in q_d among the LLL. This is due to the stability of both lateral lines Ø (16 mm) and their slope. The effect of DIC × LLL indicated significant differences in q_d at the 1% level in most cases.

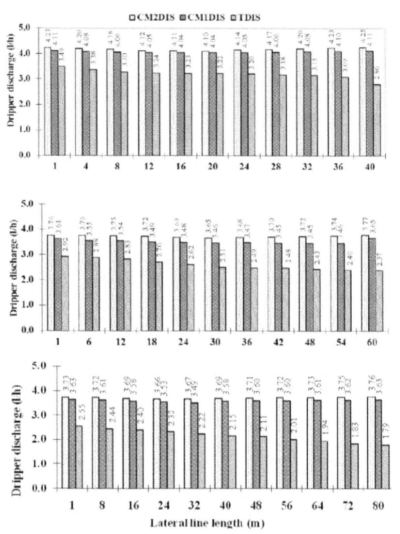

FIGURE 4.19 Effects of three irrigation closed circuit designs on emitter discharge along three lateral line lengths at operating pressure 1.0 atm and slope = 0%. Leyend: DIC = trickle irrigation circuit, LLL = lateral line length, CM2DIS = closed circuit with two manifolds separately, CM1DIS = closed circuit with one manifold, TDIS = traditional trickle irrigation system.

The maximum value of q_d was 4.18 lph and the minimum was 2.6 lph, and these were in the following interactions: CM2DIS × LLL1 and TDIS × LLL3, respectively.

4.4.1.6   Lateral Line Discharge

Data in Table 4.8 was plotted in Fig. 4.20 to illustrate the effects of DIC and LLL on discharge through the lateral line (QL). Regardless of LLL, the effects of DIC on QL can be summarized in the ascending order: TDIS < CM1DIS <CM2DIS. The ascending order, LLL3 < LLL2 < LLL1, showed that the differences in QL among LLL were significant at 1% level except between LLL1 and LLL2. Although LLL has no significant effects on q_d, the effect of LLL on QL was significant, due to the increasing number of emitters per lateral line with increasing length of lateral. Number of emitters were 133, 200 and 267 for the LLL1, LLL2, and LLL3 (40, 60, 80 m), respectively. The effects of the interaction DIC × LLL on QL was significant at the 1% level with few exceptions. The maximum value of QL (= 990 lph) and the minimum value (= 426 lph) were achieved in the interactions: CM2DIS × LLL3 and TDIS × LLL1, respectively.

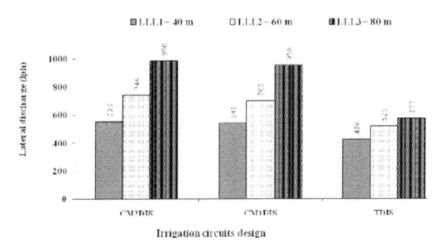

FIGURE 4.20   Effects of three irrigation closed circuit designs on lateral line discharge for three lateral line lengths at operating pressure 1.0 ATM and slope = 0%. Leyend: DIC = Trickle irrigation circuit, LLL = Lateral line length, CM2DIS = Closed circuit with two manifolds separately, CM1DIS = Closed circuit with one manifold, TDIS = Traditional trickle irrigation system; LLL1 = 40 m, LLL2 = 60 m, LLL3 = 80 m.

4.4.1.7   Uniformity Coefficient

Table 4.8 and Fig. 4.21 exhibits the role of DIC and LLL on uniformity coefficient (UC). The mean effects of DIC on UC can be put in the ascending order: TDIS ≤ CM1DIS < CM2DIS. Differences in UC among DIC were significant at the 1% level except that between CM2DIS and TDIS. Base of the effects of LLL on UC, regardless of DIC used, can be put in the ascending order: LLL3 ≤ LLL2 ≤ LLL1. Differences in UC among LLL were significant at the 1% level between LLL1 and LLL3 only.

It is worthwhile to mention that the values of UC took an opposite trend compared to that of QL, due to increasing of QL and LLL that reduced pressure head and increased friction loss. The effect of the interaction DIC × LLL on UC was significant at the 1% level, except between the interactions CM2DIS × LLL2 and CM1DIS × LLL2. The maximum value of UC was 97.74% and the minimum was 80.53% in the interaction CM2DIS × LLL1 and TDIS × LLL3, respectively.

The values of CV were acceptable in all cases except interactions below the line of acceptability as shown Fig. 4.21: (CM1DIS × LLL3), (TDIS × LLL2) and (TDIS × LLL3).

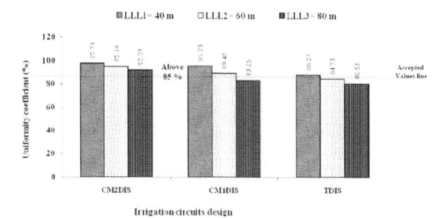

FIGURE 4.21   Effects of three irrigation closed circuit designs on uniformity coefficient (UC) for three lateral line lengths at operating pressure 1.0 atm and slope = 0%. Leyend: DIC = Trickle irrigation circuit, LLL = Lateral line length, CM2DIS = Closed circuit with two manifolds separately, CM1DIS = Closed circuit with one manifold, TDIS = Traditional trickle irrigation system; LLL1 = 40 m, LLL2 = 60 m, LLL3 = 80 m.

4.4.1.8   Coefficient of Variation for Emitter Discharge

Data on the effects of DIC and LLL on (coefficient of variation) CV are tabulated in Table 4.8 and plotted in Fig. 4.22. It is concluded that the trend of CV values was similar to that of QL, whereas it was opposite to that of UC. The effects of DIC on CV despite of LLL can be arranged in the ascending order: TDIS < CM1DIS < CM2DIS. Differences in CV among DIC were significant at the 1% level. Data indicated that the degree of CV acceptability according to [5] was excellent and good using CM2DIS and both CM1DIS and TDIS, respectively.

The Fig. 4.22 and Table 4.8 illustrates the effects of LLL on CV despite of DIC. The effects of LLL on CV can be summarized in the ascending order LLL3 < LLL2 < LLL1. The difference between any two LLL treatments was significant at the 1% level. It is concluded from the data that CV acceptability was excellent and good in LLL1 and both LLL2 and LLL3, respectively. The differences in CV values were insignificant at the 1% level among any of the following interactions: (CM2DIS × LLL1, CM2DIS × LLL2; CM1DIS × LLL1), (CM2DIS × LLL3, CM1DIS × LLL2; TDIS × LLL1) and (CM1DIS × LLL2; TDIS × LLL3). The highest value of CV was 0.28 and the lowest was 0.063 in the interactions: (TDIS × LLL3), and (CM2DIS × LLL2), respectively. Finally, the degree of CV

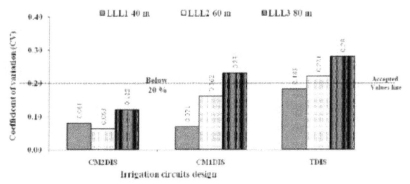

FIGURE 4.22   Effects of three irrigation closed circuit designs on coefficient of variation (CV) for three lateral line lengths at operating pressure 1.0 atm and slope = 0%. Leyend: DIC = Trickle irrigation circuit, LLL = Lateral line length, CM2DIS = Closed circuit with two manifolds separately, CM1DIS = Closed circuit with one manifold, TDIS = Traditional trickle irrigation system; LLL1 = 40 m, LLL2 = 60 m, LLL3 = 80 m.

acceptability of DIC × LLL were excellent, fair and good in the interaction: (CM2DIS × LLL1, CM2DIS × LLL2), CM1DIS × LLL1), (TDIS × LLL3) and in all the other interactions, respectively.

Through DIC and LLL, a trickle irrigation system can be managed to improve all the hydraulic characteristics under investigation. This would cause an increase in uniformity of distribution of both water and fertilizers, and subsequently in plant growth, yield, water use efficiency, fertilizer use efficiency and reduction in cost analysis. The acceptable values of CV were above line of acceptability in all cases except interactions (CM1DIS × LLL3), (TDIS × LLL2) and (TDIS × LLL3) as shown in Fig. 4.22. This may be due to the difference in pressure head in different closed circuits and along different lateral line lengths therefore reflecting on velocity head, head loss and CV values of emitter discharge.

4.4.1.9   Comparing the Practical Data of Head Loss Along the Lateral Line in the Laboratory With Those Calculated Using HydroCal Simulation Program

The discharge rates and pressures in trickle irrigation systems were measured under field conditions at three locations down the lateral lines for CM2DIS, CM1DIS, and TDIS using three different LLL (LLL1 = 40 m, LLL2 = 60 m and LLL3 = 80 m). Empirical estimates were used to validate the trickle simulation program (HydroCal Simulation program copyright 2009 developed by NETAFIM, USA). HydroCal is a computer simulation program used for planning and design of trickle or sprinkler irrigation systems. Modification of trickle irrigation closed circuit (DIC) and lateral lines lengths (LLL) depend mainly on hydraulic equations such as: Hazen-William's equations, Bernoulli's equations, etc. The data inputs provided to HydroCal are shown in Table 4.11. The empirical data depended on the laboratory measurements of emitter pressure, discharge, and uniformity of water distribution.

The predicted outputs of HydroCal simulation program (exponent (X), head loss (m) and velocity (m/s)) are shown in Tables 4.12 and 4.13 and Figs. 4.23–4.25. The differences in exponent (x) values of built-in emitters are attributed to the different closed circuits and different lateral line lengths that affects the pressure and exponent (x) values.

The predicted head loss along the lateral lines was calculated by HydroCalc simulation program for trickle irrigation systems: CM2DIS and CM1DIS compared with TDIS under different LLL of LLL1, LLL2,

TABLE 4.11 Inputs for the HydroCalc Simulation Program For Closed Circuit Designs in Trickle Irrigation Systems

Manifold		Lateral drip line		Emitters	
Inputs	**Value**	**Item**	**Value**	**Item**	**Value**
Pipe type	PVC	Tubes type	PE	Emitter type	Built in
Pipe length	—	Tubes lengths:	40, 60, and 80 m	Emitter flow (Lh-1)	4
Pipe diameter	0.05 m	Inner diameter	0.16 m	Emitters distance	0.30 m
C, Pipe roughness	150	C: Pipe roughness	150	Pressure head required (m)	10.0 m
Slope	0 m/m	Slope	0.0 m/m	Calculation method	Flow rate variation
Extra energy losses	0.064	Spacing	0.7 m	—	—

PVC: Polyvinyl chloride; PE: Polyethylene.

TABLE 4.12 Predicted Values of Exponent (x), Head Loss (m) and Velocity (m.sec^{-1}) Using the HydroCalc Simulation Program For Closed Circuits Trickle Irrigation Design

	DIC								
	CM2DIS			CM1DIS			TDIS		
LLL (m)	Exponent (x)	Head loss (m)	Velocity m/sec	Exponent (x)	Head loss (m)	Velocity m/sec	Exponent (x)	Head loss (m)	Velocity m/sec
40	0.72	0.53	1.40	0.69	0.82	1.35	0.58	1.12	0.87
60	0.65	1.27	1.08	0.61	1.69	0.98	0.55	2.19	0.71
80	0.58	1.69	0.79	0.52	2.96	0.75	0.53	3.98	0.62

DIC: Trickle irrigation circuits, L.L.L.: Lateral line lengths, CM2DIS: Closed circuit with two manifolds separately, CM1DIS: Closed circuit with one manifold, TDIS: Traditional trickle irrigation system

and LLL3. The predicted and measured head losses values are tabulated in Tables 4.12 and 4.13. The relationships among the predicted and measured head losses are shown in Figs. 4.23–4.25 that also include regression equations under CM2DIS, CM1DIS, and TDIS methods, respectively.

Based on predicted and measured values of heat loss, LLL1 and LLL3 can be ranked in the ascending order: CM2DIS <CM1DIS <TDIS. Under LLL2, the irrigation circuits can be ranked in the following ascending order: CM1DIS <CM2DIS <TDIS. The variation in the rankings may be attributed to how many emitters were built-in within lateral line length. The regression (R^2) was used to compare the significance of the predicted

TABLE 4.13   Effects of Different DIC and Different LLL on Hydraulic Parameters at Operating Pressure 1.0 atm and Slope = 0%: Calculated by HydroCalc Simulation Program

Hydraulic param- eters	CM2DIS			CM1DIS			TDIS		
	LLL1	LLL2	LLL3	LLL1	LLL2	LLL3	LLL1	LLL2	LLL3
No. of emitters	133	200	267	133	200	267	133	200	267
Emitter (q) (lH)	4.09	3.63	3.56	4.02	3.57	3.51	3.16	2.56	2.04
Total (Q) (lH)	544	726	950	535	714	937	420	512	545
Avg. flow velocity, m/Sec	0.86	1.54	1.88	0.91	1.73	1.92	0.94	1.62	1.97
Reynolds number	3238	3001	3062	3859	3753	3810	3234	3489	3612
Flow type					Turbulent				
Critical velocity, m/s	0.82	1.48	2.83	0.87	1.68	1.85	0.89	1.58	1.93
$f = \varepsilon / d$	0.23								
H_f(m)	0.53	1.07	1.75	0.83	1.09	2.57	1.34	2.31	4.28

f = ε/d = roughens Coefficient, Re > 3000 = turbulent flow, Re < 3000 = laminar flow (Hathoot, et al., 1993). LLL1: lateral line length = 40 m, LLL2: lateral line length = 60 m, LLL3: lateral line length = 80 m CM2DIS: closed circuit with two manifolds separately, CM1DIS: closed circuit with one manifold; TDIS: traditional trickle irrigation system.

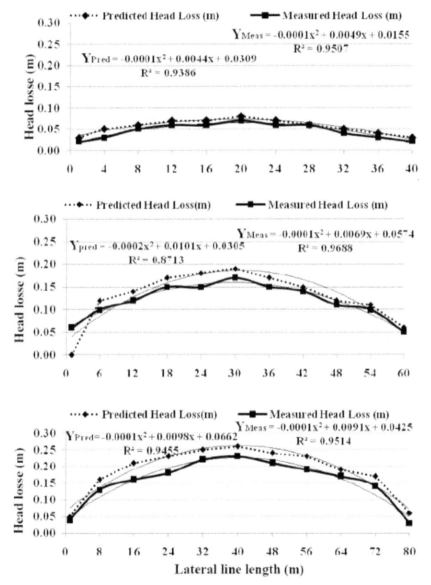

FIGURE 4.23   The relationship between different lateral line lengths (40, 60 and 80 m) and head losses (the predicted and measured) at pressure head = 1.0 atm. under CM2DIS design.

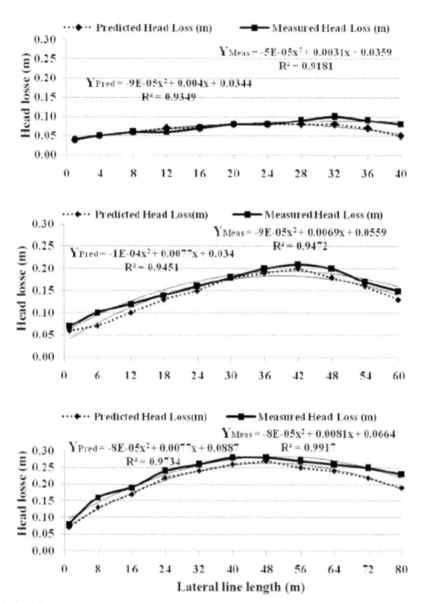

FIGURE 4.24 The relationship between different lateral line lengths (40, 60 and 80 m) and head losses (the predicted and measured) at pressure head = 1.0 atm. under CM1DIS design.

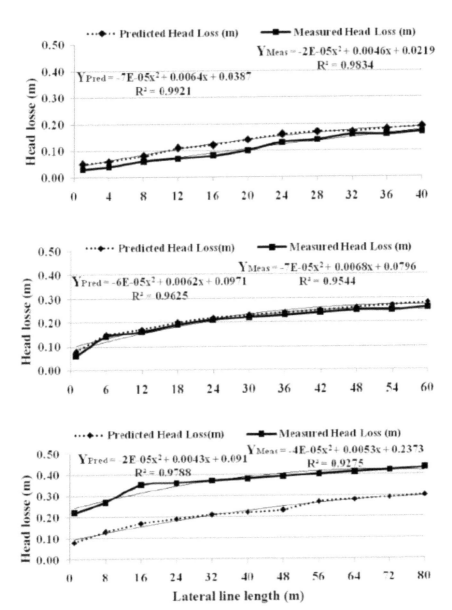

FIGURE 4.25 The relationship between different lateral line lengths (40, 60 and 80 m) and head losses (the predicted and measured) at pressure head = 1.0 atm. under TDIS design.

and measured head loss along the lateral lines for three closed circuit designs. The deviations were significant between all predicted and measured values, except the interaction TDIS × LLL3.

Generally, the values of regression coefficients between predicted and measured values were significant at 1% level, under different DIC and LLL (experimental conditions).

4.4.2 EFFECTS OF DIC AND LLL ON VEGETATIVE GROWTH AND YIELD OF CORN

Table 4.14 shows the effects of three trickle irrigation circuits (DIC) and three lateral line lengths (LLL) on vegetative growth and yield parameters of corn. Measured parameters were: average leaf area (cm²), plant height (cm), leaf length (cm), number of leaves, grain yield (tons/fed) and Stover yield (tons/fed).

4.4.2.1 Leaf Area

Table 4.14 illustrates the effects of DIC and LLL on leaf area (LA, cm²). Based on LA values, DIC can be ranked in the descending order: CM2DIS > CM1DIS > TDIS. Differences in LA among DIC were significant at the 1% level. The effect of LLL on LA can be put in the descending order: LLL1> LLL2> LLL3. Differences in LA values were significant at the 1% level. The effect of interactions, DIC × LLL, on LA were significant at 1% level. The maximum value of LA was 499.73 cm² and the minimum was 478.31 cm² were obtained in the interactions: CM2DIS × LLL1 and TDIS × LLL3, respectively.

4.4.2.2 Plant Height

Data in Table 4.14 indicates the effects of DIC and LLL on height of plant (HP, cm). Based on HP values, DIC and LLL can be written in the descending orders: CM2DIS > CM1DIS > TDIS, and LLL1> LLL2> LLL3. Differences in HP values among DIC and/or LLL treatments were significant at 1% level except that between CM2DIS and CM1DIS.

The interactions, DIC × LLL, affected HP significantly at the 1% level with the exception of the interactions: CM2DIS × LLL3, CM1DIS × LLL2, CM1DIS × LLL3 and TDIS × LLL3. The maximum value of HP was 193.78 cm and minimum was 191.45 cm in the interactions: CM2DIS × LLL1, and TDIS × LLL3, respectively.

4.4.2.3   Leaf Length

Table 4.14 shows the effects of both DIC and LLL on leaf length (LL, cm). Based on the LL values, DIC and LLL treatments can be mentioned in the descending order: CM1DIS > CM2DIS > TDIS and LLL1 ≥ LLL2 > LLL3, respectively. Differences in LL among LLL1, LLL2 and LLL3 treatments were significant at the 1% level.

The effects of interactions DIC × LLL were significant at the 1% level. The maximum value of LL was 68.15 cm and the minimum was 64.26 cm in the interactions: CM2DIS × LLL1 and TDIS × LLL3, respectively.

4.4.2.4   Number of Leaves Per Plant

The effects of DIC and LLL on a number of leaves per plant (LN) are shown Table 4.14. Based on values of LN, DIC and LLL can be stated in the descending order: CM2DIS > CM1DIS > TDIS and LLL1 > LLL2 > LLL3. Neither DIC nor LLL treatments had significant effects on LN at the 1% level. Differences in LN per plant between the means of the two factors studied were significant at the 1% level. The data illustrate that the interactions DIC × LLL treatments had significant effects on LN at the 1% level. The maximum value of LN was 15.45 and the minimum was 14.55 in the interactions: CM2DIS × LLL1 and TDIS × LLL3, respectively. The superiority of the studied growth parameters in CM2DIS, CM1DIS relative to TDIS, LLL1, and LLL2 relative to LLL3 can be noticed that this superiority was due to improving both water and fertilizer use efficiencies.

4.4.2.5   Grain Yield

Data in Table 4.14 indicates the effects of DIC and LLL treatments on corn grain yield (GY, ton per fed.). Based on values of GY, the treatments can be arranged in the ascending orders: TDIS < CM1DIS < CM2DIS and LLL3 < LLL2 < LLL1. Differences in GY among DIC and/or LLL treatments were

TABLE 4.14   Effects of Trickle Irrigation Circuits and Lateral Lines Lengths on Corn Plant Growth and Yield Parameters

DIC	LLL (m)	Leaf area (cm²)	Plant height (cm)	Leaf length (cm)	No. of leaves per plant	Grain	Stover
		— Growth and Yield Characters at Harvest (Average)				Yield (ton/fed)	
CM2DIS	40 LLL1	499.73a	193.78a	68.51a	15.45a	5.41a	3.52a
	60 LLL2	491.53d	192.21f	66.85c	15.32b	5.14c	3.47d
	80 LLL3	488.37e	192.75dc	65.25 g	15.15c	5.05ed	3.42f
CM1DIS	40	498.43b	193.30b	67.21b	14.97d	5.30b	3.50ba
	60	485.33 g	192.85c	66.34e	14.78f	5.05fe	3.44e
	80	479.83h	191.53h	64.42h	14.66h	4.99 g	3.40h
TDIS	40	496.35c	192.66e	66.58d	14.86e	5.05d	3.48cb
	60	486.78f	191.83 g	65.73f	14.72 g	4.64h	3.41 g
	80	478.31i	191.45ih	64.26i	14.55i	4.38i	3.40ih
(1) × (2)	LSD 0.01	1.27	0.11	0.14	0.09	0.03	0.02
Means, (1)	CM2DIS	492.77a	192.75a	66.87b	15.31a	5.26a	3.47a
	CM1DIS	488.29b	192.72ba	66.99a	14.80ba	5.11b	3.45b
	TDIS	487.15c	191.98c	65.52c	14.71ca	4.69c	3.43c
	LSD 0.01	4.18	0.12	0.08	1.77	0.07	0.02
Means, (2)	40	498.17a	193.25a	67.43a	15.09a	5.26a	3.50a
	60	487.88b	192.30b	66.31ba	14.94ba	4.94b	3.44b
	80	482.17c	191.91c	64.64c	14.79ca	4.81c	3.41c
	LSD 0.01	3.72	0.26	2.77	1.81	0.04	0.02

DIC: Trickle irrigation circuits, LLL: Lateral line lengths, CM2DIS: Closed circuit with two manifolds separated, CM1DIS: Closed circuit with one manifold, TDIS: Traditional trickle irrigation system, LSD = Least square difference.

Values with same letter are not significant at P = 0.01

significant at the 1% level. The effects of interaction DIC × LLL on GY were significant at the 1% level except that between any two interactions of CM2DIS × LLL3, CM1DIS × LLL2 and TDIS × LLL1. The maximum and the minimum GY were 5.14 and 4.38 ton per fed in the interactions: CM2DIS × LLL1 and TDIS × LLL3, respectively. We can observe that corn GY took the same trend of the other growth parameters due to the close correlation between vegetative growth from one side and GY from the other side.

4.4.2.6   Stover Yield

The effects of DIC and LLL treatments on Stover yield (SY, ton per fed is shown in Table 4.14. We can conclude that the change in SY took the same trend as of other growth parameters under study. Based on the values of SY, the DIC and LLL can be ranked in descending orders: CM2DIS > CM1DIS > TDIS and LLL1 > LLL2 > LLL3, respectively. The differences in SY among DIC and LLL treatments were significant at 1% level.

It is obvious that the effects of the interactions DIC × LLL treatments on SY were significant at 1% level except that between the interactions: CM1DIS × LLL3, TDIS × LLL2, CM1DIS × LLL3, TDIS × LLL2, TDIS × LLL3.

In conclusion, compared to TDIS the closed trickle irrigation circuits (CM1DIS and CM1DIS) and decreasing LLL improved selected hydraulic characteristics of the irrigation system: pressure head, friction loss, flow velocity, velocity head, uniformity, coefficient of variation. This of course improved the distribution of water and fertilizers along the lateral lines and subsequently all the growth parameters under study.

4.4.2.7   Water Use Efficiency of Grain and Stover

Table 4.15 and Figs. 4.26 and 4.27 show the effects of DIC and LLL treatments on WUE of grain and Stover (WUEg and WUEs). One can deduce that the changes in WUEg and WUEs followed the same trend as the vegetative growth parameters under study (leaf area, plant height, leaf length and number of leaves per plant). This can be due to the positive effect of DIC

TABLE 4.15   Effects of Different Irrigation Circuits Designs and Different Lateral Lines Lengths on Water Use Efficiency of Corn (WUE) For Grain and Stover Yields

DIC	LLL (m)	Applied water (m³/fed)	Grain Yield (kg/fed)	WUEg (kg/m³)	Stover Yield (kg/fed)	WUEs (kg/m³)
CM2DIS	40		5411.8a	1.33a	3522.7a	0.87a
	60		5139.0c	1.27c	3466.6d	0.85dab
	80		5049.7ed	1.24ed	3416.7f	0.84fb
CM1DIS	40	4060.14	5302.0b	1.31ba	3496.6a	0.86ba
	60		5046.5fe	1.24d	3443.4e	0.85eab
	80		4986.1 g	1.23 g	3400.1h	0.84 gb
TDIS	40		5052.3d	1.24f	3475.4c	0.86ca
	60		4634.3h	1.14i	3404.2 g	0.84hb
	80		4380.5i	1.18h	3394.5i	0.84ib
1 × 2	LSD 0.01		78.6	0.02	17.62	0.02
Means (1)	CM2DIS		5200.2a	1.28a	3468.6a	0.85a
	CM1DIS		5111.5b	1.26b	3446.7b	0.85ba
	TDIS		4689.0c	1.19c	3424.7c	0.84cab
	LSD 0.01		64.3	0.01	9.7	0.03
Means (2)	40		5255.4a	1.29a	3489.2a	0.86a
	60		4939.9b	1.22b	3438.1b	0.85ba
	80		4805.5c	1.22c	3403.8c	0.84cba
	LSD 0.01		89.4	0.03	25.4	0.02

DIC: Trickle irrigation circuits, L.L.L.: Lateral line lengths, LLL1: Lateral line length = 40 m, LLL2: Lateral line length = 60 m, LLL3: Lateral line length = 80 m CM2DIS: Closed circuit with two manifolds separately, CM1DIS: Closed circuit with one manifold; TDIS: Traditional trickle irrigation system. Values with the same letter are not significant at P = 0.01.

and LLL treatments on the vegetative growth parameters mentioned in the previous section of this chapter.

According to Waugh and WUEs values, DIC can put in the descending order: CM2DIS > CM1DIS > TDIS and CM2DIS > CM1DIS > TDIS, respectively. Differences in WUEg among DIC were significant at 1% level. With respect to the WUEg and WUEs values, the LLL can be illustrated in the descending orders: LLL1> LLL2> LLL3 and LLL1 ≥ LLL2 ≥ LLL3,

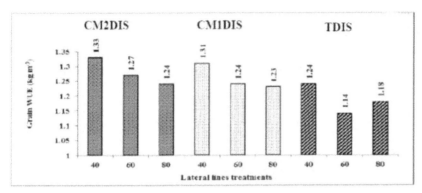

FIGURE 4.26   Effects of irrigation circuit designs and three lateral line lengths on grain WUE.

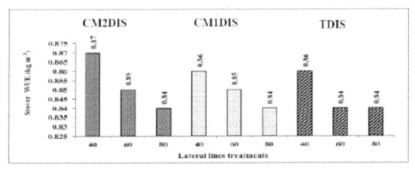

FIGURE 4.27   Effects of irrigation circuit designs and three lateral line lengths on Stover WUE.

respectively. Differences in WUEg among LLL treatments were significant at 1% level, except that between LLL2 and LLL3. On the other hand, difference in WUEs was significant at the 1% level only between LLL1 and LLL3.

The effects of the interaction DIC × LLL on WUEg were significant at 1% level, except those among the interactions CM2DIS × LLL3, CM1DIS × LLL2 and TDIS × LLL1. The effects of interaction DIC × LLL on WUEs were not significant at 1% level in most cases. The highest of WUEg and WUEs were 1.33 and 0.87 ton/fed.) and the lowest values were 1.14 and 0.84 ton/fed. were obtained in the interactions CM2DIS × LLL1 and TDIS × LLL2 or LLL3, respectively.

4.4.2.8 Fertilizers Use Efficiency (FUE)

Table 4.16 and Figs. 4.28–4.30 show the effects DIC and LLL treatments on fertilizers use efficiency (FUE) for N, P_2O_5 and K_2O fertilizers (FUEN, $FUEP_2O_5$, and $FUEK_2O$).

TABLE 4.16 Effects of Different Trickle Irrigation Circuits Designs and Lateral Lines Lengths on Fertilizer Use Efficiency (FUE)

DIC	LLL (m)	Applied fertilizers (kg/fed)			Grain yield (kg /fed)	FUE (kg yield/kg fertilizer)		
		N	P_2O_5	K_2O		NUE	PUE	KUE
CM2DIS	40 LLL1				5411.8a	89.5a	180.5a	188.1a
	60 LLL2				5139.0c	85.0c	171.4c	178.6c
	80 LLL3				5049.7ed	83.5ec	168.4e	175.5ed
CM1DIS	40	60.48	71.4	68.52	5302.0b	87.7ba	176.8b	184.2ba
	60				5046.5fe	83.4fc	168.3f	175.4fd
	80				4986.1 g	82.4 gd	166.3 g	173.3 gd
TDIS	40				5052.3d	83.5dc	168.5d	175.6d
	60				4634.3h	76.6i	154.5h	161.0h
	80				4380.5i	72.4h	146.1i	152.2i
LSD 0.01					78.6	3.2	3.5	4.1
Means	CM2DIS				5200.2a	86.0a	173.4a	180.7a
	CM1DIS				5111.5b	84.5ba	170.5b	177.6b
	TDIS				4689.0c	77.5c	156.4c	162.9c
LSD 0.01					64.3	1.6	1.8	2.2
Means	40				5255.4a	86.9a	175.2a	182.6a
	60				4939.9b	81.7b	164.7b	171.7b
	80				4805.5c	79.5cb	160.2c	167.0c
LSD 0.01					89.4	3.8	4.4	2.7

DIC: Trickle irrigation circuits, LLL: Lateral line lengths, FUE = Fertilizers use efficiency, NUE = Nitrogen use efficiency, PUE = Phosphorous use efficiency, KUE = Potassium use efficiency, LLL1: Lateral line length = 40 m, LLL2: Lateral line length = 60 m, LLL3: Lateral line length = 80 m CM2DIS: Closed circuit with two manifolds separated, CM1DIS: Closed circuit with one manifold; TDIS: Traditional trickle irrigation system. Values with same letter are not significant at P = 0.01.

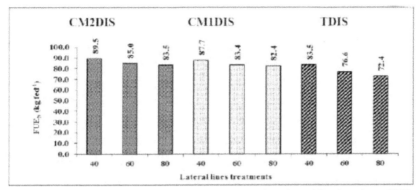

FIGURE 4.28   Effects of irrigation circuit designs and three lateral line lengths on FUE_N.

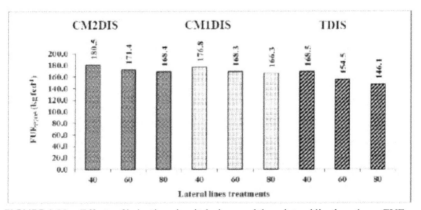

FIGURE 4.29   Effects of irrigation circuit designs and three lateral line lengths on FUE_{P2O5}.

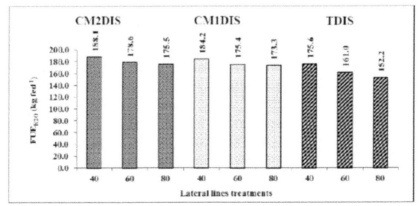

FIGURE 4.30   Effects of irrigation circuit designs and three lateral line lengths on FUE_{K2O}.

According to the FUE values for N-P-K fertilizers, the DIC and LLL treatments can be ranked in the ascending orders: TDIS < CM1DIS < CM2DIS and LLL3 < LLL2 < LLL1. Differences in FUE between any two DIC and /or LLL treatments were significant at 1% level except that between (CM2DIS, CM1DIS) and (LLL2, LLL3), in the case of FUEN. Whereas under the effects of LLL, there were significant differences at 1% level in FUE among LLL except that between LLL2 and LLL3 in FUEN.

The effects of the interactions DIC × LLL treatments on FUE were significant at 1% level among some interactions and not among the others. The highest values of FUEN, FUEP$_2$O$_5$ and FUEK$_2$O were 89.5, 180.5 and 188.1 kg of grain yield per kg of fertilizer and the lowest values were 42.5, 146.1, 152.2 in the interactions: CM2DIS × LLL1 and TDIS × LLL3, respectively. These data are supported by Baligar [9].

The results indicated that FUE took the same trend as of vegetative growth parameters, corn yield and WUE. This may be attributed to the direct relationship between WUE and FUE reported by Tayel et al. [72].

4.4.3 EFFECTS OF TRICKLE IRRIGATION CLOSED CIRCUITS (DIC) AND LATERAL LINE LENGTHS (LLL) ON COSTS ANALYSIS OF CORN PRODUCTION

Total costs of agricultural operations are major capital inputs for most farms. Tables 4.17 and 4.18 indicate effects of three closed irrigation circuits (DIC) and LLL on cost analysis of corn production (total cost, total revenue and both physical and income per unit used of irrigation water).

Table 4.17 shows that the capital costs (LE/fed) ranged from 5008 to 5658, from 5032 to 5632 and from 4962 to 5562 according to LLL treatments under CM2DIS, CM1DIS and TDIS, respectively. It was observed that the capital cost increased with decreasing LLL, due to the extra length of tubes used as manifolds and valves. The percentage of fixed cost was 40.35, 39.03, and 37.46), (40.12, 38.83, 37.45) and (39.7, 35.69, 37.0) of total cost under CM2DIS, CM1DIS, TDIS, LLL1, LLL2 and LLL3, respectively. On the other hand, the percentage operation cost was (10.04, 10.26, 10.53), (10.27, 10.5, 10.73) and (10.58, 11.29, 11.06) of the total cost in the same sequence mentioned here for fixed costs.

TABLE 4.17 Cost Analysis of Corn Production Under Different DIC and LLL (LE/fed Season)

Cost items	CM2DIS, cm			CM1DIS, cm			TDIS, cm		
	40	**60**	**80**	**40**	**60**	**80**	**40**	**60**	**80**
Capital cost (LE/fed)	5658	5358	5008	5632	5332	5032	5562	5262	4962
Fixed costs (LE/fed/season)									
1. Depreciation	396	375	351	394	373	352	389	368	347
2. Interest	226	214	200	225	213	201	222	138	198
3. Taxes and insurance	85	80	75	84	80	75	83	79	74
Sub-total = A	707	669	626	703	666	628	694	585	619
Operating costs (LE/fed/season)									
1. Electricity for pump motor		76			80			85	
2. Maintenance and Repairing		100			100			100	
Sub-total = B		176			180			185	
Total annual irrigation cost, (LE/fed/season), A + B = C	883	845	802	883	846	808	879	770	804
Total agricultural costs, D		869			869			869	
Total costs, C+D = E (LE/fed/season)	1752	1714	1671	1752	1715	1677	1748	1639	1673
Grain yield, (kg/fed.)	5412	5139	5049	5302	5046	4986	5052	4381	4381
Stover yield, (kg/fed)	3523	3467	3417	3497	3443	3400	3475	3404	3394
Grain price, (LE/fed)	3247	3083	3029	3181	3027	2992	3031	2780	2629
Stover price, (LE/fed.)	234	222	218	229	218	216	218	200	189
Total revenue, (LE/fed/season)	3481	3305	3247	3410	3245	3208	3249	2980	2818
Physical net income (kg/m³)	2.20	2.12	2.08	2.17	2.09	2.06	2.10	1.98	1.90

TABLE 4.17   (Continued)

Cost items	CM2DIS, cm			CM1DIS, cm			TDIS, cm		
	40	60	80	40	60	80	40	60	80
Net profit, (LE/ fed/season)	1740	1653	1624	1703	1621	1602	843	774	732
Net income LE/m^3	0.43	0.41	0.40	0.42	0.40	0.39	0.21	0.19	0.18

Water requirements of DIC = 4060 m^3/fed/season and fed. = 4200 m^2, CM2DIS: Closed circuit with two manifolds separated, CM1DIS: Closed circuit with one manifold; TDIS: Traditional trickle irrigation system.

Table 4.17 also presents grain yield, Stover yield, the net profit and both the physical and income per unit of irrigation water. The obtained values of these parameters were: (5412, 5139, 5049 kg/fed.), (5302, 5046, 4986 kg/ fed), (5052, 4634, 4381 kg/fed), (234, 222, 218 kg/fed), (229, 218, 216 kg/ fed) and (218, 200, 189 kg/fed), (2.20, 2.12, 2.08 kg/m^3), (2.17, 2.09, 2.06 kg/ m^3), (2.10, 1.98, 1.90 kg/m^3), (0.43, 0.41, 0.40 LE/m^3), (0.42, 0.40, 0.39 LE/m^3) and (0.21, 0.19, 0.18 LE/m^3) in the same sequence under CM2DIS, CM1DIS, TDIS and LLL1, LLL2, LLL3, respectively.

Table 4.18 presents the effects of DIC and LLL on the total cost of corn production during the crop season (LE/fed-season), total revenue (LE/fed-season), physical income (Also called production efficiency, kg/m^3) and the income (LE/m^3). Based on the effects of DIC on the parameters under consideration, the DIC can be put in the descending orders: (CM2DIS = CM1DIS > TDIS), (CM2DIS > CM1DIS > TDIS), (CM2DIS = CM1DIS > TDIS), (CM2DIS > CM1DIS > TDIS), respectively. In other wards, differences in total costs and physical income between CM2DIS and CM1DIS from one side and TDIS system from the other side were significant at 1% level, whereas the differences in both the total revenue and income per unit of irrigation water among DIC were significant at the 1% level.

In the case of the effects of LLL on all the studied parameters, LLL can be ranked in the ascending order: LLL1 < LLL2 < LLL3 except the physical income, whereas the order took the trend: LLL1 < LLL2 < LLL3. Differences in all parameters among LLL were significant at 1% level except that between LLL2 and LLL3 in the case of the physical income.

The effects of the interaction DIC × LLL on all parameters are given in Table 18. The maximum and the minimum values of total cost, total rev-

TABLE 4.18 Effects of DIC and LLL on Cost Parameters of Corn Production

DIC	LLL	Total costs (LE/fed/season)	Yield (kg/fed)		Price (LE/fed)		Total revenue, (LE fed/season)	Physical net income (kg/ m³)	Net profit, (LE/fed/season)	Net income (LE/m³)
			Grain	Stover	Grain	Stover				
CM2DIS	40	1752a	5412a	3523a	3247a	234a	3481a	2.20a	1740a	0.43a
	60	1714ed	5139c	3467d	3083c	222c	3305c	2.12c	1653c	0.41c
	80	1671hf	5049ed	3417f	3029ec	218dc	3247ec	2.08fc	1624dc	0.40dc
CM1DIS	40	1752ba	5302b	3497a	3181b	229ba	3410b	2.17ba	1703b	0.42ba
	60	1715d	5046fe	3443e	3027fc	218ec	3245fc	2.09ec	1621ed	0.40e
	80	1677f	4986g	3400h	2992g	216g	3208g	2.06gf	1602fd	0.39fe
TDIS	40	1748ca	5052d	3475c	3031dc	218fc	3249dc	2.10dc	843g	0.21g
	60	1639i	4634h	3404g	2780h	200h	2980h	1.98h	774h	0.19hg
	80	1673gf	4381i	3394i	2629i	189i	2818i	1.90i	732i	0.18ih
1x2 LSD0.01		5	79	18	60	5	64	0.05	30	0.01
Means (1)	CM2DIS	1712a	5200a	3469a	3120a	225a	3344a	2.13a	1672a	0.41a
	CM1DIS	1715ba	5111b	3447b	3067ba	221ba	3288b	2.11ba	1642b	0.40ba
	TDIS	1687c	4689c	3424c	2813c	202c	3016c	1.99c	783c	0.19c
LSD0.01		4	64	10	62	6	62	0.03	29	0.01
Means (2)	40	1751a	5255a	3498a	3153a	227a	3380a	2.16a	1429a	0.35a
	60	1689b	4940b	3438b	2963b	213b	3177b	2.06b	1349b	0.33b
	80	1674c	4805c	3404c	2883c	208cb	3091c	2.01cb	1319cb	0.32cb
LSD0.01		6	89	25	65	8	67	0.06	34	0.01

DIC: Trickle irrigation circuits, LLL: Lateral line lengths, CM2DIS: Closed circuit with two manifolds separated, CM1DIS: Closed circuit with one manifold; TDIS: Traditional trickle irrigation system. Values with same letter are not significant at P = 0.01.

enue, the physical income and the income per unit of irrigation water were achieved in the following interactions: (CM2DIS × LLL1, TDIS × LLL2), (CM2DIS × LLL1, TDIS × LLL3), (CM2DIS × LLL1, TDIS × LLL3) and (CM2DIS × LLL1, TDIS × LLL3), respectively.

This study presents effects of DIC and LLL on the investigated parameters: emitter discharge, lateral discharge, pressure head, friction loss, flow velocity, velocity head, uniformity coefficient and coefficient of variation. The positive effects of CM2DIS and CM1DIS and the shortest length of LLL on these parameters can lead to a better distribution of both water and fertilizers along the lateral lines. This has been positively reflected on corn yield per feddan and subsequently on the physical and the income per unit irrigation water. Also, it is concluded that the effects of DIC and LLL on the parameters under consideration for the fixed and operating costs were not significant.

4.5   CONCLUSIONS

Based on the results obtained in this research study on the use of closed-circuit trickle irrigation design, we can conclude as follows:

1. The problem of low water pressurized at the end of lateral lines can be tackled by using lengths of lateral lines 40 and 60 meters.
2. To avoid problem of low pressure and system breakages during the operation, when try to reduce the number of treatment units or lift head for the motor (diesel).
3. Decreased friction loss coefficient and uniformity of variation: in this study, the uniformity coefficient of emitters along the lateral was 11% and 5.5% with CM2DIS, CM1DIS compared to TDIS.
4. Fertilizer distribution uniformity, which depends on the improved regularity of the distribution of water where fertilizers added through a drip irrigation system, was improved to increase in discharge variation and uniformity of water application.
5. Maize grain yield was increased by 9.8%, 8.2% and firewood (Stover yield) was increased by 0.53%, 0.50% when using circuits CM2DIS, CM1DIS compared to TDIS.
6. WUE was increased by 7.8, 6.3% for grain yield, 1.2% of the Stover harvest under closed circuits CM2DIS, CM1DIS comparing to TDIS.

7. FUE was improved by 10.2%, 9.2% of nitrogen fertilizer, and by 9.8%, 8.2% for both fertilizers phosphorus and potassium under CM2DIS, CM1DIS compared to TDIS.
8. Net profits were 10% higher by using closed circuit compared to the traditional drip irrigation system.
9. The value of net income per unit of irrigation water (LE/m^3) was 50% higher in CM2DIS, and 51% higher in CM1DIS compared to the traditional trickle system.
10. The value of net income from the physical unit of irrigation water (kg/m^3) was increased by 6.6 and 5.2% with closed circuits CM2DIS, CM1DIS compared to TDIS.

The Author recommends using closed circuit designs in trickle irrigation system because it improved the hydraulic characteristics of the lateral lines, plant growth, corn grain yield, the physical income, WUE and FUE.

4.6 SUMMARY

The chapter discusses research studies on the effects of three trickle irrigation closed circuits (CM2DIS, CM1DIS, TDIS) and three lateral line lengths (LLL1 = 40 m, LLL2 = 60 m, and LLL3 = 80 m) on: Solution to the problem of pressure reduction at the end of lateral lines; Comparison between two types of trickle irrigation circuits with TDIS as a control; Some hydraulic parameters: Pressure head, friction loss, flow velocity and velocity head (or dynamic head); Variations in discharge, uniformity coefficient, and coefficient of variation of an emitter; Corn growth and productivity, water and fertilizers use efficiencies under field conditions; and Cost analysis of corn production, economic net income and physical net income due to modified trickle irrigation systems. Laboratory tests were conducted at the Agric. Eng. Res. Inst., ARC, MALR, Egypt. Field experiments were conducted at the Experimental Farm of Faculty of Agriculture, Southern Illinois University at Carbondale (SIUC).

According to the data on the pressure head (m), uniformity coefficient (%) and emitter discharge (lph) in DIC, these be ranked in the descending order: CM2DIS > CM1DIS > TDIS, meanwhile LLL can be ranked in descending order: LLL1 > LLL2 > LLL3. Based on data for friction loss and coefficient of variation, DIC and LLL can be ranked in the order opposite to the previous orders here. With respect to flow velocity, velocity head and lateral discharge, DIC and LLL can be ranked in the descending order: CM2DIS> CM1DIS >TDIS and LLL3> LLL2> LLL1, respectively. According to the validation of predicted and measured energy head loss, the regression coefficients between predicted and measured values were significant at the 1% level, under different DIC and LLL treatments. Base on values of one leaf area, plant height, leaf length and number of leaves per plant, DIC and LLL can be arranged in the descending order: CM2DIS > CM1DIS > TDIS and LLL3> LLL2 > LLL1, respectively. According to grain and Stover water use efficiency and fertilizers use efficiency, DIC and LLL can be arranged in the descending order: CM2DIS > CM1DIS > TDIS and LLL3> LLL2> LLL1, respectively. Cost analysis indicated that CM2DIS, CM1DIS, LLL1 and LLL2 achieved the highest values of revenue, net profits, economic net income per unit irrigation water and physical net income per unit irrigation water.

KEYWORDS

- ascending order
- check valve
- citrus
- closed circuit
- closed circuit design
- closed circuit with one manifold, CM1DIS
- closed circuit with two manifolds, CM2DIS
- coefficient of friction
- coefficient of variation
- continuity equation
- corn growth
- corn plant
- corn yield

- cost analysis
- Darcy–Weisbach equation
- discharge through lateral line, QL
- drip irrigation
- dynamic head
- Egypt
- emitter
- emitter discharge
- emitter flow
- emitter spacing
- energy cost
- energy gradient
- feddan
- fixed cost
- flow velocity
- friction coefficient
- friction loss
- growth parameters
- Hazen–Williams equation
- Hazen–Williams friction
- head loss
- HydroCal
- irrigation circuit design
- labor cost
- laminar flow
- lateral line
- lateral line length
- LE
- least square difference
- micro irrigation
- operating pressure
- operational cost
- plastic pipe

- polyethylene, PE
- power law
- pressure
- pressure head, PH
- pressure regulator
- pressurized irrigation system
- Reynolds' number, Re
- smooth pipe
- traditional drip irrigation system
- trickle irrigation
- turbulent flow
- uniformity coefficient
- vacuum relief valve
- vegetative
- velocity head
- water
- water flow

REFERENCES

1. Abdel-Aziz, A. A., 2003. Possibility of applying modern irrigation systems in the old citrus farm and economic return, *J. Agric. Sci. Mansoura Univ.,* 28 (7): 5621–5635.
2. Abou Kheira A. A., 2009. Comparison among different irrigation systems for deficit irrigated corn in the Nile Valley. *Agricultural Engineering International: the CIGR E-journal*, Manuscript LW 08 010, XI:1–25.
3. Alizadeh, A., 2001. *Principles and Practices of Trickle Irrigation.* Ferdowsi University, Mashad, Iran.
4. Allen R.G, Pereira L.S., Raes D and M. Smith, 1998. *Crop evapotranspiration-Guidelines for computing crop water requirements.* FAO Irrigation and drainage paper 56. Rome.
5. ASAE Standards, 43rd ed., 1996. EP458: Field Evaluation of Microirrigation Systems. St. Joseph, Mich. ASAE. Pages 756–761.
6. ASAE Standards, 2003. EP405.1: Design and Installation of Microirrigation Systems. ASAE, St. Joseph, MI. Pages 901–905.
7. ASAE, 1999. Design and installation of microirrigation Systems. *In: ASAE Standards, 2000, EP405.1*, ASAE, St. Joseph, MI. Pages 875–879.

8. Bacon, M. A., 2004. *Water use efficiency in plant biology*. CRC Press.
9. Baligar, V.C. and O.L. Bennett, 1986. NPK-fertilizer efficiency: A situation analysis for the tropics. *Fert. Res.*, 10:147–164.
10. Barber, S. A., 1976. Efficient fertilizer use. *Agronomic Research for Food*. Madison, WIASA Special Publication No 26. Am. Soc. Agron, Patterson, F.l., (Ed.). Pages 13–29.
11. Bazaraa, A.S., 1982. Pressurized irrigation systems, sprinkler and trickle irrigation. Faculty of Engineering, Cairo University, Egypt. Pages 64–79.
12. Berkowitz, S.J., 2001. Hydraulic performance of subsurface wastewater drip systems. In On-Site Wastewater Treatment: Proc. Ninth International Symposium on Individual and Small Community Sewage Systems. St. Joseph, MI: ASAE. Pages 583–592.
13. Bhatti, A.U., and M. Afzal, 2001. Plant nutrition management for sustained production. Deptt. Soil and Envir. Sci. NWFP Agri. Uni. Peshawar. Pages 18–21.
14. Bombardelli, F. A., M. H. Garcia, 2003. Hydraulic design of large diameter pipes. *Journal of Hydraulic Engineering*, 129(11):839–846.
15. Brady, N.C., 1974. Supply and availability of phosphorus and potassium. In: *The Nature and Properties of Soils* by R.S. Buckman. Macmillan Publishing Co. Inc., New York. Pages 456–480.
16. Camp, C.R., 1998. Subsurface drip irrigation: A review. *Transactions of ASAE*, 41(5):1353–1367.
17. Camp, C.R., E.J. Sadler and W.J. Busscher, 1997. A comparison of uniformity measure for drip irrigation systems. *Transactions of the ASAE*, 40:1013–1020.
18. Camp, C.R., E.J. Sadler and W.J. Busscher, 1989. Subsurface and alternate-middle micro irrigation for the south-eastern Coastal Plain. *Trans. of ASAE*, 32(2): 451–456.
19. Cary, J. W. and H.D. Fisher, 1983. Irrigation decision simplified with electronics and soil water sensors. *Soil Science Society of American Journal*, 47:1219–1223.
20. Cetin, B., S. Yozgan and T. Tipia, 2004. Economics of drip irrigation of olives in Turkey. *Agricultural Water Management*, 66:145–151.
21. Charlesworth, P., 2000. Soil water monitoring, CSIRO Land and Water, Australia. Pages 56–64.
22. Dagdelen, N. H. Bas, Eyilmaz, T. Gurbu, Z., and S. Akcay, 2009. Different drip irrigation regimes affect cotton yield, water use efficiency and fiber quality in Western Turkey. *Agricultural Water Management*, 96:111–120.
23. Deba, P. D. 2008. Characterization of drip emitters and computing distribution uniformity in a drip irrigation system at low pressure under uniform land slopes. M.Sc. Thesis, Biological & Agricultural Engineering Department, Texas A&M University. Pages 5–18.
24. Dhuyvetter, K. C., F. R. Lamm and D. H. Rogers, 1995. Subsurface drip irrigation (sdi) for field corn – an economic analysis. Proceedings of the 5th International Microirrigation Congress, Orlando, Florida, April 2–6, Pages 395 – 401.
25. Dioudis, P., Filintas, T. A. and H. A. Papadopoulos, 2008. Corn yield in response to irrigation interval and the resultant savings in water and other overheads. *Irrigation and Drainage Journal*, 58:96–104.
26. Doorenbos, J. and A. H. Kassam, 1986. Yield response to water. FAO Irrigation and Drainage Paper 33, FAO, Rome, Italy, pages 101–104.

27. El Amami, H., A. Zairi, L. S. Pereira, T. Machado, A. Slatni and P. Rodrigues, 2001. Deficit irrigation of cereals and horticultural crops: economic analysis. Agricultural Engineering International: the CIGR Journal of Scientific Research and Development, manuscript LW 00 007b, 3:1–11.

28. El-Awady, M. N., G. W. Amerhom and M. S. Zaki, 1976. Trickle irrigation trail on pea in condition typical of Qalubia, Egypt. *J. Hort. Sc.,* 3(1):99–110.

29. El-Shawadfy, M. A., 2008. Influence of different irrigation systems and treatments on productivity and fruit quality of some bean varieties. M.Sc. Thesis, Agricultural Engineering Department, Ain Shams University. Pages 71–73.

30. Filintas, T. A., 2003. Cultivation of maize in Greece: Increase and growth, management, output yield and environmental Sequences. University of Aegean, Faculty of Environment, Department of Environmental Studies, Mitilini, Greece. (In Greek with English abstract).

31. Filintas, T. A., 2005. Land use systems with emphasis on agricultural machinery, irrigation and nitrates pollution, with the use of satellite remote sensing, geographic information systems and models, in watershed level in Central Greece. M.Sc. Thesis, University of Aegean, Faculty of Environment, Department of Environmental Studies, Mitilini, Greece. (In Greek with English abstract).

32. Filintas, T. A., I.P. Dioudis, T.D. Pateras, N. J. Hatzopoulos, and G. L. Toulios, 2006. Drip irrigation effects in movement, concentration and allocation of nitrates and mapping of nitrates with GIS in an experimental agricultural field. Proc. of 3rd HAICTA International Conference on Information Systems in Sustainable Agriculture, Agro Environment and Food Technology (HAICTA'06), Volos, Greece, September 20–23, 2:253–262.

33. Filintas, T. A., I.P. Dioudis, T.D. Pateras, E. Koutseris, N. J. Hatzopoulos, and G. L. Toulios 2007. Irrigation water and applied nitrogen fertilizer effects in soils nitrogen depletion and nitrates GIS mapping. Proc. of First International Conference on: Environmental Management Engineering, Planning and Economics CEMEPE/SECOTOX), June 24–28, Skiathos Island, Greece, 3:201–207.

34. Filintas, T.Ag., I.P. Dioudis, T.D. Pateras, N. J. Hatzopoulos, and G. L. Toulios, 2006. Drip Irrigation effects in movement, concentration and allocation of nitrates and mapping of nitrates with GIS in an experimental agricultural field. Proc. of 3rd HAICTA International Conference on: Information Systems in Sustainable Agriculture, Agroenvironment and Food Technology, (HAICTA'06), Volos, Greece, September 20–23, 2:253–262.

35. Fulton, J. M., 1970. Relationship among soil moisture stress plant population, row spacing and yield of corn. *Can. J. Plant Sci.,* 50:31–38.

36. Gee, G. W. and J. W. Bauder, 1986. Particle-size analysis. In: Klute (ed.) *Methods of soil analysis,* Part 1. ASA and SSSA, Madison, WI. Pages 383–412.

37. Gerrish, P.J., W.H. Shayya and V.F. Bralts, 1996. An improved analysis of incorporating pipe components into the analysis of hydraulic networks. *Trans. ASAE,* 39(4):1337–1343.

38. Gill, K.S., Gajri, P.R., Chaudhary, M. R. and B. Singh, 1996. Tillage, mulch and irrigation effects on corn (*Zea mays* L.) in relation to evaporative demand. Soil & Tillage Research, 39:213–227.

39. Megh R. Goyal, 2012. Management of Drip/Trickle or Micro Irrigation. Oakville, ON, Canada: Apple academic Press Inc.,

40. Megh R. Goyal, 2015. Research Advances in Sustainable Micro Irrigation, Volumes 1 to 10. Oakville, ON, Canada: Apple academic Press Inc.,

41. Goyal, Megh R. and Eric W. Harmsen, 2013. Evapotranspiration: Principles and Applications for Water Management. Oakville, ON, Canada: Apple academic Press Inc.,

42. Hathoot, H. M., A.I. Al-Amoud, and F.S. Mohammed, 1991, Analysis of a pipe with uniform lateral flow. *Alexandria Eng.. J.*, 30(1):C49-C54.

43. Hathoot, H. M., A.I. Al-Amoud and F.S. Mohammed, 1993 Analysis and design of trickle irrigation laterals. *J. Irrig. And Drain. Div. ASAE*, 119(5):756–767.

44. Hillel, D. 1982. *Drip/Trickle irrigation in action. Volume I.* Academic Press New York. Pages 1–49.

45. Holt, D.F. and D.R. Timmons, 1968. Influence of precipitation, soil water, and plant population interactions on corn grain yields. *Agron. J.*, 60:379–381.

46. Howell, T.A., A. Yazar, A. D. Schneider, A.D., D.A. Dusek, and K.S. Copeland, 1995. Yield and water use efficiency of corn in response to LEPA irrigation. *Trans. ASAE*, 38(6):1737–1747.

47. James, L. G., 1988. *Principles of farm irrigation system design.* John Wiley and Sons,. Pages 261–263.

48. Kirnak, H., E. Dogan, S. Demir and S. Yalcin, 2004. Determination of hydraulic performance of trickle irrigation emitters used in irrigation system in the Harran Plain. *Turk. J. Agric. For.*, 28:223–230.

49. Klute, A., 1986. Moisture retention. In: A. Klute (ed.), *Methods of soil analysis*, Part 1. ASA and SSSA, Madison, WI. Pages 635–662.

50. Laboski, C.A.M., R.H. Dowdy, R.R. Allmaras and J. A. Lamb, 1998. Soil strength and water content influences on corn root distribution in a sandy soil. *Plant and Soil*, 203:239–247.

51. Larry, G. J., 1988. *Principles of farm irrigation system design*, John Wiley and Sons, Inc. Pages 260–275.

52. Maisiri, N., A. Senzanje, J. Rockstrom and S. J. Twomlow, 2005. On farm evaluation of the effect of low cost drip irrigation on water and crop productivity compared to conventional surface irrigation. *Physics and Chemistry of the Earth*, 30:783–791.

53. Mansour, H. A., 2006. The response of grape fruits to application of water and fertilizers under different localized irrigation systems. M.Sc. Thesis, Faculty of Agriculture, Ain Shams university, Egypt. Pages 78–81.

54. Mansour, Hani A., 2012. Design considerations for closed circuits of drip irrigation. Ph.D. Dissertation, Department of Agricultural Engineering, Faculty of Agriculture, Ain Shams University, Egypt.

55. Memon, K. S., 1996. Soil and Fertilizer Phosphorus. In: *Soil Science* by (eds.) Elena Bashir and Robin Bantel. National Book Fund. Islamabad. Pages 308- 311.

56. Metwally, M. F., 2001. Fertigation management in sandy soil. M. Sc. Thesis, Agric. Eng., Fac. of Agric., El Mansoura University.

57. Mizyed, N and E.G. Kruse, 1989. Emitter discharge evaluation of subsurface trickle irrigation systems. *Transactions of the ASAE*, 32: 1223–1228.

58. Modarres, A.M., R.I. Hamilton, M. Dijak, L.M. Dwyer, D.W. Stewart, D.E. Mather and D.L. Smith, 1998. Plant population density effects on maize inbred lines grown in short-season environment. *Crop Sci.*, 38:104–108.

59. Mogazhi, H. E. M., 1998. Estimating Hazen-Williams coefficient for polyethylene pipes. *J. Transp. Eng.*, 124(2):197–199.

60. Musick, J. T., F.B. Pringle, W.L. Harman, and B. A. Stewart, 1990. Long-term irrigation trends: Texas High Plains. *Applied Engineering in Agriculture*, 6:717–724.

61. Nakayama, F.S. and D.A. Bucks, 1986. *Trickle Irrigation for Crop Production: Design, Operation, and Management*. Developments in Agricultural Engineering 9. Elsevier, New York, USA. Pages 1–2.

62. Narayanan, R., D.D. Steele and T.F. Scherer, 2000. Computer model to optimize above-ground drip irrigation systems for small areas. *Applied Eng. in Agric.*, 18(4):459–469.

63. Ogola, J.B.O., T.R. Wheeler and P.M. Harris, 2002. The water use efficiency of maize was increased by application of fertilizer N. *Field Crops Research*, 78(2–3):105–117.

64. Ortega, J. F., J. M. Tarjuelo and J. A. de Juan, 2002. Evaluation of irrigation performance in localized irrigation system of semiarid regions (Castilla-La Mancha, Spain). *Agricultural Engineering international: CIGR Journal of Scientific Research and Development*, 4:1–17.

65. Rebecca, B., 2004. Soil Survey Laboratory Methods Manual. Soil Survey Laboratory Investigations Report No. 42 by Rebecca Burt Research Soil Scientist MS 41, Room 152, 100 Centennial Mall North, Lincoln, NE 68508–3866, (402):437–5006.

66. Sabreen, Kh. P., 2009. Fertigation technologies for improving the productivity of some vegetable crops. Ph.D. Thesis, Faculty of Agriculture, Ain Shams University, Egypt.

67. Safi, B., M.R. Neyshabouri, A.H. Nazemi, S. Massiha and S.M. Mirlatifi, 2007. Water application uniformity of a subsurface drip irrigation system at various operating pressures and tape lengths. *Turkish Journal of Agriculture and Forestry*, 31(5):275–285.

68. Smajstrla, A.G. and G.A. Clark, 1992. Hydraulic performance of micro irrigation drip tape emitters. Annual meeting of ASAE paper no. 92–2057. St. Joseph, Mich., ASAE.

69. Steel, R. G. D and J. H. Torrie, 1980. *Principles and Procedures of Statistics. A biometrical approach*. 2nd Ed., McGraw Hill Inter. Book Co., Tokyo, Japan.

70. Talozi, S.A., and D.J. Hills, 2001. Simulating emitter clogging in a micro irrigation subunit. *Trans. ASAE.*, 44(6):1503–1509.

71. Tayel, M. Y. and Kh. P. Sabreen, 2011. Effects of irrigation regimes and phosphorus level on two Vica Faba varieties, II: Yield, water and phosphorous use efficiency. *J. App. Sci. Res.*, 7(11):1518–1526.

72. Tayel, M. Y., I. Ebtisam, A. Eldardiry and M. Abd El-Hady, 2006. Water and fertilizer use efficiency as affected by irrigation methods. *American-Eurasian J. Agric. & Environ. Sci.*, 1(3):294–300.

73. Tayel, M. Y., I. Ebtisam, A. Eldardiry and S. M. Shaaban and Kh. P. Sabreen, 2010. Effect of injector types, irrigation, and nitrogen levels on cost analysis of garlic production. *J. App. Sci. Res.*, 6(7):822–829.

74. Tayel, M. Y, A.M. EL-Gindy and A. A. Abdel-Aziz, 2008. Effect of irrigation systems on productivity and quality of grape crop. *J. App. Sci. Res.*, 4(12):1722–1729.

75. Tayel, M.Y., S. M. Shaaban, I. Ebtisam, K. El-dardiry and Kh. P. Sabreen, 2010b. Effect of injector types, irrigation and nitrogen levels on garlic yield, water and nitrogen use efficiency. *Journal of American Science*, 6(11):38–46.

76. Tsipori, Y. and D. Shimshi, 1979. The effect of trickle line spacing on yield of tomatoes. *Soil. Sci. Soc. Am. J.*, 43:1225–1228.
77. Von Bernuth, R. D., 1990. Simple and accurate friction loss equation for plastic pipes. *J. Irrig. and Drain. Eng.*, 116(2):294–298.
78. Warrick, A.W. and M. Yitayew, 1988 Trickle lateral hydraulics, I: Analytical solution. *J. Irrig. Drain. Eng. (ASCE)*, 114(2):281–288.
79. Watters, G.Z., J. Urbina and J. Keller, 1977. Trickle irrigation emitter hose characteristics. Proceedings of International Agricultural Plastics Congress, San Diego, California, pages 67–72.
80. Weatherhead, E. K. and K. Danert, 2002. *Survey of Irrigation of Outdoor Crops in England*. Cranfield University, Bedford.
81. Wood, D. J. and A. G. Rayes, 1981. Reliability of algorithms for pipe network analysis. *J. of Hydraulics Div. ASCE*, 107(HY10):1145–1161.
82. Wu, I. P., 1992. Energy gradient line approach for direct hydraulic calculation in drip irrigation design. *Irrig. Sci.*, 13:21–29.
83. Wu, I.P. and H.M. Gitlin, 1979. Hydraulics and uniform for drip irrigation. *Journal of Irrigation and Drainage.*
84. Yazgan, S., H. Degirmenci, H. Byukcangaz and C. Demirtas, 2000. Irrigation problems in olive production in Bursa region. In: *National First Olive Symposium*. Bursa, Turkey. Pages 275–281.
85. Yildirim, G. and N. Agiralioglu, 2008. Determining operating inlet pressure head incorporating uniformity parameters for multi outlet plastic pipelines. *Journal of Irrigation and Drainage Engineering, ASCE,* 134(3):341–348.
86. Yitayew, M. and A.W. Warrick, 1988. Trickle lateral hydraulics, II: Design and examples. *J. Irrig. Drain. Engrg., ASCE*, 114(2):289–300.
87. Younis, S. M., 1986. Study on different irrigation methods in Western Nobaria to produce tomato. *Alex. J. Agric. Res.*, 31(3):11–19.
88. Zhang. H and T. Oweis, 1999. Water yield relations and optimal irrigation scheduling of wheat in the Mediterranean region. *Agricultural Water Management*, 38(3):195–211.

CHAPTER 5

PERFORMANCE OF MAIZE UNDER BUBBLER IRRIGATION SYSTEM

M. Y. TAYEL, H. A. MANSOUR, and S. K. PIBARS

CONTENTS

5.1 INTRODUCTION

Maize (*Zea Mays L.*) is cultivated in regions lying between 58°N latitude and 40°N latitude at an elevation of 3,800 meters above mean sea level. It is a grain crop that is irrigated worldwide [8, 17]. The maize irrigation requirements of maize vary from 500 to 800 m3 for maximum production [7]. Irrigation is of the utmost importance, from the appearance of the first

---

In this chapter: one feddan (Egyptian unit of area) = 4200 m².

Modified and adopted from *M. Y. Tayel, H. A. Mansour, and S. K. Pibars, 2014. Impact of bubbler irrigation system discharges and irrigation water amounts on maize: Irrigation efficiency, growth, grain and biomass production. International Journal of Advanced Research, 2(1):716–724.*

silk strands until the milky stage in the maturation of the kernels on the cob [6, 10, 13]. Once the milky stage has occurred, the appearance of black layer development on 50% of the maize kernels is a sign that the crop has fully ripened. In arid regions of Middle East, maize is generally cultivated under gravity or sprinkler irrigation and few research studies to evaluate effects of drip irrigation on performance of maize crop have been reported [6, 10, 11]. Generally, the pan evaporation method is used to calculate the amount of water needed for irrigation [22]. Drip irrigation easily be automated by using automatic irrigation controllers [3–5]. The application of fertilizers is usually by hand with low efficiency, resulting in higher costs and environmental problems [1]. Maize is one of the most important cereals in Egypt and is grown for both grain and forage.

When water supply is limited, drip irrigation results in greater partitioning of water to transpiration and less to soil evaporation, which will result in slightly less water stress. At greater irrigation needs, drip irrigation delivered water and nutrients in the root zone that may increase may crop yield [5, 16]. Payero et al. [18] investigated effects of the deficit irrigation on soybeans using surface drip irrigation at Curtis and solid-set sprinklers at North Platte, They used a greater range of irrigation requirements than at Colby, but relative performance drip and sprinkler could not be compared because these were at different locations and for different years. Bubbler irrigation is basically just a slight modification of trickle irrigation systems and therefore cost of these systems is similar to drip irrigation system.

This chapter presents research results on the performance of growth parameters of maize crop under two discharge rates of bubbler irrigation (BID) systems and three irrigation amounts, based on 100%, 75%, 50% of evapotranspiration.

5.2 MATERIALS AND METHODS

During the growing season of 2013, a field experiment was established at the Agricultural Research Farm of National Research Center (NRC), El-Nobaria, Egypt. The treatments were:

1. Two BID drippers with discharge rates of: BID8 = 8 lph and BID12 = 12 lph; and

2. Three irrigation amounts based on evapotranspiration (ET): I100 = 100% of ET, ET75 = 75% of ET, and I50 = 50% of ET.

Effects of these treatments were evaluated on vegetative growth and yield parameters of maize (*Zea mays L. cv. HF-10*): Leaf area, leaf length, leaf number of leaves per plant, plant height, grain yield and biomass yield. Soil texture at the experimental site and soil moisture retention constants were determined based on methods described by Gee and Bauder [12], Klute [14] and Rebecca [19].

The experiments setup was split-plot randomized complete block design with three replications. The total cultivated area was one feddan (= 0.42 ha). This area was divided to two parts for each of the dripper discharge of bubbler irrigation system. Each main plot was subdivided into three subplots for each of the irrigation treatments. Maize seeds were planted for a plant density of 30,000 plants/fed. were on May 3, 2013. The distance between rows was 0.7 m and 0.25 m between plants in the row. Irrigation season of maize was ended 15 days before the last harvest. Maize was harvested on September 15. According to Ministry of Agriculture in Egypt, fertilization program was followed using fertigation technique. The amounts of fertilizers NPK (20–20–10) were 80 kg/fed of (20% N), 40 kg/fed of (20% K_2O), and 65 kg/fed of (10% P2O5). For all plots, weed and pest control measures were followed based on recommendations for maize production in El-Nobaria, Egypt.

5.2.1   IRRIGATION

Source of irrigation water was from ground water. Plants were irrigated every 4 days using BID. Irrigation amount was based to compensate for ETc of maize and salt leaching requirement. The (ETc) was computed using the Class A pan evaporation method for estimating (ETo) on daily basis with a climatic data from nearest meteorological station. Pan coefficient (Kp) was assumed as 0.76 for short green cropped and medium windy area [2]. Daily pan evaporation (Epan) was estimated as 7.5 mm/day at the site. The reference evapotranspiration (ETo) was then multiplied by a crop coefficient Kc at particular growth stage to determine crop consumptive use at that particular stage of maize growth. Details of irrigation system and controls for pressure and water supply have been described by Mansour (Chapter 4 in this volume) and Safi et al. [20]. The irrigation interval was 4 days in this chapter.

5.2.2   IRRIGATION SYSTEM EFFICIENCY

Yoder and Eisenhauer [23] stated that the bubbler irrigation system efficiency (Eb) is the effectiveness of irrigation system in delivering all the water beneficially to produce the crop. It is defined as the ratio of the volume of water (Vb) that is beneficially used to the volume of irrigation water applied (Vf):

$$Eb = (Vb/Vf) \times 100 \tag{1}$$

where: Eb = bubbler irrigation efficiency, Vb = volume of water beneficially used (fed-cm), and Vf = volume of water delivered to the field (fed-cm). Overall bubbler irrigation efficiency (Eob) is calculated by multiplying the water conveyance and water application efficiencies:

$$Eob = (Ec \times Ea) \times 100 \tag{2}$$

where: Eob = overall bubbler irrigation efficiency (%), Ec = water conveyance efficiency (in decimals), and Ea = water application efficiency (in decimals). Effective bubbler irrigation efficiency (Ebe) the overall bubbler irrigation efficiency corrected for runoff and deep percolation that is recovered and reused or restored to the water source without reduction in water quality. It is expressed as:

$$Ebe = [Eob + (FR) \times (1.0 - Eob)] \times 100 \tag{3}$$

where: FR = fraction of surface runoff, seepage, and /or deep percolation that is recovered.

5.2.3   MEASUREMENTS OF PLANT GROWTH PARAMETERS

Plant measurements and observations were started 21 days after planting, and were terminated on the harvest date. Grain yield was determined by hand harvesting the 8 m sections of three adjacent center rows in each plot and was adjusted to 15.5% water content. In all treatment plots, the grain yields of individual rows were determined in order to evaluate the yield uniformity among the rows.

Treatment means for all parameters and treatments were compared using analysis of variance (ANOVA) and the least significant difference (L.S.D) at 1%, according to Steel and Torrie [21].

5.3 RESULTS AND DISCUSSION

Table 5.1 illustrates the effects of two different discharge rates of bubbler emitters on bubbler irrigation system efficiency, overall bubbler irrigation efficiency and bubbler effective irrigation efficiency.

5.3.1 LEAF AREA

Table 5.2 illustrates the effects of different BIDD and ET% on leaf area. Data can be ranked in the descending orders: 8 lph > 12 lph and I_{100} > I_{75} > I_{50}. Based on the values of leaf area, results indicate significant differences among means values of both main effects BIDD and submain effects ET%. According to the effects of interaction between both investigated

TABLE 5.1 The Effects of Two Discharge Rates of Bubbler Emitters Drippers Discharges on Irrigation System Efficiencies

BID	Bubbler Irrigation system efficiency, Eb (%)	Overall bubbler irrigation efficiency, Eob (%)	Effective bubbler irrigation efficiency, Ebe (%)
8 LPH	85.17	76.65	28.6
12 LPH	84.04	75.64	25.3

Equations were:

$Eb = (Vb/Vf) \times 100$

$Eob = (Ec \times Ea) \times 100$

$Ebe = [Eob + (FR) \times (1.0 Eob)] \times 100$

TABLE 5.2   Effects of Discharge Rates of Bubbler Emitters and Irrigation Amounts on Performance Parameters of Maize Plants

BIDD (1)	Based on ET (2)	Leaf area	Plant height	Leaf		Maize yield	
				Length	Nos.	Grain	Biomass
	%	cm²	cm	cm	per plant	Kg/feddan	
8 lph	100	492.1a	191.4a	68.7a	15.6a	5532a	4827a
	75	489.2ca	190.8ba	67.3c	15.4ba	5527ba	4823ba
	50	476.5e	189.7dc	65.4e	14.3e	5070d	4295e
12 lph	100	490.3ba	190.7c	68.2b	15.4ca	5478c	4553c
	75	487.8db	189.5ec	65.6d	14.5de	4668e	4365d
	50	472.5fe	188.7f	64.3f	14.1fe	4436f	3925f
(1) × (2)	LSD at 0.01	**4.5**	**1.5**	**0.8**	**0.5**	**12** at 0.05	**14** at 0.05
(1) Means	8 lph	485.9a	190.6a	67.1a	15.1a	5376a	4648a
	12 lph	483.5b	189.6b	66.0b	14.7b	4861b	4281b
	LSD at 0.05	**2.33**	**0.09**	**0.06**	**0.84**	88	76
(2) Means	100	491.2a	191.1a	68.5a	15.5a	5505a	4690a
	75	488.5b	190.2b	66.5b	15.0b	5098b	4594b
	50	474.5c	189.2c	64.c	14.2c	4753c	4110c
	LSD at 0.05	2.8	1.3	0.5	0.4	**68** at 0.01	**56** at 0.01

BIDD: bubbler irrigation dripper discharges rate in liter per hour (lph), ET, %: Amounts of irrigation based on evapotranspiration (Also called deficit irrigation), LSD: least significant differences at P = 0.01 or at P = 0.05.

factors, the highest and lowest values of leaf area were recorded for 8 LPH and 100%ET. Also, it was observed that under all BIDD, all highest values were observed at 100% ET. For leaf area parameter, differences were significant at 5% level among all means values of BIDD and ET%. The effects of interaction between two studied factors were not significant at 5% level, except in the interactions: 8 LPH × 100 and 12 LPH × 100, 8 LPH × 75 and 12 LPH × 75. The maximum and minimum values of leaf area were found in the interactions: 8 LPH × 100 and 12 LPH × 50, respectively.

5.3.2 PLANT HEIGHT

Table 5.2 indicates that plant height followed the trend similar to the leaf area. The effects of BIDD and ET% can be ranked in the descending orders: 8 LPH > 12 LPH and 100 > 75 > 50, respectively. Differences in plant height were significant at 5% level among all means values of BIDD and ET%. The effects of interaction between two studied factors were significant at 5%, level except in the interactions: 8 LPH × 100 and 12 LPH × 100, 8 LPH × 75 and 12 LPH × 75. The maximum and minimum values of plant height were found in the interactions of 8 LPH × 100 and 12 LPH × 50, respectively.

5.3.3 LEAF LENGTH

Table 5.2 shows the effects of BIDD and ET% on leaf length (cm). According to values of leaf length, BIDD and ET% can be ranked in the descending orders: 8 LPH > 12 LPH and 100 > 75 > 50, respectively.

Based on values of leaf length, data indicated that there is significant difference within means values of BIDD and ET%, while the highest and lowest values were observed for 8 LPH and 12 LPH, respectively. There is significant difference between interactions among BIDD and ET%. The effects of interaction among the two study factors, data indicated that there were significant differences between treatments at 5% level. The maximum and minimum values of leaf length were recorded at 8 LPH × 100 and 12 LPH × 50.

5.3.4 NUMBER OF LEAVES PER PLANT

Table 5.2 shows effects of BIDD and ET% on number of leaves per plant, which can be ranked in descending order: 8 LPH > 12 LPH and I100 > I75 > I50, respectively. Differences in number of leaves per plant between means of two factors studied were significant at 5% level. While the highest and lowest values under BIDD and ET% were achieved in 8 LPH; 12 LPH and 100;50, respectively. The maximum and minimum values of number of leaves were (significant at 5%) recorded in 8 LPH × 100 and 12

LPH × 50, respectively. The superiority of the studied vegetative growth parameters in all treatments of BIDD and ET% can be observed due to improving of water and fertilizer distribution uniformities. This was due to the treatments 100%ET and 75%ET that were gave convergent results in values which implies that the amount of water added was is difference between the 100 − 75 = 25% of ET. This amounts to excess of plants required under the current conditions of the experiment.

5.3.5 GRAIN YIELD

Table 5.2 indicates the effects of BIDD and ET% on maize grain yield (GY, kg/feddan), which can be ranked in the ascending orders: 8 LPH > 12 LPH and I100>I75>I50, respectively. With respect to the main effects of BIDD on grain yield, one can observe that the differences in grain yield were significant among BIDD treatments at 5% level. The highest and lowest grain yields were obtained in 8 LPH and 12 LPH, respectively. According to grain yield, the effects of ET% treatments on grain yield, were significantly different at 5% level among I100, I75 and I5. However, highest and lowest values were achieved in I100 and I50, respectively.

Based on the effects of BIDD × ET% on grain yield, there were significant differences at 5% level, except for the interactions: 8 LPH × I100, 8 LPH × I75. The maximum and minimum values of grain yield were obtained in 8 LPH × I100 and 12 LPH × I50, respectively. Lamm [15] indicated a range of seasonal irrigations applied relative to meeting the full irrigation requirement. Grain yield vs. seasonal irrigation were grouped for years having average or greater rainfall (1998, 1999, 2004) or significant drought (2000–2003) for simulated low-pressure precision applicators and drip irrigation systems, where yield and seasonal irrigations were averaged for each group of years. For average to wet years, grain yield with drip irrigation was slightly greater than simulated low-pressure precision applicators, but vice versa for drought years. In average to wet years, differences in grain yields were primarily due to kernel weight, but in drought years due to the number of kernels per ear (see Ref. [15] for actual yield data).

5.3.6   BIOMASS YIELD

Table 5.2 indicates the effect of both BIDD and ET% on maize biomass yield (BY, kg/feddan). We can notice that the change in maize biomass yield took the same trend as of vegetative growth parameters and thus took the trend of grain yield too. Based on the positive effects of BIDD and ET% on biomass yield, these can be ranked in descending orders: 8 LPH>12 LPH and I100 > I75 > I50. With respect to BIDD and ET% effects on biomass yield, one can notice significant differences at 1% level among all mean values of BIDD and ET%. According to the interaction effects of the investigated factors, the highest and lowest values of biomass yield were recorded under interactions: 8 LPH × I100 and 12 LPH × I50.

We can notice that maize grain and biomass yields took the same trend of other vegetative growth parameters. This finding can be attributed to the close correlation between vegetative growth from one side and grain/biomass yields from the other one; and also due to positive relations between the increasing of growth parameters and increasing of grain and biomass yields.

5.4   CONCLUSIONS

The effects of two discharge rates of BIDD were evaluated on bubbler irrigation system efficiency, overall bubbler irrigation efficiency, and effective bubbler irrigation efficiency, Data could be ranked in the descending orders: 8 LPH > 12 LPH. Based on these results, decreasing of bubbler dripper discharge will increase and give the greater efficiencies and vice versa. The improvement of the studied vegetative growth and yield parameters were observed in: leaf area, plant height, leaf length, number of leaves per plant, grain yield and biomass yield under 8 LPH; 12 LPH and 100,75 due to improving of water and fertilizer distribution uniformities in the sandy soil. Parameters under study can be ranked in descending orders: 8 LPH > 12 LPH and 100>75>50. With respect to BIDD and ET% effects on maize biomass yield, one can notice significant differences at 1% level among all mean values of BIDD and ET%. According to the interaction effects of the investigated factors, the highest and lowest values of maize biomass yield were recorded under interactions: 8 LPH × 100 and 12 LPH × 50.

Maize grain and biomass yield took the same trend as of other vegetative growth parameters, due to the close correlation between vegetative growth and grain yield.

5.5 SUMMARY

This chapter presents research results on the performance of growth and yield parameters of maize crop under two discharge rates of BID systems and three irrigation amounts, based on 100%, 75%, 50% of evapotranspiration. Vegetative growth and yield parameters included: leaf area, leaf length, number of leaves per plant, plant height, grain yield, and biomass yield. During the growing season of 2013, a field experiment was established at the Agricultural Research Farm of National Research Center (NRC), El-Nobaria, Egypt. Irrigation water was added in order to compensate for ETc of maize and salt leaching requirement.

Based on effects of two discharge rates of BIDD on bubbler irrigation system, overall bubbler irrigation, and effective bubbler irrigation efficiencies, data can be ranked in the descending orders: 8 LPH > 12 LPH. Decreasing of bubbler dripper discharge will increase and give greater values of all efficiencies and vice versa. Growth parameters under study can be ranked in descending orders: 8 LPH > 12 LPH and 100 >75>50. With respect to effects of BIDD and ET% on maize biomass yield, one can notice significant differences at 1% level among all mean values of BIDD and ET%. According to the interaction effects of the investigated factors, the highest and lowest values of maize biomass yield were recorded under interactions: 8 LPH × 100 and 12 LPH × 50. These results are due to the convergent results in treatments of 100%ET and 75%ET. This means that the amount of water added was the difference between the two: 100 – 75% = 25%ET. Therefore, it can be recommend to use water applications based on 75% of ET for saving 25% water requirements under bubbler irrigation system using 8 LPH drippers.

KEYWORDS

- biomass yield
- bubbler irrigation
- crop yield
- deficit irrigation
- discharge
- drip irrigation
- dripper
- Egypt
- evapotraspiration
- grain yield
- irrigation
- irrigation efficiency
- leaf area
- leaf length
- maize
- number of leaves per plant
- plant height
- USA
- vegetative growth

REFERENCES

1. Abou–Kheira, A. A., 2009. Comparison among different irrigation systems for deficit-irrigated transgenic and non trangenic maize in the Nile Valley. *Agricultural Engineering International: the CIGR Ejournal.* Manuscript LW08.
2. Allen, R. G., Pereira, L. S, Raes, D. and M. Smith, 1998. *Crop evapotranspiration: Guidelines for computing crop water requirements.* FAO Irrigation and Drainage Paper 56. Rome
3. Cary, J. W. and Fisher, H. D., 1983. Irrigation decision simplified with electronics and soil water sensors. *Soil science Society of American Journal,* 47:1219–1223.
4. Charlesworth, P., 2000. *Soil water monitoring.* CSIRO Land and Water, Australia.
5. Colaizzi, P. D., A. D. Schneider, S. R. Evett, and T. A. Howell, 2004. Comparison of SDI, LEPA, and spray irrigation performance for grain sorghum. *Trans. ASAE,* 47(5):1477–1492.

6. Dioudis, P., Filintas, T., H. A. Papadopoulos, 2008. Transgenic and non trangenic maize yield in response to irrigation interval and the resultant savings in water and other overheads. *Irrigation and Drainage Journal*, 58:96–104.

7. Doorenbos, J. and A. H. Kassam, 1986. *Yield response to water*. FAO Irrigation and Drainage Paper 33, FAO, Rome, Italy. Pages 101–104.

8. Filintas, T., 2003. *Cultivation of Maize in Greece: Increase and Growth, Management, Output Yield and Environmental Sequences*. University of Aegean, Faculty of Environment, Department of Environmental Studies, Mitilini, Greece.

9. Filintas, T., 2005. Land use systems with emphasis on agricultural machinery, irrigation and nitrates pollution, with the use of satellite remote sensing, geographic information systems and models, in watershed level in Central Greece. M.Sc. Thesis, University of Aegean, Faculty of Environment, Department of Environmental Studies, Mitilini, Greece.

10. Filintas, T., Dioudis, I. P., Pateras, T. D., Hatzopoulos, N. J. and G. L. Toulios, 2006. Drip irrigation effects in movement, concentration and allocation of nitrates and mapping of nitrates with GIS in an experimental agricultural field. Proc. of 3rd HAICTA international conference on: information systems in sustainable agriculture, Agro Environment and Food Technology, (HAICTA'06), Volos, Greece, September 20–23. Pages 253–262.

11. Filintas, T., Dioudis, I. P., Pateras, T. D., Koutseris, E., Hatzopoulos, N. J. and G. L. Toulios, 2007. Irrigation water and applied nitrogen fertilizer effects in soils nitrogen depletion and nitrates GIS mapping. Proc. of First International Conference on: Environmental Management, Engineering, Planning and Economics CEMEPE/SECOTOX), June 24–28, Skiathos Island, Greece, III:2201–2207.

12. Gee, G. W. and J. W. Bauder, 1986. Particle-size analysis. Pages 383–412. In: Klute (ed.) *Methods of soil analysis. Part 1*. ASA and SSSA, Madison, WI.

13. Gill, K. S., Gajri, P. R., Chaudhary, M. R. and B. Singh, 1996. Tillage, mulch and irrigation effects on transgenic and non trangenic maize (*Zea mays* L.) in relation to evaporative demand. *Soil & Tillage Research*, 39:213–227.

14. Klute, A., 1986. Moisture retention. Pages 635–662. In: A. Klute (ed.) *Methods of soil analysis. Part 1*. ASA and SSSA, Madison, WI.

15. Lamm, F. R., 2004. Comparison of SDI and simulated LEPA sprinkler irrigation for corn. CD-ROM, Irrigation Association, Annual Meeting, 14–16 Nov, Tampa, FL.

16. Lamm, F. R., H. L. Manges, L. R. Stone, A. H. Khan and D. H. Rogers, 1995. Water requirement of subsurface drip-irrigated corn in north-west Kansas. *Trans. ASAE*, 38(2):441–448.

17. Musick, J. T., Pringle, F. B., Harman, W. L. and B. A. Stewart, 1990. Long-term irrigation trends: Texas High Plains. *Applied Engineering Agriculture*, 6:717–724.

18. Payero, J. O., S. R. Melvin, and S. Irmak, 2005. Response of soybean to deficit irrigation in the semiarid environment of west-central Nebraska. *Trans. ASAE*, 48(6):2189–2203.

19. Rebecca, B., 2004. *Soil Survey Laboratory Methods Manual*. USDA – NSSC Soil Survey Laboratory Investigations Report No. 42. Room 152, 100 Centennial Mall North, Lincoln, NE 68508–3866.

20. Safi, B., M. R. Neyshabouri, A. H. Nazemi, S. Masiha and S, M. Mirlatifi, 2007. Sub-surface irrigation capability and effective parameters on onion yield and water use ef-ficiency. *Journal of Scientific Agriculture*, 1:41–53.

21. Steel, R. G. D and J. H. Torrie, 1980. *Principles and Procedures of Statistics*. A bio-metrical approach. 2nd Ed., McGraw Hill Inter. Book Co. Tokyo, Japan.

22. Weatherhead, E.K. and K. Danert, 2002. *Survey of Irrigation of Outdoor Crops in Eng-land*. Cranfield University, Bedford.

23. Yoder, R. E., and D. E. Eisenhauer, 2010. Irrigation system efficiency. In: D. R. Held-man and C.I. moraru (eds). *Encyclopedia of Agricultural, Food, and Biological Engi-neering*. Second Edition, XI:1–25.

CHAPTER 6

ENERGY AND WATER SAVINGS IN DRIP IRRIGATION SYSTEMS

H. A. MANSOUR, M. Y. TAYEL, D. A. LIGHTFOOT,
and A. M. EL-GINDY

CONTENTS

6.1   INTRODUCTION

Drip irrigation system (DIS) has many advantages and is accompanied by some clogging problems and constraints [1–9]. The problem of low pressure at the end of lateral lines has been partially tackled by the development of closed circuit designs (CCDIS) by modification to the traditional drip irrigation system by Mansour [16]. The new design of CCDIS have

In this chapter: One *feddan* (Egyptian unit of area) = 0.42 ha.

Modified and printed from, *"H.A. Mansour, M. Y. Tayel, D. A. Lightfoot, and A. M. El-Gindy, 2010. Energy and water saving by using modified closed circuits of drip irrigation system. Agricultural Sciences, 1(3):154–177. Open access at: http://www.scirp.org/journal/AS/."*

been successfully for the irrigation of fruit trees, vines, vegetable and field crops, in the Egyptian desert. The unique drip irrigation system with 0.5 to 16 lph discharge rates supplies adequate soil moisture near the root zone and row spacing between the crop remains dry throughout the season.

Sources of fossil fuel are being rapidly depleted and energy consumption is increasing at an exponential rate. The International Energy Outlook 2006 [11] projects continued growth for worldwide energy demand over the period from 2003 to 2030. The total world consumption of marketed energy expands from 421 quadrillion British thermal units (BTU) in 2003 to 563 quadrillion BTU in 2015, and then to 722 quadrillion BTU in 2030, or a 71% increase during 2003 to 2030 period (Fig. 6.1).

Pimentel et al. [19] indicated that irrigation accounts for 13% of the agricultural energy consumption. There have been some attempts to power irrigation systems with renewable energies, but most of the resulting systems are designed for large farms and the cost of such systems is usually high. Designing successful irrigation systems powered with renewable energies for small farms depends on many factors: climate, crop, crop water needs, type of irrigation system, and crop type. More accurately, it depends on the balance between the energy demand and supply. Due to the large number of factors involved in the design process of such a system, it is not easy to conduct experiments to evaluate the effect of each factor. Worldwide, various types and models of drip/trickle or micro irrigation systems are being designed, developed and marketed. Aside from the basic

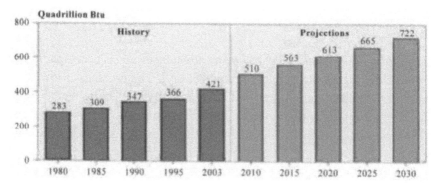

FIGURE 6.1 Global energy consumption from 1980 to 2003 and the projected consumption upto 2030 in quadrillion BTU [11].

technical differences, they differ in cost or affordability and in water distribution uniformity.

Among the most cost-effective of drip irrigation models, the drip kit has ben developed by International Development Enterprises [19]. The drip kit consists of microtube emitters inserted through plastic tape roll laterals connected to polyethylene submain pipes, which-in turn can be connected to a drum water reservoir. The system can be operated by elevating the drum reservoir at appreciable head, thereby eliminating the need for a pumping unit. Typical operating heads of the IDE drip kits range from 1.0 m to 3.0 m [13, 19]. This drip kit is suitable for developing countries because of its low cost and simplicity of design and installation.

The drip kit has gained popularity in some upland watersheds [20] for vegetable production under agroforestry systems in India, Nepal, Srilanka, Kenya and South-east Asian countries (Philippines, Vietnam and Indonesia). While distribution uniformity studies of some types of drip or trickle irrigation systems have been undertaken [4], evaluation of the performance of low-cost drip irrigation systems such as that of IDE at different heads for a given slope has not been fully explored. In fact, no rigorous study has been carried out to determine recommendable operating heads for such low-cost drip systems to generate certain levels of water distribution uniformity especially under sloping conditions. Keller and his colleagues [13] have conducted studies to determine the effects of hydraulic head and slope on the water distribution uniformity of the IDE 'Easy Drip Kit' and subsequently develop mathematical relationships to characterize the effect of slope and head on water distribution uniformity, which can serve as the basis for optimizing water use efficiency and crop productivity.

Pipelines are essential for the use of drip irrigation. They need to be operated at adequate pressures (typically 1–2 bars for drip systems) and need to be strong enough to withstand up to twice the working pressure. Energy is needed in pipe systems not only to pump water from the source to the pipe but also to overcome the energy losses due to friction as water flows down the pipe. Predicting head losses in pipes is not an exact science and it is easy to make mistakes when calculating these losses. In addition, losses can increase with the age of pipe and the increase in pipe roughness through continued use. For these reasons, the losses in the distribution system should be kept low at the design stage by choosing large pipe diameters to reduce head losses. As a guideline, energy losses in the pipe should be less than 30% of the total pumping head.

Energy is another word commonly used in everyday language in engineering. However, in hydraulic and irrigation engineering, it has a very specific meaning: Energy enables t h e useful work. In irrigation, energy is needed to lift or pump water. Water energy is supplied by a pumping device driven by human or animal power, or a motor using solar, wind or fossil fuel energy. The system of energy transfer is not perfect and energy losses occur through friction between the moving parts and are usually lost as heat energy (the human body temperature rises during exercise; an engine heats as fuel is burnt to provide power). Energy losses can be significant in pumping systems, and so can be costly in terms of fuel use [5].

6.1.1   QUALITATIVE CLASSIFICATION STANDARDS FOR EMITTERS

The emitter discharge rate (q) has been described by a power law: $q = kH^x$, where H = operating pressure, k = emitter coefficient, and x = exponent. Values of k and x depend on emitter characteristics [9, 12]. According to the manufacturer's coefficient of emitter variation (CVm), standards have been developed by ASAE [1]. CVm values below 10% are acceptable and >20% are unacceptable [1, 8]. The emitter discharge variation rate (qvar) should be evaluated as a design criterion in drip irrigation systems; qvar <10% may be regarded as good and qvar >20% as unacceptable [3, 8, 24]. Differences in emitter geometry may be caused by variation in injection pressure and heat instability during the manufacturing process, as well as by a heterogeneous mixture of materials used for the production [14]. Lamm et al. [15] evaluated the distribution uniformity of drip laterals applying wastewater from a beef lagoon. Distribution uniformities ranged from 54.3 to 97.9% for the tubing's under study. Only a small percentage of emitter plugging can reduce the application uniformity [18]. Talozi and Hills [21] have modeled the effects of emitter and lateral clogging on the discharge of water through all laterals. Results show that the discharge from laterals, that were simulated to be clogged, decreased while laterals, that were not clogged, increased. In addition to decreases in discharge for clogged emitters, the model showed an increase of pressure at the manifold inlet. Due to the increased inlet pressure, a lower discharge rate by the pump was observed. Berkowitz [2] observed reductions in emitter irrigation flow ranging from 7 to 23% at five sites. Reductions in scouring velocities were also observed from the designed 0.6 m/s (2ft/s) to 0.3 m/s (1ft/s). Lines also developed some slime built-up, as reflected by the reduction in scouring velocities, but this occurred to a less degree with higher quality effluent. He used approximate friction equations: Hazen-Williams and Scobey (who neglected the variation of the velocity head along the lateral and assumed initial uniform emitter flow). Warrick and Yitayew [22] assumed a lateral with longitudinal slots and presented design charts based on spatially varied flow. The latter solution has neglected the presence of laminar flow in a considerable length of the downstream part of the lateral. Hathoot et al. [9] provided a solution based on uniform emitter discharge but took into account the change of velocity head and the variation of Reynolds's number. They used the Darcy-Weisbach friction equation in estimating friction losses. Hathoot et al. [10] considered individual emitters with variable outflow and presented a step by step computer program for designing either the diameter or the lateral length. Watters et al. [23] considered the pressure head losses due to emitters protrusion. These losses occur when the emitter barb protrusion obstructs the water flow. Based on area on barbs, three sizes of emitter barbs are specified: small ($\leq$20 mm^2), medium (21 to 31 mm^2) and large ($\geq$32 mm^2).

This chapter presents research studies on savings in energy and water under three lateral line lengths and three DIS. Authors also compared water and energy use efficiencies, and investigated emitter discharge application uniformity and its dependence on operation pressures under these treatments.

6.2 MATERIALS AND METHODS

6.2.1 SITE LOCATION AND EXPERIMENTAL DESIGN

This laboratory experiments were conducted at Irrigation Devices and Equipment Tests Laboratory of Agricultural Engineering Research Institute, Agriculture Research Center, Cairo, Egypt, The experimental design was randomized complete block with three replications. Three irrigation lateral drip line lengths of 40, 60, 80 m were installed at a slope of 0% and were tested at ten operating pressures of 0.2, 0.4, 0.6, 0.8, 1.0, 1.2, 1.4, 1.6, 1.8, and 2.0 bars for ten minutes. Details of the drip irrigation system with pressure/water supply/valves, etc., have been described by Mansour [16] in Chapter 4 of this volume. He evaluated built-in drip-per (GR) with a discharge rate of 4 lph at one bar of nominal operating pressure and emitter spacing of 30 cm, to resolve the problem of lack of pressure at the end of lateral lines in the traditional drip irrigation system. Treatments were:

1. Three drip irrigation systems (DIS):
   • Closed circuit with two manifolds of drip irrigation system (CTMDIS);
   • Closed circuit with one manifold of drip irrigation system (COMDIS); and
   • Traditional of drip irrigation system (TDIS).
2. The lateral drip line lengths: 40, 60 and 80 m.

6.2.2 FIELD EXPERIMENTAL SITE

During one growing season of 2009–2010, the field experiment was conducted at the Experimental Farm of Faculty of Agriculture at Southern Illinois University at Carbondale (SIUC), Illinois, USA. The field is located at latitude 37°73′N, longitude of 89°16′W, and elevation of 118 m above sea level. Soil at the experimental field was silty clay loam. Physical and chemical properties were determined using methods described by Gee and Bauder [6], and Jackson [12]; and the data is presented by Mansour [16] in Chapter 4 of this volume. Ground water was the source of irrigation water. A drip irrigation system was installed in all plots to evaluate the effects of above treatments. The return manifold was used to collect the water from the laterals and carry it to the return line, which returns to the pretreatment device. Prior to connecting the return manifold to the return line, a check valve is installed to prevent water from entering the zone during the operation of other zones. The system is described in detail by Mansour [16] in Figs. 4.1–4.4 in Chapter 4 of this volume.

6.2.3   HYDRAULICS OF DRIP IRRIGATION SYSTEM

Hydraulic losses are functions of size of pipe, pipe material, pipe smoothness and roughness, pipe length, accessories/joints/valves/bends, air resistance, flow velocity, properties of fluids, field slope, topography of field, complexity of system. These losses have been determined by various investigators for laminar and turbulent flows and theoretical equations have been developed. During design of the sewerage pipelines, partially filled pipes with free-surface flow are assumed. Hydraulic calculations are performed using the formulas applicable in the case of pressure flow, when the pipe is fully filled (in other words, flow is continuous). These formulas do not take into account the resistance of air above the fluid surface, which decreases as the fluid level in pipe is reduced pipe filling is reduced. General graphs are recommended for calculation of actual pipe losses [8, 9]. In this chapter, authors used the hydraulics of pipes presented by Mansour [16] in Chapter 4 of this volume. The calculated dripper emission rates were compared with the measured values to evaluate the emitter uniformity.

6.2.4.   MEASUREMENTS PERFORMANCE PARAMETERS OF DRIP IRRIGATED MAIZE (ZEA MAYS L.)

Maize crop was harvested at maturity and the grain yield in kg/ha and kg/feddan was calculated. Equations described by Mansour [16] in Chapter 4 of this volume were used to calculate water use efficiency (WUE).

6.2.5   ESTIMATION OF ENERGY REQUIREMENTS

The amount of energy needed to pump water depends on the volume of water to be pumped and the head required and can be calculated with Eq. (1). Increasing either the volume of water or the head will directly increase the energy required for pumping. Energy use efficiency and power use efficiency [5] were calculated with Eqs. (3) and (5).

Water energy (WHP, kWh) = [volume of water (m³) × head (m)] /367   (1)

$$\text{Water energy (kWh)} = \text{water power (kW)} \times \text{operating time (h)} \quad (2)$$

$$\text{Pumping plant efficiency (\%)} = (\text{water energy/actual energy}) \times 100 \quad (3)$$

$$\text{Water power (kW)} = [9.81 \times \text{discharge (m}^3/\text{s)} \times \text{head (m)}] \quad (4)$$

$$\text{Pumping plant power efficiency (\%)} = (\text{water power/}$$
$$\text{power input}) \times 100 \quad (5)$$

6.2.6 HYDROCAL IRRIGATION SYSTEM PLANNING SOFTWARE FOR HYDRAULIC CALCULATIONS

HydroCal irrigation system planning software is designed to help the user to define the parameters of an irrigation system. HydroCal program is described in detail by Mansour [16] in Chapter 4 of this volume.

6.2.7 VALIDATION OF MEASURED DATA WITH CALCULATED DATA BY HYDROCAL

The emission rate for 10 emitters was evaluated for lateral line length at three stages (beginning, middle and end of lateral line). Hydraulic calculations for "First, middle and end of the lateral line" were calculated theoretically. The head loss due to friction and insertion of emitters was calculated and then the pressure head at every emission point was determined. The flow from emission point was calculated using the characteristic equation developed for pressure head vs. discharge for the emitter or mini-sprinkler.

6.2.8 STATISTICAL ANALYSIS

All collected data were subjected to the statistical analysis using statistical program developed by Michigan State University. Regression analysis, analysis of variance (ANOVA) and the least significant difference (L.S.D) at $P = 0.01$ were conducted and determined.

6.3. RESULTS AND DISCUSSIONS

6.3.1. EFFECTS OF DIFFERENT OPERATING PRESSURES ON DISCHARGE THROUGH LATERAL DRIP LINE AT LAND SLOPE OF 0%

Table 6.1 and Figs. 6.2–6.4 indicate that there is a direct relationship between the operating pressure and the average discharge through lateral drip lines versus lateral line length in all closed circuit designs (CM2DIS, CM1DIS and TDIS). At operating pressure of 0.8 bar in CM2DIS method, the average discharge was 4.48 lph for a lateral line length of 40 m (LLL1), compared to average discharge of 4.20 in CM1DIS, for the same length of the lateral line. With increase in operating pressure to 1.0 bar, the average discharge was 4.48 lph in CM2DIS compared to average discharge of 4.33 lph in CM1DIS, for a lateral line length of 40 m (LLL1). For LLL1/LLL2/LLL3 in the control (TDIS) and lengths 60 and 80 m (LLL2 and LLL3) in CM2DIS, CM1DIS, the average discharge did not reach the standard value of 4.00 lph at operating pressure of 1.0 bar for this type of dripper (GR built-in).

For lateral line length of 40 m (LLL1) and in CM2DIS/CM1DIS, data in Figs. 6.2–6.4 show that the average discharge was able to reach the standard value of 4.00 lph for this dripper at an operating the pressure of 0.8 bar. While for lateral line length 60 m (LLL2) in CM2DIS, the discharge was able to reach the standard value at a pressure of 1.2 bar. With comparison to TDIS under same conditions, it was not possible to reach to the standard discharge at the three lateral line lengths 40, 60 and 80 m (LLL1, LLL2, LLL3).

TABLE 6.1   Comparison Between Coefficients of Determination (R^2) For Regression Analyzes Among Pressure and Discharge Versus Lateral Line Length at land Slope of 0%

| | Coefficients of determination, R^2 | | |
| | Lateral line length, m | | |
Irrigation method	40	60	80
CM2DIS	0.9712	0.9506	0.9397
CM1DIS	0.9693	0.9414	0.9368
TDIS	0.9565	0.9354	0.9153

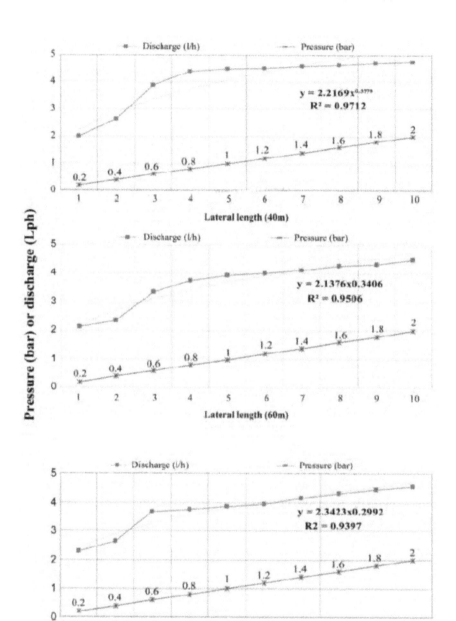

The Selected Drippers on the lateral lines of (CM2DIS)

FIGURE 6.2 Effects of different operating pressures (bar) on discharge rates of the closed circuit design with two manifolds (CM2DIS) at a land slope of 0%.

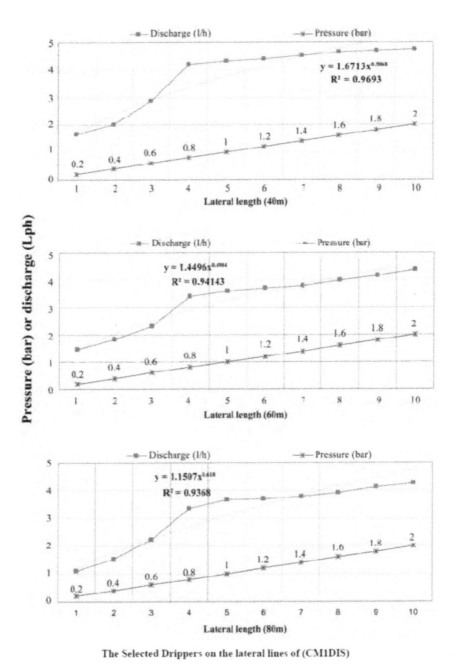

The Selected Drippers on the lateral lines of (CM1DIS)

FIGURE 6.3 Effects of different operating pressures (bar) on discharge rates of the closed circuit design with one manifold (CM1DIS) at a land slope of 0%.

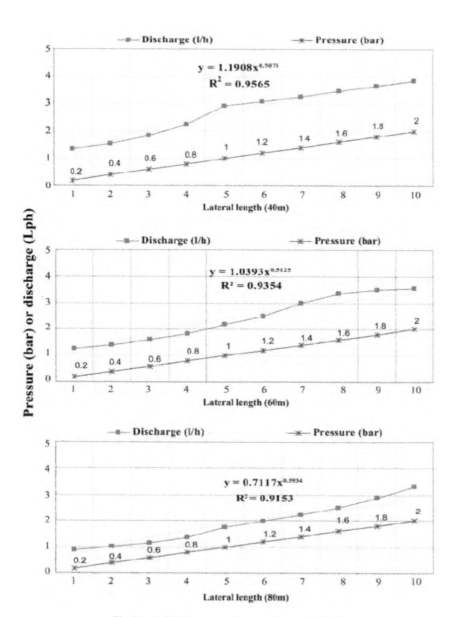

The Selected Drippers on the lateral lines of (TDIS)

FIGURE 6.4 Effects of different operating pressures (bar) on discharge for the traditional drip system (TDIS) at a land slope of 0%.

The values of coefficient of determination (R^2) are shown in Table 6.1 and Figs. 6.2–6.4, for all treatments. The regression coefficients were significant at 1% level. In CM2DIS, The R^2 values were 0.971, 0.950 and 0.939 with lateral line lengths 40, 60 and 80 m (LLL1, LLL2, LLL3), respectively. In CM1DIS, R^2 values were 0.969, 0.941 and 0.936 with LLL1, LLL2 and LLL3, respectively. While in traditional drip system (TDIS), R^2 values were 0.956, 0.935, and 0.915 with lateral lengths 40, 60 and 80 m, respectively. It can also be concluded that best R2 was for lateral line length of 40 m under CM2DIS and CM1DIS.

6.3.2 EFFECTS DIFFERENT OPERATING PRESSURES ON DISCHARGE THROUGH LATERAL DRIP LINE AT LAND SLOPE OF 2%

Table 6.2 and Figs. 6.5–6.7 indicate that there is a direct relationship between the operating pressure and the average discharge through lateral drip lines versus lateral line length in all closed circuit designs (CM2DIS, CM1DIS and TDIS). At operating pressure of 0.8 bar in CM2DIS method, the average discharge was 4.46 lph for a lateral line length of 40 m (LLL1), compared to average discharge of 4.32 in CM1DIS, for the same length of the lateral line. With increase in operating pressure to 1.0 bar, the average discharge was 4.56 lph in CM2DIS compared to average discharge of 4.45 lph in CM1DIS, for a lateral line length of 40 m (LLL1). For LLL1/LLL2/LLL3 in the control (TDIS) and lengths 60 and 80 m (LLL2 and LLL3) in CM2DIS, CM1DIS, the average discharge did not reach the standard value of 4.00 lph at operating pressure of 1.0 bar for this type of dripper (GR built-in), as shown in Table 6.2 and Figs. 6.5–6.7. For lateral line length of 40 m (LLL1) and in CM2DIS/CM1DIS, data in Figs. 6.5–6.7 show that the average discharge was able to reach the standard value of 4.00 lph for this dripper at an operating the pressure of 0.8 bar. While for lateral line length 60 m (LLL2) in CM2DIS, the discharge was able to reach the standard value at a pressure of 1.2 bar. With comparison to TDIS under same conditions, it was not possible to reach to the standard discharge at the three lateral line lengths 40, 60 and 80 m (LLL1, LLL2, LLL3).

TABLE 6.2 Coefficients of Determination For Regression Analyzes Among the Pressure and Discharge Rates Versus Lateral Line Length for All Irrigation Methods at 2% Land Slope

| | Coefficient of determination, R^2 | | |
| | Lateral line length, m | | |
Irrigation method	40	60	80
CM2DIS	0.9756	0.9618	0.9531
CM1DIS	0.9713	0.9463	0.9251
TDIS	0.9625	0.9552	0.9314

The values of coefficient of determination (R^2) are shown in Table 6.2 and Figs. 6.5–6.7, for all treatments. The regression coefficients were significant at 1% level. In CM2DIS, The R^2 values were 0.9756, 0.9618 and 0.9531 with lateral line lengths 40, 60 and 80 m (LLL1, LLL2, LLL3), respectively. In CM1DIS, R^2 values were 0.9713, 0.9463 and 0.9251 with LLL1, LLL2 and LLL3, respectively. While in traditional drip system (TDIS), R^2 values were 0.9625, 0.9552, and 0.9314 with lateral lengths 40, 60 and 80 m, respectively. It can also be concluded that best R^2 was for lateral line length of 40 m under CM2DIS and CM1DIS.

We can observe in Figs. 6.5–6.7 that the pressure value of effective more (PVEM), at land slope of 0 and 2%, make large increase in the discharge; and the discharge did not decrease after this value. In CM2DIS at all lateral line lengths of 40, 60, and 80 m, the PVEM was 0.6 bar. Under CM1DIS with all lateral line lengths of 40, 60, and 80 m, the PVEM was 0.8 bar compared to PVENM value of 1.00 bar in traditional drip method at all lateral line lengths of 40, 60, and 80 m.

6.3.3 VALIDATION OF HYDRAULIC ANALYSIS FOR LATERAL DRIP LINES USING HYDROCAL SIMULATION PROGRAM AT LAND SLOPES OF 0% AND 2%

The discharge rates and pressures in trickle irrigation systems were measured under field conditions for land slopes of 0 and 2% at three locations down the lateral lines for CM2DIS, CM1DIS, and TDIS using three dif-

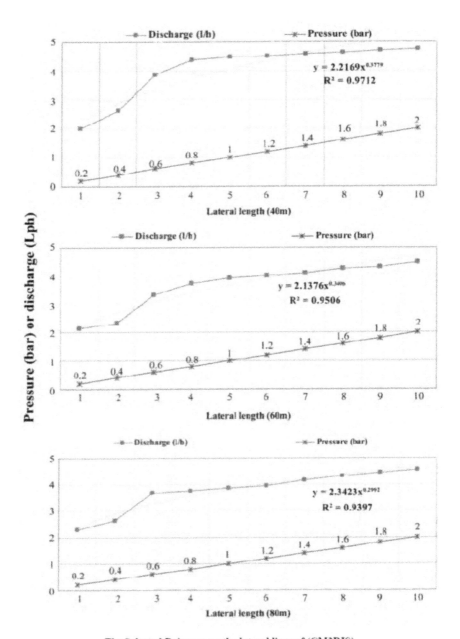

The Selected Drippers on the lateral lines of (CM2DIS)

FIGURE 6.5 Effects of different operating pressures (bar) on discharge rates of the closed circuit design with two manifolds (CM2DIS) at a land slope of 2%.

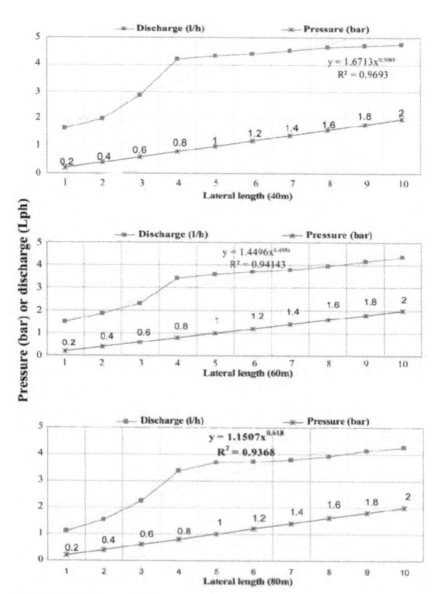

FIGURE 6.6 Effects of different operating pressures (bar) on discharge rates of the closed circuit design with one manifold (CM1DIS) at a land slope of 2%.

ferent LLL (LLL1 = 40 m, LLL2 = 60 m and LLL3 = 80 m). Empirical estimates were used to validate the trickle simulation program (Hydro-Calc Simulation program copyright 2009 developed by NETAFIM, USA).

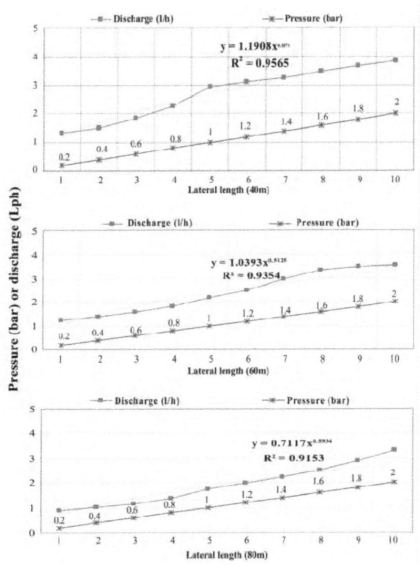

FIGURE 6.7 Effects of different operating pressures (bar) on discharge rates of the traditional drip system (TDIS) at a land slope of 2%.

HydroCalc is a computer simulation program used for planning and design of trickle or sprinkler irrigation systems. Modification of trickle irrigation closed circuit (DIC) and lateral lines lengths (LLL) depend mainly on hydraulic equations such as: Hazen-William's equations, Bernoulli's equations, etc. The data inputs provided to HydroCalc are shown in Table 6.3 for land slopes of 0 and 2%. The empirical data depended on the laboratory measurements of emitter pressure, discharge, and uniformity of water distribution.

The predicted outputs of HydroCal simulation program (exponent (X), head loss (m) and velocity (m/s)) are shown in Tables 6.4. For more details, the reader should refer to Figs. 4.23–4.25 of Chapter 4 of this volume The differences in exponent (x) values of built-in emitters are attributed to the different closed circuits and different lateral line lengths that affects the pressure and exponent (x) values.

6.3.4 PREDICTED AND MEASURED HEAD LOSS ANALYSIS ALONG THE LATERAL DRIP LINE OF CLOSED CIRCUITS AT 0% LAND SLOPE

The predicted head loss along the lateral lines was calculated by Hydro-Calc simulation program for trickle irrigation systems: CM2DIS and CM1DIS compared with TDIS under lateral drip line lengths of LLL1,

TABLE 6.3 Input Values For HydroCal Simulation Program For Closed Circuits Drip Irrigation Systems

		Lateral drip line		Emitters	
Item	Value	Name	Value	Name	Value
Pipe type	PVC	Tube type	PE	Emitter type, PE	Built-in
Pipe length	—	Tube lengths, m	40, 60, and 80 m	Emitter flow, lph	4.0
Pipe diameter	0.05 m	Inner diameter	0.0142 m	Emitter spacing	0.30 m
C: Pipe roughness	150	C: Pipe roughness	150	Pressure head required	10.0 m
Slope	0 m/m	Slope	0 or 0.02 m/m	Calculation Method	Flow rate variation
Extra energy losses	0.064	Spacing	0.7 m	—	—

TABLE 6.4 Predicted Values For Hydraulic Analysis by HydroCal Simulation Program for Closed Circuit Drip Irrigation Systems at Land Slopes of 0 and 2%

		Irrigation connection method, DIS								
		CM2DIS			CM1DIS			TDIS		
Field slope (%)	Drip line length, LLL	Exponent	Head loss	Velocity	Exponent	Head loss	Velocity	Exponent	Head loss	Velocity
%	m	x	m	m/s	x	m	m/s	x	m	m/s
0	40	0.72	0.64	1.58	0.69	0.73	1.55	0.58	1.43	1.52
	60	0.65	1.48	1.63	0.61	1.55	1.57	0.55	2.35	1.64
	80	0.58	3.00	1.92	0.52	3.11	1.88	0.53	3.58	2.18
2	40	0.76	0.45	1.51	0.71	0.76	1.51	0.63	1.38	1.51
	60	0.68	1.34	1.57	0.64	1.55	1.55	0.59	2.26	1.62
	80	0.61	2.92	1.89	0.58	3.00	1.74	0.55	3.37	1.97

DIC: Trickle irrigation circuit, LLL.: Lateral line lengths, CM2DIS: Closed circuit with two manifolds separately, CM1DIS: Closed circuit with one manifold, TDIS: Traditional trickle irrigation system.

LLL2, and LLL3 (40, 60 and 80 m). The predicted and measured head loss values are given Table 6.4. The relationships among the predicted and measured head losses at land slope of 0% are shown in Figs. 4.23–4.25 of Chapter 4 of this volume that also include regression equations under CM2DIS, CM1DIS, and TDIS methods, respectively.

Based on predicted and measured values of head loss, LLL1 and LLL3 can be ranked in the ascending order: CM2DIS < CM1DIS < TDIS. Under LLL2, the irrigation circuits can be ranked in the following ascending order: CM1DIS < CM2DIS < TDIS. The variation in the rankings may be attributed to how many emitters were built-in within lateral line length. The coefficient of determination (R^2) was used to compare the significance of the predicted and measured head loss along the lateral lines for three closed circuit designs. The deviations were significant between all predicted and measured values, except the interaction TDIS × LLL3. Generally, the values of regression coefficients were significant at 1% level, under all DIC and LLL (experimental conditions) treatments.

6.3.5 PREDICTED AND MEASURED HEAD LOSS ANALYSIS ALONG THE LATERAL DRIP LINE OF CLOSED CIRCUITS AT 2% LAND SLOPE

Figures 6.8 and 6.9 show the predicted head loss values along the lateral lines at 2% land slope that were calculated by HydroCalc simulation program for trickle irrigation systems: CM2DIS and CM1DIS compared with TDIS under lateral drip line length of LLL1, LLL2, and LLL3 (40, 60 and 80 m). The predicted and measured head losses values are given Table 6.4. The relationships among the predicted and measured head losses at land slope of 2% are shown in Figs. 6.8 and 6.9 that also include regression equations under CM2DIS, CM1DIS, and TDIS methods, respectively.

Based on predicted and measured values of head loss under LLL1 and LLL2, irrigation methods can be ranked in the ascending order: CM2DIS <CM1DIS <TDIS. Under LLL2 = 60 m, the irrigation circuits can be ranked in the ascending order: CM1DIS <CM2DIS <TDIS. The variation in the rankings may be attributed to how many emitters were built-in

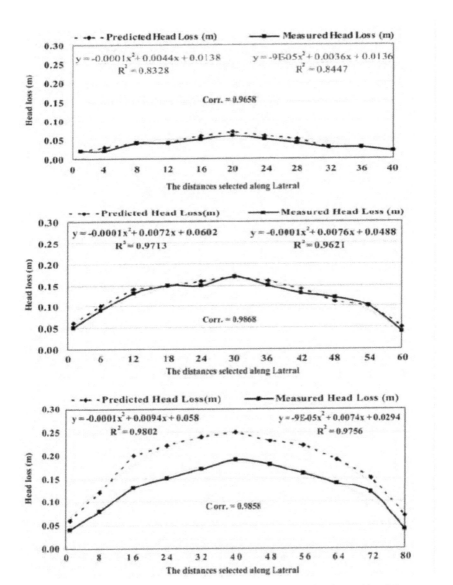

FIGURE 6.8   The relationships along predicted and measured values of head loss as a function of lateral line lengths at a land slope of 2% under closed circuit CM1DIS method.

within lateral line length. The coefficients of determination (R^2) were used to compare the significance of the predicted and measured head loss along the lateral lines for three closed circuit designs. Generally, the values of

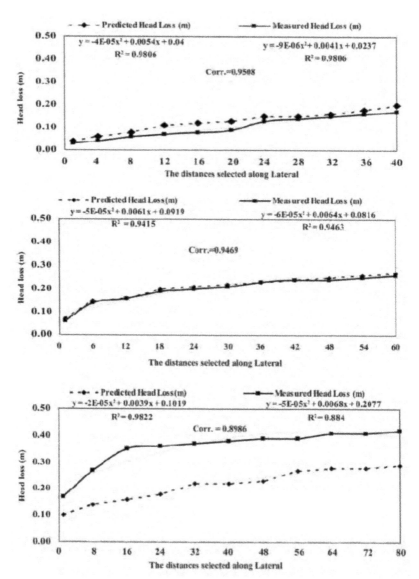

FIGURE 6.9 The relationships along predicted and measured values of head loss as a function of lateral line lengths at a land slope of 2% under traditional drip irrigation (TDIS) method.

regression coefficients were greater than 0.90. The regression coefficients were significant at 1% level, in all DIC and LLL (experimental conditions) treatments.

TABLE 6.5 Energy Saving of Closed Circuit Modified Methods Had Been Calculated by Comparing with TDIS

Field slope	Energy saving, % of irrigation method TDIS					
	CM2DIS			CM1DIS		
	Lateral line length, m					
%	40	60	80	40	60	80
0	32.27	33.21	34.37	30.84	28.96	27.45
2	31.57	33.14	34.25	30.15	28.98	27.53

6.3.6 ENERGY SAVING UNDER CLOSED CIRCUIT DESIGNS AT LAND SLOPE OF 2%

Table 6.5 indicates the values of percentage energy saving in CM2DIS and CM1DIS for three lateral line lengths of 40, 60 and 80 m at a land slope of 0 and 2%. The values were higher for a land slope of 0% compared to that for a land scope of 2%. At a land slope of 2%, percentage energy saving values were 31.57 in LLL1, 33.14 in LLL2, and 34.25 in LLL3 under CM2DIS compared to 30.15 in LLL1, 28.98 in LLL2 and 27.53 in LLL3 under CM1DIS, respectively. All values in DIS and LLL were higher than TDIS.

6.3.7 WATER USE EFFICIENCY (WUE) OF MAIZE FOR THREE CLOSED CIRCUIT DESIGNS AND THREE LATERAL LINE LENGTHS AT LAND SLOPES OF 0 AND 2%

For two land slopes, three irrigation methods (CM1DIS, CM2DIS, TDIS) and three lateral line lengths (LLL1, LLL2, LLL3), Tables 6.6 and 6.7 show values for applied water, grain yield, WUE, water requirement, actual energy, water energy, and energy use efficiency (EUE). For a 0% land slope, Table 6 indicates that WUE under CM2DIS was 1.67 in LLL1, 1.18 in LLL2, and 0.87 kg/m^3 in LLL3 compared to 1.65 in LLL1, 1.16 in LLL2, and 0.86 kg/m^3 in LLL3 under with CM1DIS. WUE under TDIS was 1.35 in LLL1, 1.04 in LLL2, and 0.75 kg/m^3 in LLL3. For a 2% land slope, Table 7 indicates that WUE under CM2DIS was 1.76, 1.29, and 0.84 kg/m^3 compared to 1.77, 1.30, and 0.87 kg/m^3 with CM1DIS

TABLE 6.6 Effect of Closed Circuits Drip Irrigation Methods on WUE and EUE When Slope Level 0%

Irrigation method, DIS	Lateral line length, LLL m	Applied water m³/ha	Yield kg/ha	WUE kg/m³	Water demand m³	Actual energy kwh	Water energy kwh	EUE %
CM2DIS	40	7725.16	12885.27	1.67	9879.73	255.74	199.97	78.19
	60	10338.91	12235.62	1.18	13583.81	322.01	245.09	76.11
	80	13757.42	12023.18	0.87	18686.05	366.59	269.90	73.62
CM1DIS	40	7638.29	12623.69	1.65	9973.74	250.02	191.48	76.58
	60	10382.71	12015.51	1.16	14509.10	328.13	234.81	71.56
	80	13782.14	11871.72	0.86	20693.90	388.50	258.74	66.60
TDIS	40	8932.25	12029.28	1.35	16865.39	407.16	215.64	52.96
	60	10652.88	11034.12	1.04	20954.56	444.78	226.12	50.85
	80	15212.70	11429.77	0.75	31484.54	514.73	248.71	48.32

DIS: Trickle irrigation circuit, LLL: Lateral line lengths, CM2DIS: Closed circuit with two manifolds separately, CM1DIS: Closed circuit with one manifold, TDIS: Traditional trickle irrigation system.

TABLE 6.7 Effect of Closed Circuit Drip Irrigation Methods on WUE and EUE at Field Slope of 2%

Irrigation method, DIS	Lateral line length, LLL m	Applied water m³/ha	Yield kg/ha	WUE kg/m³	Water demand m³	Actual energy kwh	Water energy kwh	EUE %
CM2DIS	40	7488.73	13152.71	1.76	9558.78	250.04	195.89	78.34
	60	9823.52	12641.23	1.29	12872.84	305.86	233.41	76.31
	80	14893.68	12551.34	0.84	20172.39	390.26	288.13	73.83
CM1DIS	40	7515.22	13291.25	1.77	9791.56	248.12	190.44	76.75
	60	9664.75	12538.78	1.30	13451.66	311.55	223.84	71.85
	80	13123.36	11423.16	0.87	19591.78	371.02	248.52	66.98
TDIS	40	8897.93	12512.87	1.41	16597.52	401.60	215.30	53.61
	60	10322.34	11521.87	1.12	20230.36	431.07	219.95	51.02
	80	14985.81	11318.13	0.76	30869.30	511.40	248.27	48.55

DIS: Trickle irrigation circuit, LLL: Lateral line lengths, CM2DIS: Closed circuit with two manifolds separately, CM1DIS: Closed circuit with one manifold, TDIS: Traditional trickle irrigation system.

and 1.41, 1.12, and 0.76 kg/m³ under TDIS (for lateral lengths 40, 60, and 80 m, respectively). Based on WUE values for both slopes, the descending order of irrigation methods was CM2DIS-CM1DIS-TDIS. Highest values were observed in CM2DIS for both slopes.

6.4 CONCLUSIONS

The effective pressure value was 0.6 bar in CM2DIS compared to 0.8 bar in CM1DIS and 1.00 bar in TDIS. Irrigation systems at 40, 60, 80 m can be arranged according to EUE and WUE, in the ascending order: TDIS < CM1DIS < CM2DIS. According to friction losses of lateral lines, irrigation systems can be arranged in ascending order: CM2DIS < CM1DIS < TDIS. At 0% land slope under CM2DIS, percentage of EUE were 32.27, 33.21, and 34.37% compared to 30.84, 28.96, and 27.45% under CM1DIS. For slope of 2%, EUE were 31.57, 33.14, and 34.25 with CM2DIS with CM2DIS compared to 30.15, 28.98, and 27.53 under CM1DIS for lateral line lengths of 40, 60 and 80 m, respectively. For 0% slope, WUE were 1.67, 1.18, and 0.87 kg/m³ under CM2DIS compared to 1.65, 1.16, and 0.86 kg/m³ with CM1DIS and 1.35, 1.04, and 0.75 kg/ m³ with TDIS, for three line lengths. With level slope 2%, WUE were 1.76, 1.29, and 0.84 kg/m³ under CM2DIS compared to 1.77, 1.30, and 0.87 kg/m³ with CM1DIS and 1.41, 1.12, and 0.76 kg/m³ under TDIS (for lateral lengths 40, 60, and 80 meters, respectively). Percentage of water saving varied widely within individual lateral lengths and between circuit types relative to TDIS.

6.5 SUMMARY

This chapter presents research studies on savings in energy and water under three lateral line lengths and three drip irrigation systems (DIS). Authors also compared water and energy use efficiencies, and investigated emitter discharge application uniformity and its dependence on operation pressures under these treatments. The treatments were: One manifold for lateral lines in closed circuit drip irrigation system (CM1DIS); two manifolds for lateral lines in closed circuit drip irrigation system (CM2DIS);

Traditional drip irrigation system (TDIS) as a control. Three lengths of lateral drip lines were: 40, 60, and 80 meters, at 0 and 2% of land slope. Experiments were conducted at the Agric. Res. Fields., Soil and Plant & Agric. System Department, Agric. Faculty, Southern Illinois University, Carbondale (SIUC), Illinois, USA.

The effective pressure value was 0.6 bar in CM2DIS compared to o.8 bar in CM1DIS and 1.00 bar in TDIS. Irrigation systems at 40, 60, 80 m can be arranged according to EUE and WUE, in the ascending order: TDIS < CM1DIS < CM2DIS. According to friction losses of lateral lines, irrigation systems can be arranged in ascending order: CM2DIS < CM1DIS < TDIS. At 0% land slope under CM2DIS, percentage of EUE were 32.27, 33.21, and 34.37% compared to 30.84, 28.96, and 27.45% under CM1DIS. For slope of 2%, EUE were 31.57, 33.14, and 34.25 with CM2DIS with CM2DIS compared to 30.15, 28.98, and 27.53 under CM1DIS for lateral line lengths of 40, 60 and 80 m, respectively. For 0% slope, WUE were 1.67, 1.18, and 0.87 kg/m^3 under CM2DIS compared to 1.65, 1.16, and 0.86 kg/m^3 with CM1DIS and 1.35, 1.04, and 0.75 kg/m^3 with TDIS, for three line lengths. With level slope 2%, WUE were 1.76, 1.29, and 0.84 kg/m^3 under CM2DIS compared to 1.77, 1.30, and 0.87 kg/m^3 with CM1DIS and 1.41, 1.12, and 0.76 kg/m^3 under TDIS (for lateral lengths 40, 60, and 80 meters, respectively). Water saving percent varied widely within individual lateral lengths and between circuit types relative to TDIS. Under slope 0%, water saving values were 19.26, 12.48, and 14.03% under CM2DIS compared to 18.51, 10.50, and 12.78% under CM1DIS (for three lateral line lengths). For a land slope of 2%, water savings were 19.93, 13.26, and 10.38% under CM2DIS compared to 20.49, 13.96, and 13.23% under CM1DIS (for lateral lengths 40, 60, 80 meters, respectively). The energy use efficiency and water saving were observed under CM2DIS and CM1DIS when using the shortest lateral length of 40 meters, while the lowest value was observed when using lateral length of 80 meters. These results depend on the physical and hydraulic characteristics of the emitters, lateral line uniformity, and friction losses. CM2DIS gave higher values of energy use efficiency and water saving compared to CM1DIS or TDIS.

KEYWORDS

- ascending order
- check valve
- citrus
- closed circuit
- closed circuit design
- closed circuit with one manifold, CM1DIS
- closed circuit with two manifolds, CM2DIS
- coefficient of determination
- coefficient of friction
- coefficient of variation
- continuity equation
- Darcy – Weisbach equation
- discharge through lateral line, QL
- drip irrigation
- dynamic head
- Egypt
- emitter
- emitter discharge
- emitter flow
- emitter spacing
- energy cost
- energy gradient
- energy use efficiency, EUE
- feddan
- flow velocity, FV
- friction coefficient
- friction loss, FL
- growth parameters
- Hazen–Williams equation
- head loss
- HydroCal
- irrigation circuit design, ICD

- laminar flow
- lateral line
- lateral line length, LLL
- least square difference, LSD
- maize
- micro irrigation
- operating pressure
- plastic pipe
- polyethylene, PE
- power law
- pressure
- pressure head, PH
- pressure regulator
- pressurized irrigation system
- Reynolds' number, Re
- smooth pipe
- Southern Illinois University
- traditional drip irrigation system, TDIS
- trickle irrigation
- turbulent flow
- uniformity coefficient
- vacuum relief valve
- vegetative growth
- velocity head, VH
- water
- water flow
- water use efficiency, WUE

REFERENCES

1. ASAE Standards, 2003. EP405.1, Design and installation of microirrigation systems. ASAE, St. Joseph.
2. Berkowitz, S. J., 2001. Hydraulic performance of subsurface wastewater drip systems. In: *On-Site Wastewater Treatment*. Proceedings of 9th International Symposium on Individual and Small Community Sewage Systems, ASAE, St. Joseph, 583–592.

3. Camp, C. R., Sadler, E. J. and Busscher, W. J., 1997. A comparison of uniformity measure for drip irrigation systems. *Transactions of the ASAE*, 40:1013–1020.
4. Capra, A. and Scicolone, B., 1998. Water quality and distribution uniformity in drip/ trickle irrigation systems. *Journal of Agriculture Engineering Research*, 70:355- 365.
5. FAO, 1992. Small scale pumped irrigation: energy and cost. FAO land and Water Development Division, pages 5–40.
6. Gee, G. W. and Bauder, J. W., 1986. Particle size analysis. In: *Methods of soil analysis, Part 1.* Agronomy Monograph 9, 2nd Edition, ASA and SSSA, Madison, pages 383–412.
7. Goyal, Megh R. (Ed.), 2015. *Research Advances in Sustainable Micro Irrigation*, Volumes 1 to 10. <appleacademicpress.com>
8. Megh R. Goyal, 2012. *Management of Drip/Trickle or Micro Irrigation*. Oakville, ON: Apple Academic Press Inc.,
9. Hathoot, H. M., Al-Amoud, A. I. and Mohammed, F. S., 1991. Analysis of a pipe with uniform lateral flow. *Alexandria Engineering Journal,* 30(1):C49- C54.
10. Hathoot, H. M., Al-Amoud, A. I. and Mohammed, F. S., 1993. Analysis and design of trickle irrigation laterals. *Journal of the Irrigation and Drainage Division of ASCE*, 119(5):756–767.
11. International Energy Annual, 2003. *EIA: Projection, System for the Analysis of Global Energy Markets.*
12. Jackson, M. L., 1967. *Soil chemical analysis*. Prentice Hall, Inc., Englewood Cliffs.
13. Keller, J., 2002. Evolution of drip/microirrigation: Traditional and nontraditional uses. Paper Presented as Keynote Address at the International Meeting on Advances in Drip/ Micro Irrigation, 2 to 5 December, Puerto de la Cruz, Tenerife.
14. Kirnak, H., Dogan, E., Demir, S. and Yalcin, S., 2004. Determination of hydraulic performance of trickle irrigation emitters used in irrigation system in the Harran Plain. *Turkish Journal of Agriculture and Forestry*, 28:223- 230.
15. Lamm, F. R., Trooien, T. P., Clark, G. A., Stone, L. R., Alam, M., Rogers, D. H. and Chlgel, A. J., 2002. Using beef lagoon effluent with SDI. In: Proceedings of Irrigation Association in International Irrigation Technical Conference, 24–26 October, New Orleans.
16. Mansour, Hani A., 2012.
17. Mizyed, N. and Kruse, E. G., 1989. Emitter discharge evaluation of subsurface trickle irrigation systems. *Transactions of the ASAE*, 32:1223–1228.
18. Nakayama, F. S. and Bucks, D. A., 1981. Emitter clogging effects on trickle irrigation uniformity. *Transactions on ASAE*, 24(1):77–80.
19. Pimentel, D. and Giampietro, M., 1994. *Food, Land, Population and the U.S. Economy*. Carrying Capacity Network. <http://www.dieoff.com/page55.htm>
20. Reyes, M. R., 2007. Agroforestry and sustainable vegetable production in South-east Asian watersheds. Annual Report, SANREM-CRSP, North Carolina A&T State University.
21. Talozi, S. A. and Hills, D. J., 2001. Simulating emitter clogging in a microirrigation subunit. *Transactions on ASAE*, 44(6):1503–1509.
22. Warrick, A. W. and Yitayew, M., 1988. Trickle lateral hydraulics. I: Analytical solution. *Journal of Irrigation and Drainage Engineering*, ASCE, 114(2):281–288.
23. Watters, G. Z. and Keller, J., 1978. Trickle irrigation tubing hydraulics. ASAE Technical Publication, St. Joseph, MI.
24. Wu, I. P. and Gitlin, H. M., 1979. Hydraulics and uniform for drip irrigation. *Journal of the Irrigation and Drain- age Division, ASCE*, 99(IR3):157–168.

CHAPTER 7

AUTOMATION OF MINI-SPRINKLER AND DRIP IRRIGATION SYSTEMS

H. A. MANSOUR, H. M. MEHANNA, M. E. EL-HAGAREY, and A. S. HASSAN

CONTENTS

7.1 INTRODUCTION

The continuous increasing food demand requires rapid improvement in food production technology. In a countries like India and Egypt,

---

One *feddan* (Egyptian unit for area) = 4200 m².

This chapter is combined version of *H. A. Mansour, H. M. Mehanna, M. E. El-Hagarey, and A. S. Hassan 2013. Using automation controller system and simulation program for testing closed circuits of mini-sprinkler irrigation system. Open Journal of Modeling and Simulation, 1:14–23. http://www. scirp.org/.*

Mansour, H. A., 2013. Using simulation program and automation controller under closed circuits of drip irrigation system. International Journal of Automation and Control Engineering, 2(3):128–136. www.seipub.org/ijace.

where the economy is mainly based on agriculture and the climatic conditions are isotropic, we have not been able to make full use of agricultural resources, due to the lack of rains fall and scarcity of land reservoir water. The continued extraction of water from groundwater is lowering the water level. Also a significant amount of water goes waste due to unplanned use of water. In the modern closed circuits of mini-sprinkler irrigation systems, the most significant advantage is that water is supplied near the root zone of the plants by mini-sprinkler due to which a large quantity of water can be saved. At the present era, the farmers have been using an irrigation technique with manual control gate valves to irrigate the land at the regular intervals. This process sometimes consumes more water or sometimes the water reaches late due to which the crops suffer water stress. Water deficiency can be detrimental to plants before visible wilting occurs, thus causing slower growth rate and low quality fruits. This problem can be overcome with the use of automatic irrigation controller so that field is irrigated only when necessary.

Irrigation system uses gate valves to turn the irrigation ON and OFF. These valves can be easily automated by using controllers and solenoid valves [3]. Automating farm or nursery irrigation allows farmers to apply the right amount of water at the right time, regardless of the availability of labor to turn valves on and off. In addition, farmers using automation equipment are able to reduce runoff from over watering saturated soils, avoid irrigating at the wrong time of day. Automatic closed circuit for mini-sprinkler irrigation system (MSIS) is a valuable tool to maintain adequate soil moisture control in highly specialized greenhouse vegetable production. It is simple and precise method for irrigation. It also helps in: saving time, elimination of human error in adjusting available soil moisture levels and to maximize the net profits.

The entire automation of irrigation system procedure involves two steps: To study the basic components of the irrigation system thoroughly; and then to design and implement the control circuit for automation. Therefore, the basis of closed circuit of MSIS is discussed in the following sub-sections.

7.1.1 DEFINITION OF IRRIGATION

Irrigation is the artificial application of water to the soil usually to grow crops. In crop production, it is mainly used in dry areas, in periods of short-falls of rainfall, and to protect plants against frost [3].

7.1.2 TYPES OF IRRIGATION SYSTEMS

Surface irrigation, localized irrigation (drip, mini-sprinkler, bubbler, etc.), closed circuit of MSIS, sprinkler irrigation, and hydroponic. Closed circuit MSIS, also known as mini-sprinkler irrigation system or microirrigation, is a sprinkler irrigation method which minimizes the use of water and fertilizer by allowing water through the mini-sprinklers slowly to the plant roots, either onto the soil surface or directly into the root zone, through a network of valves, pipes, tubing, and emitters [5].

7.1.3 MODERN IRRIGATION SYSTEMS

The conventional irrigation methods like overhead sprinkler and gravity irrigation systems usually wet the lower leaves and stem of the plants [3]. The entire soil surface is saturated and often stays wet long enough after the irrigation cycle is completed. Such conditions promote infestation by insects and diseases. The flood irrigation consumes large amount of water with low irrigation efficiency and the area between crop rows remains wet. On the contrary, the closed circuit MSIS is a type of modern irrigation technique that slowly applies small amounts of water near the root zone. Closed circuit of drip irrigation system (DIS) was developed by Hani Mansour [4] in 2010. Water is supplied frequently, often daily, to maintain favorable soil moisture conditions and prevent moisture stress in the plant. Closed circuit DIS saves water because only the root zone receives moisture. Minimum amount of water is lost due to deep percolation if the proper amount is applied. Closed circuits MSIS are popular because of increased crop yield and decreased crop water requirements and labor cost.

Closed circuit of drip irrigation system (CCDIS) requires about half of the water needed by other irrigation systems. Low operating pressures and flow rates can result in reduced energy costs. A high degree of water control is attainable. Plants can be supplied with more precise amounts of water. Disease and insect damage is reduced because plant foliage stays dry [4]. Operating cost is usually reduced. Farmers can easily do farm operations because rows between plants remain dry. Fertilizers can be fertigated resulting in reduction of fertilizer amounts and costs. When compared with overhead sprinkler systems, closed circuit of Mini-sprinkler irrigation system (CCMSIS) lead to less soil and wind erosion. CCMSIS can be used under a wide range of field conditions (Fig. 7.1). A typical closed circuits of drip irrigation system assembly is shown in Chapter 4 of this book [5].

This chapter discusses us of automation controller system and simulation program to evaluate closed circuit of mini-sprinkler irrigation system under Egyptian conditions.

7.2   MATERIALS AND METHODS

The field experiment was conducted at the Experimental Farm of Faculty of Agriculture, Southern Illinois University (SIUC), USA. The experimental

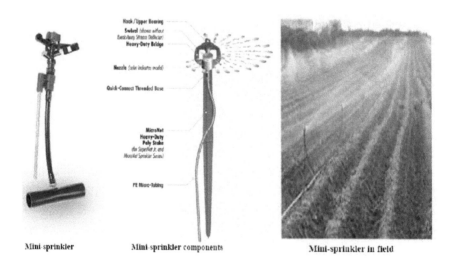

FIGURE 7.1   Mini-sprinkler irrigation system (MSIS).

design was a split-plot randomized complete block design with three replications. The field trials were conducted carried using lateral line length of
60 m, and:
1. Two mini-sprinkler irrigation circuits (MSIC):
   • One manifold for lateral lines and closed circuit mini-sprinkler
     irrigation system (CM1 MSIS); and
   • Two manifolds for lateral lines and closed circuit mini-sprinkler
     irrigation system (CM2 MSIS).
2. Two drip irrigation closed circuits (DIC):
   • One manifold for lateral lines and closed circuit drip irrigation
     system (CM1DIS); and
   • Two manifolds for lateral lines and closed circuit drip irrigation
     system (CM2DIS).
Irrigation networks are described in detail in Chapter 4 of this volume.
The components of CCMSIS include: supply lines, control valves, supply
and return manifolds, mini-sprinkler lateral lines, mini-sprinklers, check
valves and air relief valves, accessories and fittings [3]. In the CCDIS sys-
tems, emitters were used instead of mini-sprinklers.

7.2.1   IRRIGATION SCHEDULING

Irrigation intervals (I, days) in were calculated with Eq. (1) equations [3].
Net irrigation depth applied in each irrigation (d, mm) was calculated with
Eq. (2). Available soil moisture (AW in v/v, %) was determined with Eq. (3).
Irrigation interval in this study was 4 days for both CM1 MSIS and CM2
MSIS treatments.

$$I = d/(ETc) \tag{1}$$

$$d = AMD \times ASW \times Rd \times P \tag{2}$$

$$AW \text{ (v/v \%)} = ASW \text{ (w/w \%)} \times BD \tag{3}$$

where: d = net water depth applied per each irrigation (mm), and ETc = crop
evapotranspiration (mm/day) according to Goyal [3], AMD = allowable soil
moisture depletion (%), ASW = available soil water, (mm water/m depth),

Rd = effective root zone depth (m), or irrigation depth (m), and p = percentage of soil area wetted (%), and BD = Soil bulk density (gm.cm^{-3}).

7.2.2 DESIGN OF AUTOMATIC CONTROLLER BASED CLOSED CIRCUIT OF MINI-SPRINKLER IRRIGATION SYSTEM ACCORDING TO ASHOK ET AL. [1]

The following key elements should be considered while designing a mechanical model:

1. **Pressure (The force pushing the flow):** Most products operate best between 1.0 and 1.5 bars of operating pressure. Normal household pressure is 10 m (1.0 bar).
2. **Water supply and quality:** City and well water are easy to filter for CCMSIS. Pond, ditch and some well water may require special filtration equipment's. The quality and source of water will dictate the types of filtration system [3].
3. **Flow:** We can measure the output of water supply with a one or five gallon bucket and a stopwatch. Record time to fill the bucket and use the data to calculate the flow per hour: Gallons per minute × 60 = Gallons per hour.
4. **Elevation:** Variations in elevation in the field can cause a change in water pressure within the system. Pressure changes by one pound for every 2.3 foot change in elevation. Pressure-compensating emitters are designed to work in areas with large changes in elevation.
5. **Soil type and root structure:** The soil type will determine the area of coverage by a regular mini-sprinkler. Sandy soil requires closer emitter spacing as water percolates vertically at a faster rate and slower horizontally. In clay soil, water tends to spread horizontally, giving a wide distribution pattern. Emitters can be spaced further apart with clayey soil. A loamy soil will produce a more even percolation dispersion of water. Deep-rooted plants can handle a wider spacing of emitters, while shallow rooted plants are most efficiently watered (low gap emitters) with closer spacing of emitters. In clayey soil or on a hillside, short cycles repeated frequently are best. On sandy soil, applying water with

high gap emitters allows better horizontal water spread than a low gap emitter.

6. **Timing:** Watering in a regular scheduled cycle is essential. On clayey soils or hillsides, short cycles repeated frequently are best to prevent runoff, erosion and waste of water. In sandy soils, slow watering using low output emitters is recommended. Timers can help to prevent the too-dry/too-wet cycles that can cause stress and retard the plant growth. They also allow for watering at optimum times such as early morning or late evening.

7. **Watering needs:** Plants with different water needs may require their own watering circuits. For example, orchards that get watered weekly need a different circuit than a garden that must be watered daily. Plants, that are drought tolerant, will need to be watered differently than plants requiring a lot of water.

8. **The components of an automatic controller unit** in CCMSIS and CCDIS are: (a) Control head station, (b) Flow meter, (c) Control, flushing valves, (d) Chemical injection unit (Fertigation unit), (e) Manifolds and mini-sprinkler lines with emitters, (f) Moisture and temperature sensors, and (g) An automatic controller (the brain of the system). The signal sent by the sensor is boosted unto the required level by corresponding amplifier stages. Then the amplified signal is fed to A/D converters of desired resolution to obtain digital form of sensed input for use in an automatic controller (Figs. 7.2 and 7.3).

9. **Sensor**: LCD module can be used in the system to monitor current readings of all the sensors and the current status of respective valves. The solenoid valves are controlled by an automatic controller though relays [3]. A Chemical injection unit is used to mix required amount of fertilizers, pesticides, and nutrients with water, whenever required. Varying speed of pump motor can control the water pressure. Pumping unit can also be shut off and on with the help of an automatic controller unit. A flow meter is used to know total volume of water consumed. The required readings can be transferred to the centralized computer for further analytical studies, through the serial port on an automation controller unit (Figs. 7.2 and 7.3). For automation of large fields, more than one automatic controller unit can be

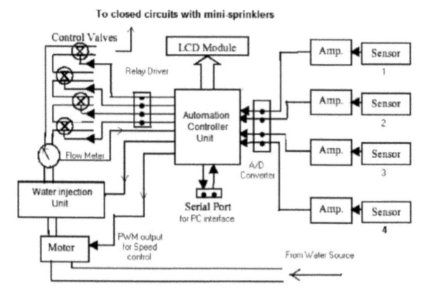

FIGURE 7.2   Circuit for automatic controller unit.

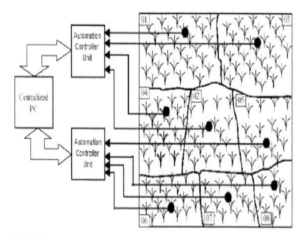

FIGURE 7.3   Field irrigation controllers.

interfaced to the centralized computer. An automatic controller unit
has a built-in timer, which operates parallel to the sensor system. In
case of sensor failure, the timer turns off the valves after a threshold
value, which can prevent further disaster. Automatic controller unit

can also warn the pump failure or insufficient amount of water input with the help of a flow meter.

7.2.3 DESCRIPTION OF AUTOMATIC CONTROLLER UNIT

The automated control system consists of moisture sensors, temperature sensors, signal conditioning circuit, digital to analog converter, LCD module, relay driver, and solenoid control valves (Fig. 7.2). The important parameters to be measured for automation of an irrigation system are soil moisture and temperature. The entire field is first divided into small sections so that each section should have one moisture sensor and a temperature sensor. RTD – PT100 can be used as a temperature sensor while Densitometer can be used as the moisture sensor to detect soil moisture content (Fig. 7.3). These sensors are buried in the ground at a required depth. Once the soil has reached desired moisture level, the sensors send a signal to the micro controller to turn off the relays, which control the valves [3].

7.2.4 USING HYDROCAL SIMULATION PROGRAM FOR HYDRAULIC CALCULATIONS

HydroCal irrigation system planning software is designed to help the user to define the parameters of an irrigation system. HydroCal program is described in detail in Chapter 4 of this volume.

7.2.5 VALIDATION OF MEASURED DATA WITH CALCULATED DATA BY HYDROCAL

The emission rate for 10 emitters or mini-sprinklers was evaluated for lateral line length at three stages (beginning, middle and end of lateral line). Hydraulic calculations for "First, middle and end of the lateral line" were calculated theoretically. The head loss due to friction and insertion of emitters or mini-sprinklers was calculated and then the pressure head at every emission point was determined. The flow from emission point was calculated using the characteristic equation developed for pressure head vs. discharge for the emitter or mini-sprinkler.

7.2.6 STATISTICAL ANALYSIS

MSTATC program (Michigan State University) was used to carry out the statistical analysis. Treatments mean were compared using the technique of analysis of variance (ANOVA) and the least significant difference (L.S.D.) among the systems at 1% were determined for the randomized complete block design according to Steel et al. [9].

7.3 RESULTS AND DISCUSSIONS

7.3.1 VALIDATION OF LATERAL LINES HYDRAULIC ANALYSIS BY HYDROCAL SIMULATION PROGRAM FOR LATERAL LINES WITH 0% AND 7% DOWNWARD LAND SLOPE: MINI-SPRINKLERS

The discharge rates and pressures for the mini-sprinkler heads were measured under field conditions at three sites along the lateral lines (start, middle and end) for CM2 MSIS, CM1 MSIS, and TDIS with lateral line length of 60 m and for two different slopes of lateral line (0 and 7%). Empirical measurements were used to validate the mini-sprinkler simulation program (HydroCal Simulation program copyright 2009 developed by NETAFIM, USA). Modification of closed circuit mini-sprinkler lateral lines depended on hydraulic equations such as, Hazen-William's and Bernolli equations, etc. The data inputs for HydroCal are shown in Table 7.1. The empirical data depended on the laboratory measurements of pressures and discharge, as well as the field uniformity.

    Table 7.2 and Figs. 7.4–7.7 show the predicted outputs of HydroCal simulation program (exponent X, pressure head loss in m, velocity in m/s, and pressure along the lateral mini-sprinkler lines).

TABLE 7.1 Inputs for the HydroCal Simulation Program for Closed Circuit Design

Closed circuit design for mini-sprinkler irrigation systems					
Manifold		Mini-sprinkler line		Mini-sprinkler	
Name	Value	Name	Value	Name	Value
Pipe type	PVC	Tube type	PE	Mini-sprinkler type	Online
Pipe length	—	Tube length	60 m	Mini-sprinkler flow, lph	12.0 lph
Pipe diameter	0.05 m	Inner diameter	0.0235 m	Mini-sprinkler distance	0.50 m
C, Pipe roughness	150	C, Pipe roughness	150	Pressure head required m	10.0 m
Slope	0 m/m	Slope	0 or 0.03 m/m	Calculation method	Flow rate variation
Extra energy losses	0.064	Spacing	0.7 m	—	—

Closed circuit design drip irrigation system (CCDIS)					
Manifold		Drip line		Emitters	
Pipe type	PVC	Tube type	PE	Emitter type	Built-in
Pipe length	—	Tube length	60 m	Emitter flow	4 lph
Pipe diameter	0.05 m	Inner diameter	0.0142 m	Emitters distance	0.30 m
C, Pipe roughness	150	C. Pipe roughness	150	Pressure head required	10.0 m
Slope	0 m/m	Slope	0 or 0.05 m/m	Calculation method	Flow rate variation
Extra energy losses	0.064	Spacing	0.7 m	—	—

7.3.2 PREDICTED AND MEASURED HEAD LOSS ANALYSIS ALONG THE LATERAL LINE FOR CLOSED CIRCUIT DESIGNS WITH 0% LAND SLOPE: MINI-SPRINKLERS

The predicted head loss analysis along the lateral Mini-sprinkler's line was done by HydroCal simulation program for closed circuits mini-sprinkler

TABLE 7.2   Predicted Exponent (x), Head Loss (m) and Velocity (m/s) by the HydroCal simulation Program For Closed Circuit Mini-Sprinkler Irrigation Design With Slopes of 0 and 7%

Field slope (%)	Irrigation connection design					
	CM2 MSIS			CM1 MSIS		
	Exponent (x)	Head loss (m)	Velocity (m/s)	Exponent (x)	Head loss (m)	Velocity (m/s)
0	0.72	0.64	1.58	0.69	0.73	1.55
	0.65	1.48	1.63	0.61	1.55	1.57
	0.58	3.00	1.92	0.52	3.11	1.88
7	0.76	0.45	1.51	0.71	0.76	1.51
	0.68	1.34	1.57	0.64	1.55	1.55
	0.61	2.92	1.89	0.58	3.00	1.74

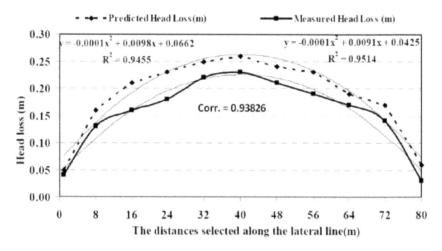

FIGURE 7.4   The relationship between the predicted and measured head losses, for lateral line length of 60 m and land slope of 0%, under the CM2 MSIS design: Mini-sprinklers.

irrigation systems (CM2 MSIS and CM1 MSIS with zero land slope and with lateral line length 60 m. Table 7.3 and Figs. 7.4 and 7.5 show the relationships among the predicted and measured head losses for CM2 MSIS and CM1 MSIS methods with no slope 0%. The correlation (Corr.) coef-

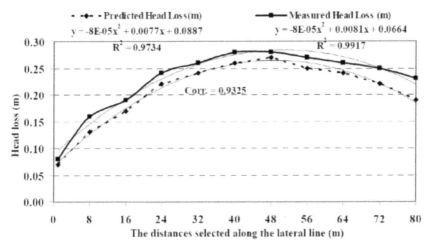

FIGURE 7.5   The relationship between the predicted and measured head losses, for lateral line length of 60 m and land slope of 0%, under the CM1 MSIS design: Mini-sprinklers.

ficients were obtained to compare the significance level of the predicted and measured head loss along the lateral lines of the two closed circuits designs. Generally, the values of correlation coefficients were >0.90 for 0% field slope.

Data in Table 7.3 show that the irrigation closed circuit designs can be ranked in the ascending order: CM1 MSIS < CM2 MSIS, according to the values of the pressure head. Friction losses for the TDIS method were higher than closed circuit systems. The $LSD_{0.01}$ values concluded that there was no significant difference between both start and end values, in CM1 MSIS. The $LSD_{0.01}$ values indicate that under CM2 MSIS there is no significant difference between start and end values of pressure head but there are significant differences between middle value and both start and end pressure head values. While under CM1 MSIS, there are significant differences between the all pressure head values of start, middle and end of lateral line length of 60 m. The interaction between irrigation methods indicates that at the start no significant differences were found between CM2 MSIS and CM1 MSIS. These results are in agreement with those reported by other investigators [2, 6, 7, 8, 10–13].

TABLE 7.3 Pressure Head Analysis Along The Lateral Lines In Closed Circuit Mini-Sprinkler Irrigation Systems (CM2 MSIS and CM1 MSIS) For Land Slope of 0%

| Distance along lateral (m) | Mini-sprinkler irrigation circuits | | | |
| | CM2 MSIS | | CM1 MSIS | |
	Predicted	Measured	Predicted	Measured
1	0.93	0.93	0.94	0.90
6	0.88	0.91	0.93	0.88
12	0.83	0.90	0.90	0.86
18	0.83	0.85	0.87	0.84
24	0.82	0.82	0.85	0.82
30	0.81	0.79	0.82	0.80
36	0.83	0.82	0.81	0.81
42	0.83	0.86	0.80	0.83
48	0.85	0.88	0.82	0.85
54	0.89	0.90	0.84	0.87
60	0.94	0.92	0.87	0.89
Average	**0.85**	**0.87**	**0.86**	**0.86**
$LSD_{0.01}$	**0.03**	**0.07**	**0.05**	**0.02**

7.3.3 PREDICTED AND MEASURED HEAD LOSS ANALYSIS ALONG THE LATERAL LINE FOR CLOSED CIRCUITS WITH A 7% DOWNWARD LAND SLOPE FROM THE MANIFOLD TO THE TERMINAL END: MINI-SPRINKLERS

Table 7.4 and Figs. 7.6 and 7.7 show the relationships between predicted and measured head losses as well as the correlation coefficients, for lateral line length of 60 m at 7% land slope downward under two closed circuit mini-sprinkler irrigation systems (CM2 MSIS and CM1 MSIS). Irrigation methods with lateral line length of 60 m can be ranked in the ascending order according the values of the predicted and measured head losses: CM1 MSIS < CM2 MSIS.

While with lateral line length of 60 m, the values of the predicted and measured head losses under irrigation methods can be ranked in the ascending orders: CM2 MSIS < CM1 MSIS. This may be attributed to the differ-

TABLE 7.4   Pressure Head Loss Analysis Along The Lateral Line in Closed Circuit Mini-Sprinkler Irrigation (CM2 MSIS and CM1 MSIS) Method, For a Land Slope of 7% Downward

| Distance along laterals (m) | Mini-sprinkler irrigation circuits | | | |
| | CM2 MSIS | | CM1 MSIS | |
	Predicted	Measured	Predicted	Measured
1	0.94	0.95	0.94	0.91
6	0.90	0.91	0.93	0.89
12	0.86	0.87	0.90	0.88
18	0.85	0.85	0.87	0.84
24	0.84	0.85	0.85	0.81
30	0.83	0.83	0.82	0.78
36	0.84	0.85	0.81	0.79
42	0.86	0.87	0.80	0.80
48	9.89	0.88	0.82	0.81
54	0.90	0.90	0.84	0.82
60	0.93	0.96	0.87	0.83
Average	**0.88**	**0.88**	0.86	0.83
$LSD_{0.01}$	**0.09**	**0.04**	**0.02**	**0.04**

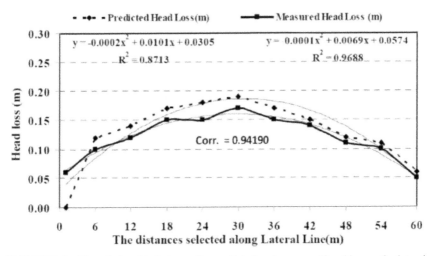

FIGURE 7.6   The relationship between the predicted and measured head losses, for lateral line length of 60 m and land slope of 7%, under the CM2 MSIS design: Mini-sprinklers.

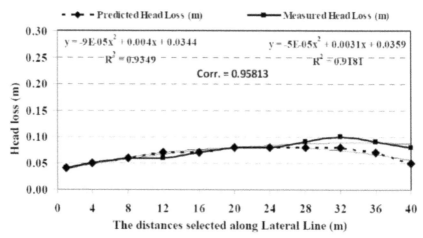

FIGURE 7.7 The relationship between the predicted and measured head losses, for lateral line length of 60 m and land slope of 7%, under the CM1 MSIS design: Mini-sprinklers.

ent number of mini-sprinklers or how many mini-sprinklers were built-in in each lateral line.

The correlation coefficients were used to compare the predicted and measured head losses along the lateral lines of all closed circuits designs. Generally, the values of correlation coefficients > 0.90 were obtained with 7% field slope and 60 m length of lateral line for all closed circuits.

Data in Table 7.4 and Figs. 7.6 and 7.7 show that the head loss along the lateral lines of the different closed circuit designs (CM2 MSIS and CM1 MSIS). According to lateral line length of 60 m, the values of the pressure head under irrigation methods can be arranged in the ascending orders: CM1 MSIS < CM2 MSIS. This may be attributed to the decreased head loss along lateral line length by using the modified method CM2 MSIS and CM1 MSIS. The $LSD_{0.01}$ values in Table 7.4 show that under CM2 MSIS and CM1 MSIS there is no significant difference between start and end values of pressure head but there are significant differences between middle values and both start and end pressure head values. The interaction between irrigation methods indicates that at the start there are significant differences between CM2 MSIS and CM1 MSIS. While at both of end and middle there are significant differences between all irrigation methods. These results are in agreement with those reported by other investigators [2, 6–8, 10–13].

7.3.4 VALIDATION OF HYDRAULIC ANALYSIS
FOR LATERAL LINE BY HYDROCALC SIMULATION
PROGRAM FOR LAND SLOPE OF 0% AND 5% DOWNWARD:
DRIP IRRIGATION SYSTEM

The discharge rates and pressures at the drip head were measured under field conditions at three locations along the lateral lines (start, middle and end) for CM2DIS, CM1DIS, and TDIS with lateral line length of 60 m and for two different land slopes (0 and 5%). Empirical measurements were used to validate the HydroCal drip simulation program (HydroCalc Simulation program copyright 2009 developed by NETAFIM, USA).

The data inputs to HydroCalc are shown in Table 7.1. The empirical data depended on the laboratory measurements of pressures and discharge, and the field uniformity. The predicted outputs of HydroCalc simulation program (exponent, X, pressure head loss, velocity, and pressure along the lateral line drippers) are shown in Table 7.5.

7.3.5 PREDICTED AND MEASURED HEAD LOSS ANALYSIS
ALONG THE LATERAL DRIP LINE OF CLOSED CIRCUIT
DESIGN WITH 0% SLOPE: DRIP IRRIGATION SYSTEM

The predicted head loss values along the lateral drip line were calculated by HydroCalc simulation program for closed circuits drip irrigation systems

TABLE 7.5 Predicted Exponent (x), Head Loss (m) and Velocity (m/s) by the HydroCal Simulation Program For Closed Circuit Drip Irrigation Design With Different Slopes (0 and 5%)

	Irrigation connection design					
	CM2DIS			CM1DIS		
Field slope (%)	Exponent (x)	Head loss (m)	Velocity (m/s)	Exponent (x)	Head loss (m)	Velocity (m/s)
0	0.72	0.64	1.58	0.69	0.73	1.55
	0.65	1.48	1.63	0.61	1.55	1.57
	0.58	3.00	1.92	0.52	3.11	1.88
5	0.76	0.45	1.51	0.71	0.76	1.51
	0.68	1.34	1.57	0.64	1.55	1.55
	0.61	2.92	1.89	0.58	3.00	1.74

(CM2DIS and CM1DIS) with no slope (0%) and with lateral line length of 60 m. Those predicted values are tabulated in Table 7.6. Figures 7.8 and 7.9 and Table 7.6 show the relationships among the predicted and measured head losses and the correlations under CM2DIS and CM1DIS methods with 0% slope.

The irrigation methods can be ranked in the ascending order: CM1DIS < CM2DIS< TDIS. The correlation coefficients were obtained to compare the significance level of the predicted and measured head loss along the lateral lines of the two closed circuits designs. Generally, the values of correlation coefficients were >0.90.

Data in Table 7.6 show the pressure head loss along the lateral lines of the two closed circuit designs (CM2DIS and CM1DIS). Clearly the irrigation closed circuits designs under study can be ranked in the ascending order: CM1DIS < CM2DIS according to the values of the pressure head loss. Possibly this was due to increased friction losses for the traditional TDIS method. LSD at P = 0.01 values in Table 7.6 show no significant

TABLE 7.6   Pressure Head Loss Values Along the Lateral Drip Lines in Drip Irrigation Closed Circuit (CM2DIS and CM1DIS) Methods at 0% Slope

| Distance along laterals (m) | Drip irrigation circuits | | | |
| | CM2DIS | | CM1DIS | |
	Predicted	Measured	Predicted	Measured
1	0.93	0.93	0.94	0.91
6	0.88	0.91	0.93	0.89
12	0.83	0.90	0.90	0.88
18	0.83	0.85	0.87	0.84
24	0.82	0.82	0.85	0.81
30	0.81	0.79	0.82	0.78
36	0.83	0.82	0.81	0.79
42	0.83	0.86	0.80	0.80
48	0.85	0.88	0.82	0.81
54	0.89	0.90	0.84	0.82
60	0.94	0.92	0.87	0.83
Average	0.858	0.87	0.859	0.833
LSD0.01	0.05	0.09	0.03	0.04

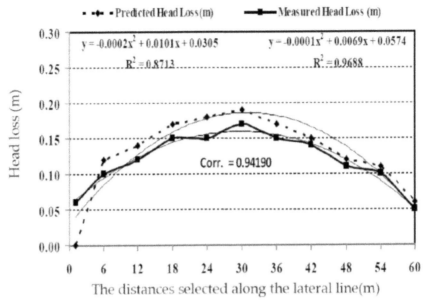

FIGURE 7.8 The relationship between lateral line length of 60 m and both the predicted and measured head losses at land slope of 0% with the CM2DIS design.

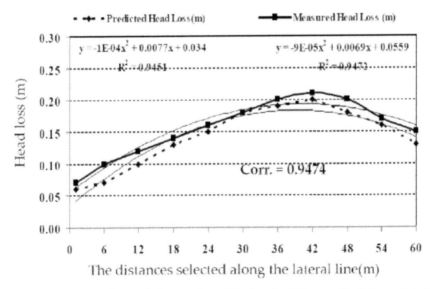

FIGURE 7.9 The relationship between lateral line length of 60 cm and both the predicted and measured head losses at land slope of 0% with the CM1DIS design.

difference between both start and end values of head loss. LSD at P = 0.01 values indicate that under CM2DIS there is no significant difference between both start and end values of pressure head but there are significant differences between middle value and both start and end pressure head values.

While under CM1DIS there are significant differences between the all pressure head loss values at start, middle and end lateral length 60 m. The interaction between irrigation methods at the start showed no significant differences between CM2DIS and CM1DIS. These data are in agreement with Mansour et al. [4–6], Tayel et al. [10–13], and Mizyed and Kruse [7].

7.3.6 PREDICTED AND MEASURED HEAD LOSS VALUES ALONG THE LATERAL DRIP LINE OF CLOSED CIRCUITS AT A LAND SLOPE OF 5% DOWNWARD FROM THE MANIFOLD TO THE FARTHEST END: DRIP IRRIGATION SYSTEM

The predicted head loss for a land slope of 5% along the lateral drip line was calculated by HydroCalc simulation program for closed circuits drip irrigation systems (CM2DIS and CM1DIS). Figures 7.10 and 7.11 and Table 7.7 show the relationship between predicted and measured head losses and the correlation coefficients. Irrigation methods under study with lateral length of 60 m can be ranked in the ascending order according the values of the predicted and measured head losses: CM1DIS < CM2DIS.

While by using lateral length of 60 m, the values of the predicted and measured head losses under irrigation methods can be ranked in the ascending orders; CM2DIS <CM1DIS. This may be attributed to the different number of drippers or how many drippers were built-in in each lateral line length. This may also be attributed to the decreased head loss in lateral line length by using the modified method CM2DIS and CM1DIS. The correlation coefficients were used to compare the predicted and measured head losses along the lateral lines for all the closed circuits designs. Generally, the values of correlation analysis > 0.90 were obtained with 5% field slope for 60 m length (experimental conditions) for all closed circuits.

LSD0.01 values in Table 7.7 show that under CM2DIS and CM1DIS there is no significant difference between both start and end values of

TABLE 7.7 Pressure Head Losses Along the Lateral Drip Lines in Drip Irrigation Closed Circuits (CM2DIS and CM1DIS) Method at 5% Slope

Distance Along Laterals (m)	Drip irrigation circuits			
	CM2DIS		CM1DIS	
	Predicted	Measured	Predicted	Measured
1	0.94	0.95	0.94	0.90
6	0.90	0.91	0.93	0.88
12	0.86	0.87	0.90	0.86
18	0.85	0.85	0.87	0.84
24	0.84	0.85	0.85	0.82
30	0.83	0.83	0.82	0.80
36	0.84	0.85	0.81	0.81
42	0.86	0.87	0.80	0.83
48	9.89	0.88	0.82	0.85
54	0.90	0.90	0.84	0.87
60	0.93	0.96	0.87	0.89
Average	**0.878**	**0.884**	0.859	0.855
LSD0.01	0.11	0.14	0.12	0.11

pressure head but there are significant differences between middle value and both start and end pressure head values. The interactions, between irrigation methods at the start, are significant between CM2DIS and CM1DIS. Also at the end and middle, there are significant differences between all irrigation methods. These data are in agreement with Mansour et al. [4–6], Tayel et al. [10–13], and Mizyed and Kruse [7].

7.4 CONCLUSIONS

7.4.1 CCMSIS

The automatic controller for CCMSIS is a real time feedback control system, which monitors and controls all activities of the irrigation system efficiently. Automation can save manpower, water to increase production and ultimately profit. Irrigation methods with lateral line length of 60 m can ranked in the ascending order according the values of the predicted

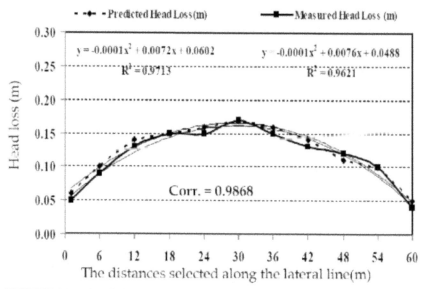

FIGURE 7.10   The relationship between lateral line length of 60 m and both the predicted and measured head loss at a land slope of 5% downward with the CM2DIS design.

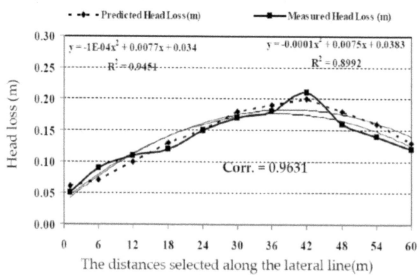

FIGURE 7.11   The relationship between lateral line length of 60 m and both the predicted and measured head losses at a land slope of 5% downward with the CM1DIS design.

and measured head losses: CM1 MSIS < CM2 MSIS. Generally, the values of correlation coefficients were >0.90 for all closed circuits. The interaction between irrigation methods gave significant differences between CM2 MSIS and CM1 MSIS.

7.4.2 CCDIS

Irrigation methods under study with lateral line length of 60 m can be ranked in the ascending order according the values of the predicted and measured head losses: CM1DIS < CM2DIS. Generally, the values of correlation coefficients > 0.90 were obtained at 0% field slope with 60 m length (experimental conditions) for all closed circuits. The interaction between irrigation methods at the start are significant between CM2DIS and CM1DIS.

7.5 SUMMARY

7.5.1 CCMSIS

The field experiments were conducted at the experimental farm of Faculty of Agriculture, Southern Illinois University SIUC, USA. The study uses the automated irrigation automated. The use of low cost sensors and the simple circuit makes it a low cost product, which can be bought even by a poor farmer. This research work is best suited in water scarcity regions. This chapter presents a model to modernize the agriculture at a mass scale with optimum expenditure. In the field of agricultural engineering, use of sensor method of irrigation operation is important and it is well known that CCMSIS are very economical and efficient. Closed circuits are considered one of the modifications of mini-sprinkler irrigation system, which can relieve problem of low operating pressures at the end of the lateral lines. In the conventional closed circuits of mini-sprinkler irrigation system, the farmer has to keep watch on irrigation timetable, which is different for different crops.

Irrigation methods can be ranked in the ascending order according the values of the predicted and measured head losses: CM1 MSIS < CM2

MSIS. Generally, the values of correlation coefficients were > 0.90 with 0% field slope for all closed circuits. The interaction between irrigation methods was significantly different between CM2 MSIS and CM1 MSIS.

7.5.2. CCDIS

Irrigation methods can be ranked in the following ascending order according the values of the predicted and measured head losses: CM1DIS < CM2DIS. Generally, the values of correlation coefficients >0.90 were obtained at 0% field slope with 60 m length (experimental conditions) for all closed circuits. The interaction between irrigation methods at the start are significant between CM2DIS and CM1DIS.

KEYWORDS

- automatic controller
- CM1DIS
- CM2DIS
- CM1 MSIS
- CM2 MSIS
- closed circuit
- drip irrigation
- Egypt
- evapotranspiration
- gravity irrigation
- head loss
- HydroCal
- India
- irrigation
- irrigation interval
- mini-sprinkler

- **operating pressure**
- **sensor**
- **simulation program**
- **sprinkler irrigation**
- **surface irrigation**

REFERENCES

1. Ashok P. and K. Ashok, 2010. Microcontroller based drip irrigation system. International Journal & Magazine of Engineering, Technology, Management and Research, http://www.yuvaengineers.com/.

2. Burt, C. M., A. J. Clemens, T. S. Strelkoff, K. H. Solomon, R. D. Blesner, L. A. Hardy, and T.A. Howell, 1997. Irrigation performance measures: Efficiency and uniformity. *J. Irrig. and Drain. Div.*(ASCE), 123(6):423–442.

3. Megh R. Goyal, 2012. *Management of Drip/Trickle or Micro Irrigation*. Apple Academic Press Inc., Oaksville, ON, Canada.

4. Mansour, H. A., M. Y. Tayel, David A. Lightfoot, A. M. El-Gindy, 2010. Energy and water saving by using closed circuits of mini-sprinkler irrigation systems. *Agriculture Science Journal*, 1(3):1–9.

5. Mansour, H. A., 2012. *Design considerations for closed circuits of drip irrigation system*. PhD Thesis, Faculty of Agriculture, Agric., Ain Shams university, Egypt.

6. Mansour, H.A. and Abdullah S. Aljughaiman, 2012.Water and fertilizers use efficiency of corn crop under closed circuits of drip irrigation system. *Journal of Applied Sciences Research*, 8(11):5485–5493.

7. Mizyed, N. and E. G. Kruse. 1989. Emitter discharge evaluation of subsurface trickle irrigation systems. *Transactions of the ASAE*, 32:1223–1228.

8. Smajstrla, A. G. and G. A. Clark, 1992. Hydraulic performance of microirrigation mini-sprinkler tape emitters. ASAE paper no 92–2057. St Joseph, MI:ASAE.

9. Steel, R. G. D and J. H. Torrie, 1980. *Principles and Procedures of Statistics*. A biometrical approach. 2nd Ed., McGraw Hill Inter. Book Co. Tokyo, Japan.

10. Tayel, M. Y., A. M. El-Gindy and H. A. Mansour, 2012. Effect of drip irrigation circuit design and lateral line lengths, III- on dripper and lateral discharge. *Journal of Applied Sciences Research*, 8(5):2725–2731.

11. Tayel, M. Y., A. M. El-Gindy and H. A. Mansour, 2012. Effect of drip irrigation circuit design and lateral line lengths, IV- on uniformity coefficient and coefficient of variation. *Journal of Applied Sciences Research*, 8(5):2741–2748.

12. Tayel, M. Y., H. A. Mansour and David A. Lightfoot, 2012. Effect of drip irrigation circuit design and lateral line lengths I- on Pressure head and friction loss. *Agric. Sci.*, 3(3):392–399.

13. Tayel, M. Y., H. A. Mansour and David A. Lightfoot, 2012. Effect of drip irrigation circuit design and lateral line lengths, II- on flow velocity and velocity head. *Agric. Sci.*, 3(4):531–537.

APPENDIX I: TYPICAL PHOTOS OF IRRIGATION AUTOMATION

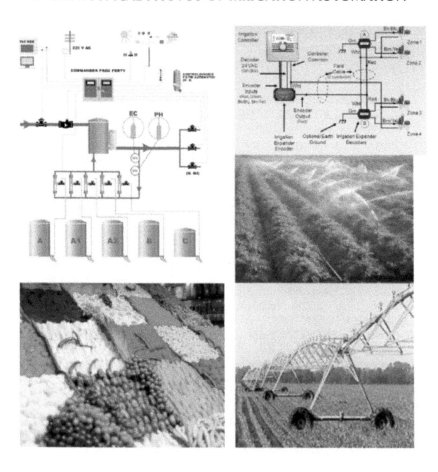

PART III

APPLICATIONS IN SANDY SOILS

CHAPTER 8

WATER AND FERTILIZER USE EFFICIENCIES FOR DRIP IRRIGATED MAIZE

H. A. MANSOUR, M. A. EL-HADY, and C.S. GYURCIZA

CONTENTS

8.1   INTRODUCTION

Water use efficiency (WUE) of Maize is a function of physiological characteristics of maize, genotype, soil water holding capacity, meteorological conditions and agronomic practices. To improve WUE, integrative measures should aim to optimize cultivar selection and agronomic practices. The most

---

One *feddan* (Egyptian unit of area) = 4200 m².

Modified and printed from *H. A. Mansour, M. A. El-Hady and C.S. Gyurciza, 2013. Effect of localized irrigation systems and humic compost fertilizer on water and fertilizer use efficiency of maize in sandy soil. International Journal of Agricultural Science Research, 2(10, October):292–297. Open access article available at http://academeresearchjournals.org/journal/ijasr.*

important management interaction in many drought-stressed maize environments is between soil fertility management and water supply. In areas subject to drought stress, many farmers are reluctant to economic loss risk by applying fertilizer, strengthening the link between drought and low soil fertility [1]. Ogola et al. [16] reported that the WUE of maize was increased by application of nitrogen. They added that maize plants are especially sensitive to water stress because of sparse root system.

Humic compost is a final component of organic matter decomposition, and it can hold more moisture, which will increase water use efficiency in the amended sandy soil when compared to the unamended one. This may be due to the phenomenon of swelling and retention of water by the amended soil. On the other hand, humic substances are able to attach to metal ions, which will decrease nutrients leaching with irrigation water, and increase fertilizer use efficiency (FUE). Humic substances are relatively stable products of organic matter, and accumulate in the environmental systems to increase moisture retention and nutrient supply potentials of sandy soils. Laboski et al. [11] found that maize yield responded to amount of water applied by trickle irrigation.

Increasing the plant population density usually increases maize grain yield until an optimum number of plants per unit area is reached [7]. Fulton [5] also reported that higher plant densities of maize produce higher grain yields. Plant densities of 90,000 plants/ha for maize are common in many regions of the world [15]. The FUE of plant nutrients depends upon various aspects of fertilizer application like rate, method, time, type of fertilizer, crop and soil in addition to other factors. Proper method and time of fertilizer application is inevitable to reduce the losses of plant nutrients and is important for a fertility program to be effective. Nitrogenous fertilizers should be applied in split doses for the long season crops. Similarly nitrogen should not be applied to sandy soil in a single dose, as there are more chances for nitrate leaching [3]. Phosphate fertilizers applications are also of great concern. When applied to soil, they are often fixed or rendered unavailable to plants, even under the most ideal field conditions. In order to prevent rapid reaction of phosphate fertilizer with the soil, the materials are commonly banded. To minimize the contact with soil, pelleted or aggregated phosphate fertilizers are also recommended [4]. Much of the phosphate is available to the plant early stage for row crops. Similarly, data collected on the yield of maize showed that application of all phosphorus at sowing was better than its late application.

Memon [14] concluded that phosphorus uptake by plant roots depend upon the phosphorus uptake properties of roots and the phosphorus supplying properties of soil. He also added that maximizing the uniformity of water application is one of the easier ways to save water, at the farm level. The evaluation of the emission uniformity (EU) of the trickle system should be done periodically. In comparison studies between different irrigation systems Mansour [12] found that the increases in both WUE and water utilization efficiency at the 2nd season relative to the 1st one were maximum under drip irrigation system (42 and 43%, on ranking), followed by the low head bubbler irrigation system (40.7 and 37%), while the minimum ones were (30.6 and 32%, on ranking) under gated pipe irrigation system. Also he found that the increases in FUE of N, P_2O_5, and K_2O at 2nd season relative to the 1st one were (24, 23, 28%), (22%, 21%, 27%) and (9%, 8%, 14%) under drip irrigation system, low head bubbler irrigation system and gated pipe irrigation system, on ranking.

This chapter discusses the effects of localized irrigation systems (mini-sprinkler irrigation system, MSIS; bubbler irrigation system, BIS; drip irrigation system, DIS) and humic fertilizer treatments (HF: HF100 = 100 kg/fed, HF50 = 50 kg/fed and HF0 = 0 kg/fed) on WUE and FUE of maize crop under Egyptian desert conditions.

8.2   MATERIALS AND METHODS

Field experiment was conducted at the Experimental Farm of Agricultural Division, National Research Center, El-Noubaria Governor, Egypt. Maize crop (Zea mays, L cv. Gemizza 9) was grown in sandy soil during the growing season (2012/2013). Tables 8.1–8.3 show general characteristics of soil and irrigation water. The treatments were:
1.   Localized irrigation systems (LIS) were:
   •   Mini-sprinkler irrigation system (MSIS),
   •   Bubbler irrigation system (BIS), and
   •   Drip irrigation system (DIS).

TABLE 8.1 Some Physical Properties of the Soil

| Depth | Particle Size distribution, % | | | | Texture class | θS % on weight basis | | | HC | BD g/cm³ | P cm³ voids/cm³ soil |
cm	C. Sand	F. Sand	Silt	Clay		FC	WP	AW	cm/h		
0–15	8.4	77.6	8.5	5.5	Sandy	14.0	6.0	8.0	6.68	1.69	0.36
15–30	8.6	77.7	8.3	5.4	Sandy	14.0	6.0	8.0	6.84	1.69	0.36
30–45	8.5	77.5	8.8	5.2	Sandy	14.0	6.0	8.0	6.91	1.69	0.36
45–60	8.8	76.7	8.6	5.9	Sandy	14.0	6.0	8.0	6.17	1.67	0.37

* Particle Size Distribution after Gee and Bauder [6] and Moisture retention after Klute [10].

FC: Field Capacity, WP: Wilting Point, AW: Available Water, HC: Hydraulic conductivity (cm/h), BD: Bulk density (g/cm³) and P: Porosity (cm³ of voids/cm³ of soil); C = Coarse, F = Fine.

TABLE 8.2   Some Chemical Properties of the Soil

Depth, cm	pH 1:2.5	EC dS/m	Soluble Cations, meq/L				Soluble Anions, meq/L			
			Ca++	Mg++	Na+	K+	CO_3-	HCO_3-	SO_4-	Cl–
0–15	8.3	0.35	0.50	0.39	1.02	0.23	0	0.11	0.82	1.27
15–30	8.2	0.36	0.51	0.44	1.04	0.24	0	0.13	0.86	1.23
30–45	8.3	0.34	0.56	0.41	1.05	0.23	0	0.12	0.81	1.23
45–60	8.4	0.73	0.67	1.46	1.06	0.25	0	0.14	0.86	1.22

*Chemical properties after Rebecca [17].

TABLE 8.3   Some Chemical Properties of Irrigation Water

pH	EC dS/m	Soluble cations, meq/L				Soluble anions, meq/l				SAR
		Ca++	Mg++	Na+	K+	CO_3-	HCO_3-	SO_4-	Cl–	
7.3	0.37	0.76	0.24	2.6	0.13	0	0.9	0.32	2.51	4.61

2. Humic fertilizer treatments (HF) were:
    • HF100 = 100 kg/fed,
    • HF50 = 50 kg/fed, and
    • HF0 = 0 kg/fed.

The experiment design was split plot complete randomized design with three replications. Treatment means were compared using the technique of analysis of variance (ANOVA) and the least significant difference (L.S.D) at 1% according to Steel and Torrie [18].

The total experimental area was 504 m². Each of the localized irrigation system (LIS) was 168 m². For each humic fertilizers (HF) treatment, plot area was 56 m². The complete description of irrigation system was given by Mansour [13] and Tayel et al. [19–22]. Maize seeds were sown 12nd of May at row-to-row spacing of 0.7 m and hill-to-hill spacing of 0.25 m down the row. Plant density was 24000 plant/fed. Each row was drip irrigated by a single straight lateral line according to the daily reading of Class A pan evaporation. Irrigation frequency was 4 days. The amount of irrigation water need per irrigation was calculated according to the following equation:

$$IWA = 4.2 \times [\{(ETo \times Kc \times Kr)/IE\} + LR] \qquad (1)$$

where: IWA = irrigation water applied ($m^3fed^{-1}irrigation^{-1}$), ET_o = Potential evapotranspiration using Class A pan evaporation (mm day^{-1}), Kc = crop coefficient, Kr = reduction factor [9], I = irrigation intervals (day), IE = irrigation efficiency (90%), and LR = leaching requirement = 10% of the total water delivered to the treatment.

The amounts of the recommended fertilizers used were: 70.5, 84.9 and 75.8 kg fed^{-1} of N, K_2O and P_2O_5. The fertilizers were fertigated in doses according to the stage of growth via irrigation water. All plots were weeded and pest controlled according to the recommendation of Agriculture Ministry in Egypt. Maize was harvested on the 5th of September but irrigation season was ended 15 days before. The air-dried weights of both grains and Stover ($Kg.fed^{-1}$) were calculated. WUE was calculated after Howell et al. [8] using the following equations:

WUEg = [(grain yield, kg/fed)/(total water applied, m^3/fed)]   (2)

WUEs = [(Stover yield, kg/fed)/(total water applied, m^3/fed)]   (3)

8.3   RESULTS AND DISCUSSION

8.3.1   GRAIN AND STOVER WATER USE EFFICIENCY (WUE_G AND WUE_S)

Table 8.4 indicates effects localized irrigation systems (LIS) and humic fertilizer (HF) treatments on grain water use efficiency (WUE_g) and Stover water use efficiency (WUE_s). We can conclude that the changes in WUE_g and WUE_s took the same trend as vegetative growth parameters (leaf area, plant height, leaf length and number of leaves per plant). This may be due to the positive effect of LIS and HF treatments on the vegetative growth parameters mentioned here.

Based on WUE_g and WUE_s values, LIS can be ranked in the descending orders: BIS > MSIS > DIS and BIS > MSIS > DIS. Differences in WUE_g only among LIS were significant at 1% level.

With respect to the WUE_g and WUE_s values, the HF can be illustrated in the descending orders: (HF100) > (HF50) > (HF0) and (HF100) ≥ (HF50) ≥ (HF0). Differences in WUE_g among HF treatments were significant at 1%

TABLE 8.4 Effects of Three Irrigation Systems and Three Humic Fertilizer Treatments on Maize Yield and WUE

LIS (1)	Applied HF kg/fed (2)	Applied water m³/fed.	Grain yield kg/fed·	WUE$_g$ kg.m^{-3}	Stover yield Kg/fed·	WUE$_s$ k.gm^{-3}
	100		4832.6a	1.58a	4943.5a	1.61a
BIS	50	3066.2	4459.8c	1.45c	4787.4c	1.56c
	0		4370.5e	1.43e	4637.5e	1.51e
	100		4623.7b	1.51b	4817.4b	1.58b
MSIS	50	3054.8	4357.3f	1.43fe	4564.2f	1.49f
	0		4206.9 g	1.38 g	4520.9 g	1.48h
	100		4373.1d	1.44d	4696.2d	1.55d
DIS	50	3035.4	3855.1h	1.27h	4525.0h	1.49 gf
	0		3701.3i	1.22i	4415.3i	1.45i
1 X 2	LSD $_{0.01}$		**5.52**	**0.02**	**4.54**	**0.01**
	BIS		4554.3a	1.49a	4789.5a	1.56a
Means (1)	MSIS		4396.0b	1.44b	4634.2b	1.52b
	DIS		3976.5c	1.31c	4545.5c	1.50cb
	LSD $_{0.01}$		**6.41**	**0.03**	**4.48**	**0.03**
	100		4609.8a	1.51a	4819a	1.58a
Means (2)	50		4224.1b	1.38b	4625.5b	1.52b
	0		4092.9c	1.34c	4524.5c	1.48c
	LSD $_{0.01}$		**6.53**	**0.05**	**3.37**	**0.03**

LIS: Localized irrigation system, **HF:** Humic Fertilizer added, **(HF100)**: Humic amount added = 100 kg/fed, **(HF50)**: Humic amount added = 50 kg/fed, **(HF0)**: Humic amount added = 0 kg/fed; **BIS**: Bubbler irrigation system, **MSIS**: Mini-sprinkler irrigation system, **DIS**: Drip irrigation system, **WUE$_g$**: grain water use efficiency, and **WUE$_s$**: Stover water use efficiency; LSD = Least square difference.

level. On the other hand, differences in WUE$_s$ were significant at 1% level only between HF100 and HF0.

The effects of the interaction, LIS × HF on WUE$_g$, were significant at 1% level, except those among the interactions: BIS × (HF0), MSIS × (HF50) and DIS × (HF100).

The effects of interaction, LIS X HF on WUE$_s$, were not significant at 1% level in most cases. The highest WUE$_g$ and WUE$_s$ values of 1.58,

1.61 kg.m^{-3}) and the lowest values of 1.22 and 1.45 kg.m^{-3} were obtained in the interactions: BIS × (HF100) and DIS × (HF0).

8.3.2 FERTILIZERS USE EFFICIENCY (FUE)

Table 8.5 shows the effects of LIS and HF treatments on (N, P$_2$O$_5$, K$_2$O) fertilizers use efficiency (FUE$_N$, FUE$_{P2O5}$, FUE$_{K2O}$). According to the FUE

TABLE 8.5   Effects of Three Irrigation Systems and Humic Fertilizer Treatments on FUE.

LIS	HF kg/fed	N	P2O5	K2O	Grain yield kg/fed	FUEN	FUE-P2O5	FUEK2O
			Applied fertilizers kg/fed				FUE kg yield per kg fertilizer	
	100				4832.6a	68.6a	56.9a	64.0a
BIS	50				4459.8c	63.3c	52.5c	59.1c
	0				4370.5e	62.0d	51.5d	57.9d
	100				4623.7b	65.6b	54.5b	61.2b
MSIS	50	70.5	84.9	75.8	4357.3f	61.8f	51.3f	57.7f
	0				4206.9 g	59.7 g	49.6 g	55.7 g
	100				4373.1d	62.0ed	51.5ed	57.9ed
DIS	50				3855.1h	54.7h	45.4h	51.1h
	0				3701.3i	52.5i	43.6i	49.0i
LSD at P = 0.01					5.5	2.8	1.4	2.5
Means	BIS				4554.3a	64.6a	53.6a	60.3a
	MSIS				4396.0b	62.4b	51.8b	58.2b
	DIS				3976.5c	56.4c	46.8c	52.7c
LSD at P = 0.01					6.4	2.2	1.5	1.9
Means	100				4609.8a	65.4a	54.3a	61.1a
	50				4224.1b	59.9b	49.8b	55.9b
	0				4092.9c	58.1c	48.2c	54.2c
LSD at P = 0.01					6.5	1.5	1.3	2.4

LIS: Localized irrigation system, HF: Humic Fertilizer added, FUE = Fertilizers use efficiency, (FUE)$_N$ = Nitrogen use efficiency, (FUE)$_{P2O5}$ = Phosphorous use efficiency, (FUE)$_{K2O}$ = Potassium use efficiency, (HF100): Humic amount added = 100 kg/fed, (HF50): Humic amount added = 50 kg/fed, (HF0): Humic amount added = 0 kg/fed BIS: Bubbler irrigation system, MSIS: Mini-sprinkler irrigation system, DIS: Drip irrigation system.

values of the three fertilizers used, the LIS and HF treatments can be ranked in the ascending orders, DIS < MSIS < BIS and (HF0) < (HF50) < (HF100). Differences in FUE between any two LIS treatments and /or HF ones were significant at 1% level except that between (BIS; MSIS) and (HF50; HF0) in the case of $(FUE)_N$.

The effects of the interactions, LIS X HF treatments on FUE, were significant at 1% level among some interactions. The highest values of nitrogen use efficiency, FUE_N, phosphate use efficiency FUE_{P2O5}, and potassium use efficiency, FUE_{K2O} were 68.6, 56.9 and 64.0 kg of yield per kg fertilizer and the lowest were 52.5, 43.6, and 49.0 kg yield per kg of fertilizer in the interactions: BIS × (HF100) and DIS × (HF0). These results are supported by Baligar and Bennett [2].

The obtained results indicated that FUE took the same trend as of vegetative growth parameters, yield and WUE. This may be attributed to the direct linear relationship between WUE and FUE found by Tayel et al. [23].

8.4   CONCLUSIONS

At present, the world is facing challenges of food insecurity and malnutrition widespread due to the limited water resources, population growth, adverse climate changes, and environmental pollution. Water stress is one of the most important factors that lead to poor crop yield. In order to avoid the occurrence of drought and water stress to crops we must use modern irrigation methods, such as LIS.

In addition, we must use organic fertilizers such as HF. Finally, it can be recommended to use HF = 100 kg/feddan in maize under BIS. It was concluded that the impact of HF was positive on the WUE and FUE, and maize crop productivity.

8.5   SUMMARY

Irrigation water shortage, traditional irrigation systems and poor soils in arid regions and some other factors have negative impacts on crops production, energy for processing, exportation and importation of fertilizers. Field experiments were carried out during the growing season (2012/2013) in a sandy soil at the Experimental Farm of National Research Center (NRC), El-Noubaria Governor, Egypt, to study the effects of some LIS and HF on WUE, and FUE of maize crop. Three localized irrigation systems were used: MSIS, BIS, and DIS.

The humic fertilizer treatments were: (HF100), (HF50), (HF0): 50, 25, 0 kg per feddan. The N, P2O5 and K2O were applied via irrigation water (Fertigation) at the rate of 60.71 and 69 kg/feddan in doses according to growth stage. The research work concludes that the agriculture activities should choose irrigation system and humic fertilizer to increase the maize yield. Data obtained indicated that the Bubbler irrigation system and humic fertilizer treatment (HF100) can positively affect maize productivity parameters: grain yield, Stover yield, water use efficiency, and fertilize use efficiency.

KEYWORDS

- arid regions
- bubbler irrigation system
- crop production
- drip irrigation system, DIS
- Egypt
- fertilizer
- fertilizer use efficiency, FUE
- humic fertilizer, HF
- irrigation
- irrigation water
- localized irrigation system, LIS
- maize
- mini-sprinkler irrigation system, MSIS
- nitrate leaching
- nitrogen
- phosphorous
- poor soil
- potassium

- sandy soil
- traditional irrigation system
- water
- water shortage
- water use efficiency, WUE

REFERENCES

1. Bacon, M. A., 2004. *Water use efficiency in plant biology.* CRC Press.
2. Baligar, V. C. and O. L. Bennett, 1986. NPK-fertilizer efficiency-A situation analysis for the tropics. *Fert. Res.,* 10:147–164.
3. Bhatti, A. U. and M. Afzal, 2001. *Plant nutrition management for sustained production.* Deptt. Soil and Envir. Sci., NWFP Agri. Uni. Peshawar, Pakistan. Pages 18–21.
4. Brady, N.C., 1974.Supply and availability of phosphorus and potassium. In: *The Nature and Properties of Soils.* (Ed.) R.S. Buckman. Macmillan Publishing co., Inc., New York. Pages 456–480.
5. Fulton, J. M., 1970. Relationship among soil moisture stress plant population, row spacing and yield of Maize. *Can. J. Plant Sci.,* 50:31–38.
6. Gee, G. W. and J. W. Bauder, 1986. Particle-size analysis. In: *Methods of soil analysis. Part 1.* ASA and SSSA, Madison,WI. Pages 383–412.
7. Holt, D.F. and D.R. Timmons, 1968. Influence of precipitation, soil water, and plant population interactions on Maize grain yields. *Agron. J.,* 60:379–381.
8. Howell, T. A., Yazar, A., Schneider, A. D., Dusek, D. A. and K. S. Copeland, 1995. Yield and water use efficiency of Maize in response to LEPA irrigation. *Trans. ASAE,* 38(6):1737–1747.
9. Keller, J. and D. Karmeli, 1975. *Trickle irrigation design.* 1stEdition. Rain Bird Sprinkler Manufacturing Corporation, Glendora, California 91740, USA.
10. Klute, A., 1986. Moisture retention. In: *Methods of soil analysis. Part 1.* ASA and SSSA, Madison, WI. Pages 635–662.
11. Laboski, C. A. M., Dowdy, R. H., Allmaras, R.R. and J. A. Lamb, 1998. Soil strength and water content influences on Maize root distribution in a sandy soil. *Plant and Soil,* 203:239–247.
12. Mansour, H. A., 2006. The response of grape fruits to application of water and fertilizers under different localized irrigation systems. M.Sc.: Thesis, Faculty of Agriculture, Ain Shams University, Egypt. Pages 78–81.
13. Mansour, H. A., 2012. Design considerations for closed circuits of drip irrigation system. PhD. Thesis, Faculty of Agriculture, Agric., Ain Shams University, Egypt. Pages 74–82.
14. Memon, K. S., 1996. Soil and Fertilizer Phosphorus. In: *Soil Science.* (Ed.) Elena Bashir and Robin Bantel. National Book Foundation. Islamabad. Pages 308- 311.

15. Modarres, A. M., Hamilton, R. I., Dijak, M., Dwyer, L. M., Stewart, D. W., Mather, D. E. and D.L. Smith, 1998. Plant population density effects on maize inbred lines grown in short-season environment. *Crop Sci.,* 38:104–108.
16. Ogola, J. B. O., T. R. Wheeler and P.M. Harris, 2002. The water use efficiency of maize was increased by application of fertilizer N. *Field Crops Research,* 78 (2–3):105–117.
17. Rebecca, B., 2004. *Soil Survey Laboratory Methods Manual.* Soil Survey Laboratory Investigations Report No. 42,, MS 41, Room 152, 100 Centennial Mall North, Lincoln, NE 68508–3866.
18. Steel, R. G. D and J. H. Torrie, 1980. *Principles and Procedures of Statistics.* A biometrical approach.2nd Ed., McGraw Hill Inter. Book Co. Tokyo, Japan.
19. Tayel, M. Y., H.A. Mansour, and A.M. El-Gindy, 2012. Effect of different closed circuits and lateral line lengths on: III- dripper and lateral discharge. *Journal of Applied Sciences Research,* 8(5):2725–2731.
20. Tayel, M. Y., H.A. Mansour, and A.M. El-Gindy, 2012. Effect of different closed circuits and humic fertilizer treatments on: IV-uniformity coefficient and coefficient of variation. *Journal of Applied Sciences Research,* 8(5):2741–2748.
21. Tayel, M. Y., H.A. Mansour and David A. Lightfoot, 2012. Effect of different closed circuits and humic fertilizer treatments on: I – Pressure head and friction loss. *Agric. Sci.,* 3(3):392–399.
22. Tayel, M. Y., H.A. Mansour and David A. Lightfoot, 2012. Effect of different closed circuits and humic fertilizer treatments on: II- flow velocity and velocity head. *Agric. Sci.,* 3(4):531–537.
23. Tayel, M. Y., K. Ebtisam, I. Eldardiry and M. Abd El-Hady, 2006. Water and fertilizer use efficiency as affected by irrigation methods. *American-Eurasian J. Agric. & Environ. Sci.,* 1(3):294–300.

CHAPTER 9

PERFORMANCE OF DRIP IRRIGATED YELLOW CORN: KINGDOM OF SAUDI ARABIA

H. A. MANSOUR and Y. EL-MELHEM

CONTENTS

9.1 INTRODUCTION

Kingdom of Saudi Arabia (KSA) has plans to use its limited water resources efficiently, to overcome the gap between supply and demand. Using of modern irrigation techniques is very urgent for agricultural activities at the desert region at KSA. The application of fertilizers is usually by hand

In this chapter: one feddan (Egyptian unit of area) = 4200 m².

Modified and adopted from *H. A. Mansour and Y. El-Melhem, 2013. Impact the automatic control of closed circuits drip irrigation system on yellow corn growth and yield. International Journal of Advanced Research, 1(10):33–42.*

with low efficiency, resulting in higher costs and environmental problems [1]. Yellow corn (*Zea Mays* L.) is one of the most important cereals for human and animal consumption, in Egypt and is grown for both grain and forage. The questions often arise, "What is the minimum irrigation capacity for irrigated transgenic yellow corn? And what is the suitable irrigation system for irrigating yellow corn?" These are very hard questions to answer because these greatly depend on weather, yield goal, soil type, land conditions and the economic conditions necessary for profitability. Yellow corn (*Zea Mays* L.) is cultivated in areas lying between 58°N latitude and 40°S latitude up to an elevation of 3,800 meters above sea level. It is a crop, which is irrigated worldwide. USA is the main maize producing country [7, 14].

The irrigation water requirements of maize vary from 500 until 800 m³ to achieve maximum production with a variety of medium maturity of seed [6]. On a coarse textured soil, maize production was increased with a combination of deep tillage and the incorporation of hay deposits in soil, together with a general increase in irrigation [12]. Filintas et al. [9, 10] and Dioudiset et al. [5] have conducted extensive irrigation research on the cultivation of maize, and they found that the irrigation is of utmost importance, from the appearance of the first silk until the milky stage in the maturation of the corn kernels. Once the milky stage has occurred, the appearance of black layer development on 50% of the maize kernels is a sign that the crop has fully ripened. The aforementioned criteria were used in the research that is presented in this chapter.

Most research studies in KSA on the effects of irrigation on yellow corn yield are for sprinkler irrigation or furrow irrigation. In contrast, the research is limited on maize cultivation under drip irrigation [5, 9, 10]. The evaporation pan method was to calculate the amount of water needed for irrigation. This method was used in England for irrigation scheduling upto 45% of the irrigated areas of the country in outdoor cultivation [18]. Also, an additional advantage of drip irrigation is that, there are many tools available for soil moisture measurement [3, 8]. Electronic irrigation controller and electro-hydraulic systems have successfully been used for automation of irrigation networks [4, 8].

This chapter discusses research results on performance of drip irrigated yellow corn under KSA conditions. Authors studied the effects of automatic drip irrigation circuits (DIC) on vegetative growth and yield parameters of yellow corn.

9.2 MATERIALS AND METHODS

During the growing season of 2012, the experiment was conducted in a sandy loam soil at the Experimental Farm, Faculty of Agriculture, King Faisal University, Al-Hasa Governorate, KSA. The soil at the experimental site has a water field capacity of 0.22 v/v%, wilting point 0.11% and soil bulk density of 1.44 gm/cm³. Soil texture was determined according to procedure by Gee and Bauder [11]. Soil moisture retention constants were determined based on methods by Klute [13]. Chemical characteristics of soil saturation extract paste and irrigation water were determined according to methods by Rebecca [15] and the values are shown in Tables 9.1–9.3.

The experimental design was split-plot randomized complete block design with three replications. Laboratory tests were also carried out. Treatments were:

TABLE 9.1 Soil Physical Properties of the Experimental Site in KSA

Depth cm	Particle Size distribution, %				Texture class	θS % on volume basis			HC cm/h	BD g/cm³
	C. Sand	F. Sand	Silt	Clay		FC	WP	AW		
0–15	3.7	54.5	25.2	16.6	SL	0.22	0.11	0.11	1.11	1.45
15–30	3.8	55.8	24.6	15.8	SL	0.22	0.11	0.11	1.28	1.43
30–45	4.6	53.7	26.0	15.7	SL	0.22	0.11	0.11	1.28	1.43
45–60	4.6	55.9	25.5	14.0	SL	0.21	0.10	0.11	1.53	1.42

* Particle Size Distribution after Gee and Bauder [11] and Moisture retention after Klute [13].

SL: Sandy loam, FC: Field Capacity, WP: Wilting Point, AW: Available Water, HC: Hydraulic conductivity (cm/h), BD: Bulk density (g/cm³), and C = Coarse, F = Fine.

TABLE 9.2 Chemical Analysis of the Soil At The Site in KSA

Depth cm	pH 1:2.5	EC dS/m	Soluble Cations, meq/L				Soluble Anions, meq/L			
			Ca++	Mg++	Na+	K+	CO3–	HCO3–	SO4–	Cl–
0–15	8.10	1.97	6.43	4.89	185.0	18.84	0	5.64	58.7	6.65
15–30	8.13	2.98	11.53	6.49	237.1	25.01	0	5.21	62.6	10.53
30–45	8.11	3.61	12.15	7.97	279.1	26.63	0	3.68	64.0	11.48
45–60	8.03	3.76	12.56	4.17	307.1	32.28	0	3.62	66.9	5.6

*Chemical properties after Rebecca [15].

TABLE 9.3 Chemical Analysis of Irrigation Water

pH	EC dS/m	Soluble cations, meq/L				Soluble anions, meq/l			
		Ca++	Mg++	Na+	K+	CO3–	HCO3–	SO4–	Cl–
7.48	2.0	0.7	1.72	128	13	0.0	3.4	67	1.8

1. **Three irrigation lateral lines** (LLL): LLL1 = 40 m; LLL2 = 60 m; and LLL3 = 80 m.
2. **Three drip irrigation circuits** (DIC): one manifold for lateral lines for closed circuit of drip irrigation system (CM1DIS); two manifolds for lateral lines for closed circuit drip irrigation system (CM2DIS); and traditional drip irrigation system (TDIS) as a control.

9.2.1 EXPERIMENTAL SETUP

The irrigation network for this study is described in detail by Mansour in Figures 4.1–4.4 in Chapter 4 of this volume. Details of the pressure and water supply controls have been described by Safi et al. [16]. The irrigation network in this research study was automated using an automatic irrigation controller that is described in detail in the chapter titled, "*Automation of Mini-Sprinkler Irrigation and Drip Irrigation Systems by H. A. Mansour, H. M. Mehanna, M. E. El-Hagarey, and A. S. Hassan*" of this volume. Irrigation systems were evaluated to solve the problem of lack of pressure head at the end of lateral lines in the TDIS. Irrigation scheduling and estimation of crop water requirements of yellow corn were based on the methods by Allen at al. [2] and the procedure is described in detail in Chapter 4 of this volume. Irrigation Interval of four days was used for all irrigation methods in this chapter. The crop evapotranspiration (ETc) was computed using the Class Pan evaporation method for estimating (ETo) on daily basis [2] and the values are given in Table 9.4.

Yellow corn (*Zea mays L.,* cv. Ghota-82) was planted on April, 2, 2012. The row-to-row spacing was 0.7 m and plant-to-plant spacing was 0.25 m down the row. Plant density was 42,000 plants per feddan. Each row was

TABLE 9.4 Water Requirements for Transgenic Yellow Corn Grown at the Experimental Site, KSA

Item	Apr.	May	Jun.	Jul.	Aug.	Sep.
			Month			
Epan (mm/day)	6.58	6.34	7.85	9.43	9.23	7.28
Kp			0.71			
Kc	1.05	1.08	1.15	1.17	1.22	1.25
Kr	0.45	0.90	0.95	1.00	1.00	1.00
ETo (mm/day)	4.66	4.52	5.57	6.70	6.59	5.13
ETc (mm/day)	2.20	4.39	6.08	7.84	8.04	6.41
Ks			100% (1.00)			
Eu			90% (1.11)			
Lr			10%			

Item	Planting establish-ment	Vegetative growth		Flowering		Ribbing and harvesting	
			Growth stage				
Length	2–21 Ap.	21 Ap-1 Jun		2 Jun-5 Jul		6 Jul-5 Aug.	
Number of days for irrigation	19	42		34		31	
IRn (mm/month)	41.8	184.4		170.2	47.0	257.5	88.0
IRg (mm/month)	51.3	227.2		209.7	57.9	209.0	64.1

irrigated by a single straight lateral line in all the plots. The total experimental area was 4536 m². In each plot of drip irrigation circuits (DIC), plot areas for lateral line lengths were 168, 252 and 336 m² under LLL1 = 40 m, LLL2 = 60 m and LLL3 = 80 m, respectively. Irrigation season of transgenic yellow corn was ended 11 days before the last harvest. Transgenic yellow corn was harvested on September 15, 2012. Fertilizers were fertigated through the irrigation. The amounts of NPK (20–20–10) fertilizers were: 74.6 kg/fed of (20% N), 33.0 kg/fed of (20% K_2O), and 60.5 kg/fed of (10% P_2O_5). In all plots, weed and pest control measures were followed based on the recommendations for transgenic yellow corn by Al-Hasa, KSA.

9.2.1.1 Measurements of Yellow Corn Plant Growth and Yield

Measurements of parameters for growth and yield of yellow corn were: plant height (cm), leaf length (cm), leaf area (cm²), number of leaves per plant, total grain weight (kg/fed) and Stover yield (kg/fed). Plant measurements and observations were started 21 days after planting, and were terminated on the harvest date. All plant samples were dried at 65 °C until the constant weight was achieved. Grain yield was determined by hand harvesting the 8 m sections of three adjacent center rows in each plot and was adjusted to 15.5% water content. In all treatments, the grain yields of individual rows were determined in order to evaluate the yield uniformity among the rows.

MSTATC program (Michigan State University) was used to carry out statistical analysis. Treatments mean were compared using analysis of variance (ANOVA) and the least significant difference (L.S.D) at P = 0.01 among the treatments [17].

9.3 RESULTS AND DISCUSSION

Table 9.5 and Figs. 9.1–9.3 show effects of three DIC and three LLL on some vegetative growth and yield parameters of yellow corn: leaf area, plant height, leaf length, number of leaves, grain yield (ton/fed) and Stover yield (ton/fed).

9.3.1 LEAF AREA (LA)

Table 9.5 and Fig. 9.1a illustrate the effects of different DIC and LLL on LA. Data can be ranked in the descending order: CM2DIS > CM1DIS > TDIS. Values of LA indicated significant differences among mean values of main effects (DIC) and submain effects of LLL. Base on the effects of interaction between both investigated factors, the highest and lowest values

TABLE 9.5 Effects of Three Irrigation Circuit Designs and Three Lateral Line Lengths on Performance Parameters of Yellow Corn During 2012, KSA

		Growth and yield characteristics (average values)					
		Leaf area	Plant height	Leaf		Yield	
	LLL			Length	Nos. per	Grain	Straw
DIC	m	cm²	cm	cm	plant	ton/fed.	
CM2DIS	40	498.52a	192.52a	69.32a	16.63a	5.66a	3.81a
	60	495.37c	191.91b	67.25d	15.35dc	5.43c	3.54d
	80	491.13e	190.35e	66.43f	15.13f	5.23f	3.32 gf
CM1DIS	40	497.27b	191.34c	68.75b	15.52b	5.47b	3.72b
	60	489.67f	190.28f	66.38e	15.13 gf	5.28e	3.45e
	80	476.42h	189.67h	65.17h	14.95h	5.07 g	3.19h
TDIS	40	495.23d	190.97d	68.12c	15.36c	5.37d	3.61c
	60	487.78 g	189.85 g	65.18 g	15.23e	4.76h	3.32f
	80	472.85i	188.71i	64.92i	14.81i	4.53i	3.01i
(1) × (2)	LSD0.01	0.86	0.07	0.11	0.06	0.02	0.01
(1) Means	CM2DIS	495.01a	191.59a	67.66a	15.70a	5.26a	3.56a
	CM1DIS	487.78b	190.43b	66.76b	15.20ba	5.27ba	3.45b
	TDIS	485.29c	189.84c	66.07c	15.13cba	4.88c	3.31c
	LSD0.01	2.33	0.09	0.06	0.84	0.04	0.01
(2) Means	40	497.01a	191.61a	68.73a	15.84a	5.50a	3.71a
	60	490.94b	190.68b	66.27b	15.24ba	5.16b	3.44b
	80	480.13c	189.58c	65.51c	14.96cab	4.94c	3.17c
	LSD0.01	1.92	0.14	1.17	1.81	0.02	0.01

DIC: Irrigation circuit design; LLL: Lateral line length; CM2DIS: Closed circuit with tow manifolds separately; CM1DIS: Closed circuits with one manifold; TDIS: Traditional drip irrigation system.

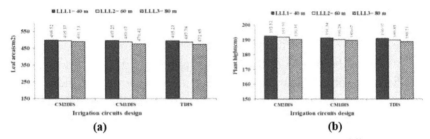

FIGURE 9.1 Effects of three irrigation circuit designs and three lateral line lengths on leaf area and plant height of yellow corn: Solid bars = 40 m, hatched bars = 80 cm and bar diagrams in the middle = 60 m.

of LA were observed in CM2DIS and LLL1. Also, data indicate that under all DIC, highest values were observed in LLL1.

9.3.2 PLANT HEIGHT

Data in Table 9.5 is plotted in Fig. 9.1b. The data indicated that plant height (ph, cm) followed the same trend as of LA. The effect of DIC and LLL can be ranked in the descending orders: CM2DIS > CM1DIS > TDIS and LLL1> LLL2 > LLL3, respectively. Differences in ph were significant at 1% level among all means values of LLL. Also, differences within values of DIC treatments were significant at 1% level except that between CM2DIS and CM1DIS. The effects of interaction between two studied factors were significant at 1% level except in the following interactions: CM2DIS × LLL3, CM1DIS × LLL2, CM1DIS × LLL3 and TDIS × LLL3. The maximum and minimum values of ph were observed in the interactions of LLL1 × CM2DIS and LLL3 × TDIS, respectively.

9.3.3 LEAF LENGTH

Data of Table 9.5 and Fig. 9.2a illustrates the effects of different DIC and LLL on leaf length (LL, cm). According to LL, DIC and LLL can be ranked in the descending orders: CM1DIS > CM2DIS > TDIS and LLL1 ≥ LLL2 > LLL3, respectively.

With respect to the LL, data indicated that there is significant difference within main effects (DIC), while the highest and the lowest values were

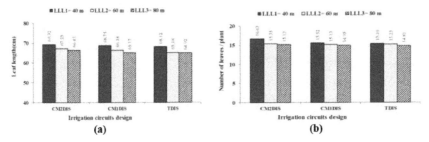

FIGURE 9.2 Effects of three irrigation circuit designs and three lateral line lengths on leaf length and number of leaves per plant of yellow corn: Solid bars = 40 m, hatched bars = 80 cm and bar diagrams in the middle = 60 m.

recorded in CM1DIS and TDIS, respectively. There is significant difference within LLL treatments except between LLL1 and LLL2 at 1% level. The highest value was recorded in LLL1 and the lowest one was recorded in LLL3 treatment. The effects of interaction among the two studied factors indicated that there were significant differences between treatments at 1% level. The maximum and minimum values of LL were recorded in CM2DIS × LLL1 and TDIS × LLL3.

9.3.3.1 Number of Leaves Per Plant

Table 9.5 and Fig. 9.2b indicates the effect of DIC and LLL on number of leaves (LN per plant), which can be ranked in descending order: CM2DIS > CM1DIS > TDIS. Differences in LN per plant, between means of the two factors studied, were significant at 1% level. The highest and lowest values under DIC and LLL were obtained in CM2DIS; TDIS and LLL1; LLL3, respectively. The maximum and minimum values of LN were significant at 1% in CM2DIS x LLL3 and TDIS x LLL1, respectively. The superiority of the studied growth parameters under (CM2DIS; CM1DIS relative to TDIS) and (LLL1; LLL2 relative to LLL3) can be noticed due to improving both water and fertilizer distribution uniformities.

9.3.3.2 Grain Yield

Data in Table 9.5 and Fig. 9.3a indicate the effects of DIC and LLL on yellow corn grain yield (GY, ton/fed), both of these can be ranked in the ascending orders: TDIS < CM1DIS < CM2DIS and LLL3 < LLL2 < LLL1, respectively. With respect to the main effects of DIC on GY, one can notice that the differences in GY were significant among all DIC at 1% level. The highest and lowest values of GY were obtained in CM2DIS and TDIS, respectively. According to effects of LLL on GY, there are significant differences at 1% level among LLL1, LLL2 and LLL3. Highest and lowest values were achieved in LLL1 and LLL3, respectively. With respect to the effects of DIC × LLL on GY, there were significant differences at 1% level, except in the interactions: CM2DIS × LLL3, TDIS × LLL1, CM2DIS × LLL3 and CM1DIS × LLL2. The maximum and minimum values of GY were obtained in CM2DIS × LLL1 and TDIS × LLL3, respectively.

We can notice that yellow corn GY took the same trend as other vegetative growth parameters, due to the close correlation between vegetative growth from one side and grain yield from the other side.

9.3.3.3   Stover Yield

Table 9.5 and Fig. 9.3b indicates the effects of DIC and LLL on Stover yield of yellow corn (SY, ton/fed). We can observe that the change in SY took the same trend as of vegetative growth parameters.

Based on the positive effects of DIC and LLL on SY, these can be ranked in descending orders: CM2DIS > CM1DIS > TDIS and LLL1 > LLL2 > LLL3. With respect to effects of DIC and LLL on the SY, one can

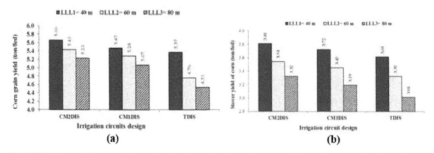

FIGURE 9.3   Effects of three irrigation circuit designs and three lateral line lengths on grain and Stover yields of yellow corn: Solid bars = 40 m, hatched bars = 80 cm and bar diagrams in the middle = 60 m.

notice significant difference at 1% level between all mean values of DIC and LLL.

9.4 CONCLUSIONS

Following conclusions can be drawn based on this research study in KSA:
1. Based on values of vegetative growth and yield parameters (leaf area, plant height, leaf length, number of leaves, grain and Stover yields of yellow corn, DIC and LLL can be ranked in the ascending orders: TDIS < CM1DIS < CM2DIS and LLL3 < LLL2 < LLL1, respectively for all studied parameters.
2. The effects of interaction DIC × LLL on vegetative growth and yield parameters were significant at 1% level with few exceptions. The highest values of leaf area, plant height, leaf length, number of leaves, grain and Stover yield were 498.52 cm^2, 192.52 cm, 69.32 cm, 16.63 per plant, 5.66 ton/fed, and 3.81 ton/fed; and the lowest values were 472.85 cm2, 188.71 cm, 64.92 cm, 14.81 per plant, 4.53 ton/fed and 3.01 ton /fed. These can be observed in the interactions: CM2DIS × LLL1 and TDIS × LLL3, respectively.

9.5 SUMMARY

This chapter discusses research results on performance of drip irrigated yellow corn under KSA conditions. Authors studied the effects of automatic DIC on vegetative growth and yield parameters of yellow corn. During the growing season of 2012, the experiment was conducted in a sandy loam soil at the Experimental Farm, Faculty of Agriculture, King Faisal University, Al-Hasa Governorate, KSA. The soil at the experimental site has a water field capacity of 0.22 v/v%, wilting point 0.11% and soil bulk density of 1.44 gm/cm^3. The field experiment was carried out under automatic irrigation system for three irrigation lateral line lengths of 40, 60, 80 m for three DIC of: one manifold for lateral lines closed circuit drip irrigation system (CM1DIS); closed circuit with two manifolds for lateral lines (CM2DIS); and traditional drip irrigation system (TDIS) as a control. Irrigation water was added in order to compensate for ETc and salt leaching requirement.

Based on values of vegetative growth and yield parameters (leaf area, plant height, leaf length, number of leaves, grain and Stover yields of yellow corn, DIC and LLL can be ranked in the ascending orders: TDIS < CM1DIS < CM2DIS and LLL3 < LLL2 < LLL1, respectively for all studied parameters. The effects of interaction DIC × LLL on vegetative growth and yield parameters were significant at 1% level with few exceptions. The highest values of leaf area, plant height, leaf length, number of leaves, grain and Stover yield were 498.52 cm^2, 192.52 cm, 69.32 cm, 16.63 per plant, 5.66 ton/fed, and 3.81 ton/fed; and the lowest values were 472.85 cm^2, 188.71 cm, 64.92 cm, 14.81 per plant, 4.53 ton/fed and 3.01 ton/fed. These can be observed in the interactions: CM2DIS × LLL1 and TDIS × LLL3, respectively.

KEYWORDS

- automation controller
- closed circuit
- closed circuit irrigation design
- drip Irrigation
- Egypt
- evapotranspiration
- fertigation
- fertilizer
- grain yield
- irrigation depth
- irrigation interval
- K_2O
- Kingdom of Saudi Arabia
- lateral line
- N
- NPK
- P_2O_5
- plant height

- **straw yield**
- **USA**
- **water use efficiency**
- **yellow corn**

REFERENCES

1. Abou-Kheira, A. A., 2009. Comparison among different irrigation systems for deficit- irrigated transgenic and non trangenic yellow corn in the Nile Valley. *Agricultural Engineering International: the CIGR Ejournal*, 11:1–25.
2. Allen, R. G, Pereira L. S., Raes, D. and Smith, M., 1998. *Crop evapotranspiration- Guidelines for computing crop water requirements*. FAO Irrigation and Drainage paper 56. Rome.
3. Cary, J. W. and Fisher, H. D., 1983. Irrigation decision simplified with electronics and soil water sensors. *Soil science Society of American Journal*, 47:1219–1223.
4. Charlesworth, P., 2000. *Soil water monitoring*. CSIRO Land and Water, Australia.
5. Dioudis, P., Filintas, T., Papadopoulos, H. A., 2008. Transgenic and nontransgenic yellow corn yield in response to irrigation interval and the resultant savings in water and other overheads. *Irrigation and Drainage Journal*, 58:96–104.
6. Doorenbos, J. and Kassam, A. H., 1986. *Yield response to water*. FAO Irrigation and Drainage Paper 33. FAO, Rome, Italy. Pages 101–104.
7. Filintas, T., 2003. *Cultivation of maize in Greece: increase and growth, management, output yield and environmental sequences*. University of Aegean, Faculty of Environment, Department of Environmental Studies, Mitilini, Greece.
8. Filintas, T., 2005. *Land use systems with emphasis on agricultural machinery, irrigation and nitrates pollution, with the use of satellite remote sensing, geographic information systems and models, in watershed level in Central Greece*. M.Sc. Thesis, University of Aegean, Faculty of Environment, Department of Environmental Studies, Mitilini, Greece.
9. Filintas, T., Dioudis, I. P., Pateras, T. D., Hatzopoulos, N. J. and Toulios, G. L., 2006. Drip irrigation effects in movement, concentration and allocation of nitrates and mapping of nitrates with GIS in an experimental agricultural field. Proc. of 3rd HAICTA International conference on: *information systems in sustainable agriculture*, Agro Environment and Food Technology, (HAICTA'06), Volos, Greece, September 20–23, pages 253–262.
10. Filintas, T., Dioudis, I. P., Pateras, T. D., Koutseris, E., Hatzopoulos, N. J. and Toulios, G. L., 2007. Irrigation water and applied nitrogen fertilizer effects in soils nitrogen depletion and nitrates GIS mapping. Proc. of First International Conference on: Environmental Management, Engineering, Planning and Economics CEMEPE/SECOTOX), June 24–28, Skiathos Island, Greece, 3:2201–2207.
11. Gee, G. W. and Bauder, J. W., 1986. Particle-size analysis. Pages 383–412. In: Klute (ed.) *Methods of soil analysis*, Part 1. ASA and SSSA, Madison,WI.

12. Gill, K. S., Gajri, P. R., Chaudhary, M. R. and Singh, B., 1996. Tillage, mulch and ir-rigation effects on transgenic and non trangenic yellow corn (*Zea mays* L.) in relation to evaporative demand. *Soil & Tillage Research*, 39:213–227.

13. Klute, A., 1986. Moisture retention. Pages 635–662. In: A. Klute (ed.) *Methods of soil analysis. Part 1*. ASA and SSSA, Madison, WI.

14. Musick, J. T., Pringle, F. B., Harman, W. L. and Stewart, B. A., 1990. Long-term irriga-tion trends: Texas High Plains. *Applied Engineering Agriculture*, 6:717–724.

15. Rebecca, B., 2004. *Soil Survey Laboratory Methods Manual*. USDA – NSSC Soil Sur-vey Laboratory Investigations Report No. 42. Room 152, 100 Centennial Mall North, Lincoln, NE 68508–3866.

16. Safi, B., Neyshabouri, M. R., Nazemi, A. H., Masiha, S. and Mirlatifi, S. M., 2007. Subsurface irrigation capability and effective parameters on onion yield and water use efficiency. *Journal of Scientific Agricultural*, 1:41–53.

17. Steel, R. G. D. and Torrie, J. H., 1980. *Principles and Procedures of Statistics*. A bio-metrical approach. 2nd Ed., McGraw Hill Inter. Book Co. Tokyo, Japan.

18. Weatherhead, E. K. and Danert, K., 2002. *Survey of Irrigation of Outdoor Crops in England*. Cranfield University, Bedford.

CHAPTER 10

WATER AND FERTILIZER USE EFFICIENCIES FOR DRIP IRRIGATED CORN: KINGDOM OF SAUDI ARABIA

H. A. MANSOUR and A. S. ALJUGHAIMAN

CONTENTS

10.1   INTRODUCTION

Water is one of the most important natural resources. Population growth and higher living standards have caused ever-increasing demands for good water quality in the future, exerting an extreme pressure on water resources. Water is essential for supplying domestic, municipal, industrial, and agriculture needs. Furthermore, while growing populations and

In this chapter: one *feddan* (Egyptian unit of area) = 4200 m².

Modified and printed from *H. A. Mansour and A. S. Aljughaiman, 2012. Water and fertilizers use efficiency of corn crop under closed circuits of drip irrigation system. Journal of Applied Sciences Research, 8(11): 5485–5493. Open access article at: http://www.scirp.org/journal/jasr/.*

increasing water requirements are a certainty, it is not known how climates will change and at what extent they will be affected by man's activities.

The Kingdom of Saudi Arabia (KSA), with a total area of about 2.15 million km², is by far the largest country in the Arabian Peninsula. It is bordered in the north by Jordan, Iraq and Kuwait, in the east by the Persian Gulf with a coastline of 480 km, in the south-east and south by Qatar, the United Arab Emirates (UAE), Oman and Yemen, and in the west by the Red Sea with a coastline of 1,750 km. KSA falls in the tropical and subtropical desert region. The winds reaching the country are generally dry, and almost all the area is arid. Because of the aridity, and hence the relatively cloudless skies, there are great extremes of temperature, but there are also wide variations between the seasons and regions. Average precipitation is 59 mm/year. The cultivable area has been estimated at 52.7 million-ha, which was almost 25% of the total area. Land under cultivation has grown from under 0.25 million-ha (1,600 km²) in 1976 to more than 5 million-ha (32,000 km²) in 1993. In 1992, the cultivated area was 1.61 million-ha, of which 1.51 million-ha consisted of annual crops and 95,500 ha consisted of permanent crops.

Al-Hassa is one of the largest oases in KSA, and its water originates from an underground source through a number of artesian springs. The water from these free-flowing springs has been used for centuries to irrigate about 20,000 ha of arable land. In recent years, the area has been troubled by salinization of cultivated land due to inefficient water management practices and deterioration of the natural drainage caused by shifting sand dunes. The arable land has been reduced to less than 50% of its original area of 20,000 ha. In pursuance of the Government's policy of using oil revenues to diversify the country's economy, the Ministry of Agriculture and Water embarked on an ambitious reclamation scheme (HIDA) during the 1960's and established the Hofuf Agricultural Research Centre (HARC), with a mission to revitalize the once flourishing oasis by improving water management practices.

Water, of course, is the key to agriculture in KSA, which has implemented a multifaceted program to provide the vast supplies of water necessary to achieve the spectacular growth of the agricultural sector. A network of dams has been built to trap and use precious seasonal floods. Vast underground water reservoirs have been tapped through deep wells.

Desalination plants have been built to produce fresh water from the sea for urban and industrial use, thereby freeing other sources for agriculture. Facilities have also been put into place to treat urban and industrial run-off for agricultural irrigation. These efforts collectively have helped transform vast tracts of the desert into fertile farmland. Agriculture in KSA consumes about 90% of the water used in the country. The efficient use of finite water resources is essential for attaining sustainability of agriculture and protection of the fragile environment KSA.

All agriculture is irrigated and in 1992 the water managed area was estimated at about 1.6 million-ha, all equipped for full/partial control irrigation. Surface irrigation was practiced on the old agricultural lands (cultivated since before 1975), which represented about 34% of the irrigated area. Sprinkler irrigation was practiced on about 64% of the irrigated areas. The central pivot sprinkler system covered practically all the lands cropped with cereals. Normally, pumped groundwater from one deep well supplied one or two central pivots. The irrigation application efficiency of this method was estimated at between 70 and 85%. Vegetables and fruit trees were in general irrigated by drip and bubbler methods, respectively. The average cost for irrigation development was about 1,093, 372 and 251 US$/ha for micro irrigation, sprinkler irrigation and surface irrigation systems respectively.

Available irrigation water is the main limiting factor for crop production in KSA. Therefore, irrigation techniques adapted to conserve crop water use are a must in order to face water shortages under local conditions. Drip irrigation potentially provides the opportunity for more efficient water use.

Drip irrigation has become a well-established method for irrigating high-value crops in KSA. In KSA, the use of drip has increased five-folds since the 2010 s. Due to the rapid increase in the use for drip irrigation system in the fields and greenhouses in recent years in most agricultural regions in KSA, there is wide variation in the availability of many drip irrigation components and products in the local market made by different manufacturers. These devices differ in their qualities and standards. Polyethylene drip line is one the most economical choices for water delivery in crops and plant irrigation systems. Polyethylene drip lines offer many benefits: Resistant to chemicals and fertilizers; Excellent environmental stress crack resistance; Easily stabilized with carbon black to prevent ultra-violet degradation; Coiled for ease of shipping and field installation; Variety of diameters, coil lengths and emitter spacing; Factory installed emitters (integral); Lightweight and flexible; and Environmentally friendly.

Water use efficiency (WUE) of corn depends physiological characteristics of corn, genotype, soil moisture constants, climatic conditions and agronomic practices. To improve WUE, integrative measures should aim to optimize cultivar selection and agronomic practices. The soil fertility management and water supply are most important management practices in drought-stressed corn regions [2]. Ogola et al. [16] reported that WUE of corn was increased by application of nitrogen. They added that corn plants are especially sensitive to water stress because of sparse root system. Laboski et al. [11] found that corn yield responded to amount of water applied by trickle irrigation. Increasing the plant population density usually increases corn grain yield until an optimum number of plants per unit area is reached. Fulton [6] also reported that higher plant densities of corn produced higher grain yields. Corn plant density of 90,000 plants/ha is common in many regions of the world [15].

The nutrient use efficiency of plants depends upon fertilizer application rate, method, time, type of fertilizer, crop and soil properties. Proper method and time of fertilizer application are inevitable to reduce the losses of plant nutrients and are important for a fertility program to be effective. Nitrogenous fertilizers should be applied in split doses for the long season crops. Similarly nitrogen should not be applied in sandy soil in a single dose, as there are more chances for nitrate leaching [4]. Phosphate fertilizers are often fixed or rendered unavailable to plants, even under the most ideal field conditions. In order to prevent rapid reaction of phosphate fertilizer with the soil, the materials are commonly placed in localized bands. To minimize the contact with soil, pelleted or aggregated phosphate fertilizers are also recommended by Brady [5].

Brady [5] also reported that much of the phosphate is used early stage for row crops. Similarly, data on the corn yield showed that application of all phosphorus at sowing was better than its late application. Memon [14] concluded that phosphorus uptake by plant roots depended on the phosphorus uptake properties of roots and the phosphorus supplying properties of soil. He also added that maximizing the uniformity of water application is one of the easier ways to save water. It is not frequently considered by the irrigators. The evaluation of the emission uniformity of the trickle system should be done periodically.

In comparison studies between different irrigation systems, Mansour [12, 13] found that the increases in water use efficiency and water utilization efficiency were maximum under drip irrigation system (42 and 43%, respectively), followed by the low head bubbler irrigation system (40.7 and 37%), while the minimum values were (30.6 and 32%, respectively) under gated pipe irrigation system. Also, he found that the increases in fertilizers use efficiency of N-P-K were (24, 23 and 28%), (22, 21 and 27%) and (9, 8 and 14%) under drip irrigation system, low head bubbler irrigation system and gated pipe irrigation system, respectively.

This chapter discusses research studies to evaluate effects on water and fertilizer use efficiencies of corn crop by three closed circuit drip irrigation designs and three lateral line lengths, in Kingdom of Saudi Arabia.

10.2   MATERIALS AND METHODS

During the growing season of 2012, the experiment was conducted in a sandy loam soil at the Experimental Farm of Irrigation and Drainage Authority Project, Al-Hassa Governorate, Al Hassa City, Hufof State, KSA (Fig. 10.1). Soil texture was determined according to procedure by Gee and Bauder [7]. Soil moisture retention constants were determined based on methods by Klute [10]. Chemical characteristics of soil saturation extract paste and irrigation water were determined according to methods by Rebecca [17]. Tables 10.1–10.3 indicate the physical and chemical properties of soil and irrigation water at the site.

The experimental design was split-plot randomized complete block design with three replications. Laboratory tests were also carried out. Treatments were:

1. **Three irrigation lateral lines** (LLL): LLL1 = 40 m; LLL2 = 60 m; and LLL3 = 80 m.

2. **Three drip irrigation circuits** (DIC): one manifold for lateral lines for closed circuit of drip irrigation system (CM1DIS); two manifolds for

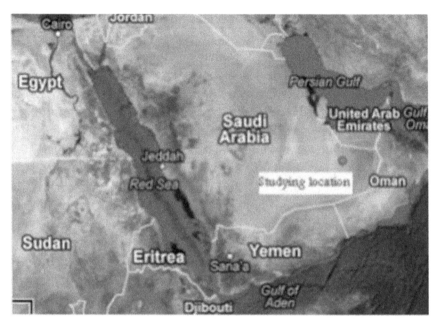

FIGURE 10.1   Site Location at Al-Hassa City in KSA.

TABLE 10.1   Soil Physical Properties At the Experimental Site, KSA

Depth cm	Particle size distribution, %				Texture class	θ_s % on volume basis		
	C. Sand	F. Sand	Silt	Clay		FC	WP	AW
0–15	0.8	75.2	8.9	15.1	SL	0.22	0.11	0.11
15–30	0.7	76.5	7.5	15.3	SL	0.22	0.11	0.11
30–45	0.6	78.6	6.0	14.8	SL	0.22	0.11	0.11
45–60	0.6	77.1	7.8	14.5	SL	0.21	0.10	0.11

Particle Size Distribution after Gee and Bauder [7] and Moisture retention after Klute [10].

SL: Sandy loam, FC: Field capacity, WP: Wilting point, AW: Available water, and C = Coarse, F = Fine.

lateral lines for closed circuit drip irrigation system (CM2DIS); and traditional drip irrigation system (TDIS) as a control.

The irrigation network for this study is described in detail by Mansour in Figs. 4.1–4.4 in Chapter 4 of this volume [12, 13]. Details of the

TABLE 10.2   Soil Chemical Properties At the Experimental Site, KSA

Depth cm	pH 1:2.5	EC dS/m	Soluble Cations, meq/L				Soluble Anions, meq/L			
			Ca++	Mg++	Na+	K+	CO_3-	HCO_3-	SO_4-	Cl-
0–15	7.7	4.26	9.4	2.8	25.4	1.9	0.00	3.3	7.8	28.4
15–30	7.6	4.23	9. 6	2.4	24.5	1.8	0.00	3.6	5.8	28.9
30–45	7.4	4.25	9.5	2.5	25.6	1.2	0.00	3.9	6.2	28.7
45–60	7.2	4.27	9.8	2.3	24.6	1.6	0.00	3.8	5.9	28.6

Chemical properties after Rebecca [17].

TABLE 10.3   Chemical Properties of Irrigation Water Used, KSA

pH	EC dS/m	Soluble cations, meq/L				Soluble anions, meq/l			
		Ca++	Mg++	Na+	K+	CO_3-	HCO_3-	SO_4-	Cl–
7.7	1.37	2.81	3.30	4.16	1.23	0.00	3.30	2.00	4.70

pressure and water supply controls have been described by Safi et al. [18]. Irrigation systems were evaluated to solve the problem of lack of pressure head at the end of lateral lines in the TDIS. Irrigation depth in each plot was based on the method described by Mansour [8, 9, 12, 13]. Irrigation scheduling and estimation of crop water requirements of yellow corn were based on the methods by Allen at al. [1] and the procedure is described in detail in Chapter 4 of this volume [8, 9, 13]. Irrigation Interval of four days was used for all treatments in this chapter. The crop evapotranspiration (ETc) was computed using the Class Pan evaporation method for estimating (ETo) on daily basis [1] and the values are given in Table 10.4.

Corn (*Zea mays L.,* cv. Giza-155) was planted on April 9, 2012. The row-to-row spacing was 0.7 m and plant-to-plant spacing was 0.25 m down the row. Plant density was 35,700 plants per feddan. Each row was irrigated by a single straight lateral line in all the plots. The total experimental area was 4536 m^2. In each plot of drip irrigation circuits (DIC), plot areas for lateral line lengths were 168, 252 and 336 m^2 under LLL1 = 40 m, LLL2 = 60 m and LLL3 = 80 m, respectively. The fertilizers were fertigated in doses according to the growth stage. The

TABLE 10.4   Water Requirements For Corn Grown at the Experimental Site

Item	Month					
	Apr.	May	Jun.	Jul.	Aug.	Sep.
Epan (mm/day)	6.56	6.36	7.84	9.44	9.28	7.23
Kp				0.71		
Kc	1.05	1.08	1.15	1.17	1.22	1.25
Kr	0.45	0.90	0.95	1.00	1.00	1.00
ETo (mm/day)	4.66	4.52	5.57	6.70	6.59	5.13
ETc (mm/day)	2.20	4.39	6.08	7.84	8.04	6.41
Ks				100% (1.00)		
Eu				90% (1.11)		
Lr				10%		

Item	Growth stage					
	Planting establishment	Vegetative growth	Flowering		Ribbling and harvesting	
Length	2–21 Ap.	21 Ap-1 Jun	2 Jun-5 Jul		6 Jul-5 Aug.	
Number of days for irrigation	19	42	34		31	
IRn (mm/month)	41.8	184.4	170.2	47.0	257.5	88.0
IRg (mm/month)	51.3	227.2	209.7	57.9	209.0	64.1

amounts of NPK (20–20–10) fertilizers were: 68.5 kg/fed of (20% N), 81.5 kg/fed of (20% K_2O), and 77.5 kg/fed of (10% P_2O_5). In all plots, weed and pest control measures were followed based on the recommendations for corn by Al-Hassa, KSA. Irrigation season of corn was ended 11 days before the last harvest. Corn was harvested on September 15, 2012. The air-dried weights of grains and Stover were recorded and yields were calculated in Kg/fed. Water use efficiency (WUE) for the grain and Stover was calculated using methods described by Mansour in Chapter 4 of this volume [8, 12, 13]. Fertilizer use efficiency (FUE) was also determined. The values of WUE and FUE are listed in Tables 10.5 and 10.6.

MSTATC program (Michigan State University) was used to carry out statistical analysis. Treatment means were compared using analysis of variance (ANOVA) and the least significant difference (L.S.D) at P = 0.01 among the treatments [19].

TABLE 10.5 Effects of Different Irrigation Circuit Designs And Different Lateral Lines Lengths (Operating   Pressure = 1 atm and slope = 0%) on WUE of Corn For Grain and Stover Yields

DIC	LLL (m)	Applied water (m³/fed)	Grain Yield (kg/fed)	WUEg (kg/m³)	Stover Yield (kg/fed)	WUEs (kg/m³)
	40		5661.2a	1.51a	3809.3a	1.02a
	60		5429.8c	1.45cb	3537.9d	0.95d
CM2DIS	80		5231.1f	1.40fe	3318.1 g	0.89f
	40	3745.56	5469.2b	1.46b	3723.3b	0.99b
	60		5275.6e	1.41e	3447.2e	0.92e
CM1DIS	80		5068.8 g	1.35 g	3187.9h	0.85h
	40		5383.3d	1.43d	3608.6c	0.96c
	60		4757.1h	1.27h	3323.4f	0.89 gf
TDIS	80		4527.3i	1.21i	3007.2i	0.80i
1 × 2	**LSD 0.01**		18.6	0.02	15.62	0.03
	CM2DIS		5440.7a	1.45a	3555.1a	0.95a
	CM1DIS		5271.2b	1.41b	3452.8b	0.92b
Means	TDIS		4889.2c	1.30c	3313.1c	0.88c
(1)	**LSD 0.01**		**24.5**	**0.03**	**88.7**	**0.02**
	40		5504.6a	1.47a	3713.7a	0.99a
	60		5154.2b	1.38b	3436.2b	0.92b
	80		4942.4c	1.32c	3171.1c	0.85c
Means (2)	**LSD 0.01**		**93.6**	**0.05**	**95.4**	**0.04**

DIC: Trickle irrigation circuits, L.L.L.: Lateral line lengths, LLL1: Lateral line length = 40 m, LLL2: Lateral line length = 60 m, LLL3: Lateral line length = 80 m CM2DIS: Closed circuit with two manifolds separately, CM1DIS: Closed circuit with one manifold; TDIS: Traditional trickle irrigation system. WUEg: Grain water use efficiency, WUEs: Stover water use efficiency. Values with the same letter are not significant at P = 0.01.

TABLE 10.6 Effects of Three Closed Circuit Drip Irrigation Designs and Three Lateral Line Lengths on FUE of Corn

DIC	LLL (m)	Applied fertilizers (kg/fed)			Grain yield (kg /fed)	FUE (kg yield per kg fertilizer)		
		N	P₂O₅	K₂O		N U E	P U E	K U E
	40 LLL1				5661.2a	116.7a	109.9a	121.7a
CM2DIS	60 LLL2				5429.8c	111.9cb	105.4cb	116.8cb
	80 LLL3				5231,1f	107.85f	101.5fe	112.5te
	40				5469.2b	112.7b	106.2b	117.6b
CM1DIS	60	48.5	51.5	46.5	5275.6e	108.7ed	102.4e	113.5e
	80				5068.8 g	104.5 g	98.4 g	109.0 g
	40				5383.3d	110.9dc	104.5d	115.7d
TDIS	60				4757.1h	98.1h	92.3h	102.3h
	80				4527.3i	93.3i	87.9i	97.3i
LSD 0.01					18.6	2.5	1.8	2.3
	CM2DIS				5440.7a	112.2a	105.6a	117.0a
	CM1DIS				5271.2b	108.7ba	102.4b	113.4ba
Means	TDIS				4889.2c	100.8c	94.9c	105.1c
LSD 0.01					24.5	6.2	2.1	4.3
	40				5504.6a	113.5a	106.9a	118.4a
	60				5154.2b	106.3b	100.1b	110.8b
Means	80				4942.4c	101.9c	96.0c	106.3c
LSD 0.01					93.6	4.1	3.6	3.1

DIC: Trickle irrigation circuits, LLL: Lateral line lengths, FUE = Fertilizers use efficiency, NUE = Nitrogen use efficiency, PUE = Phosphorous use efficiency, KUE = Potassium use efficiency, LLL1: Lateral line length = 40 m, LLL2: Lateral line length = 60 m, LLL3: Lateral line length = 80 m CM2DIS: Closed circuit with two manifolds separated, CM1DIS: Closed circuit with one manifold; TDIS: Traditional trickle irrigation system. Values with same letter are not significant at P = 0.01.

10.3 RESULTS AND DISCUSSION

10.3.1 GRAIN YIELD

Table 10.5 indicates the effects of DIC and LLL on corn grain yield (GY, kg/fed.). The treatments can be ranked in the orders: CM2DIS > CM1DIS > TDIS and LLL1 > LLL2 > LLL3. With respect to the main effects of DIC on GY, one can notice that the differences in GY were significant among all DIC at 1% level. The highest and lowest GY were obtained in CM2DIS and TDIS, respectively. With respect to the effects of LLL on GY, there are significant differences at 1% level among LLL1, LLL2 and LLL3. Highest and lowest values were achieved under LLL1 and LLL3, respectively. Concerning the effects of DIC × LLL on GY, there were significant differences at 1% level, except at the following interactions: CM2DIS × LLL3, TDIS × LLL1, CM2DIS × LLL3 and CM1DIS × LLL2. The maximum and minimum values of GY were obtained in CM2DIS × LLL1 and TDIS × LLL3, respectively. These findings can be attributed to the close correlation between vegetative growth from one side and grain yield from the other one.

10.3.2 STOVER YIELD

Table 10.5 indicates the effects of DIC and LLL on Stover yield (SY, kg/fed.). We can observe that the change in SY took the same trend as of GY. Concerning the positive effects of DIC and LLL on SY, these can be ranked in orders: CM2DIS > CM1DIS > TDIS and the LLL1 > LLL2 > LLL3. With respect to DIC and LLL effects on the SY, one can notice significant differences at 1% level between all means values of DIC and LLL. According to the interaction effects of the investigated factors, the highest value of SY was obtained in CM2DIS × LLL1. While the lowest one was achieved in TDIS × LLL3, respectively.

10.3.3 GRAIN AND STOVER WATER USE EFFICIENCIES (WUEG AND WUES)

Table 10.5 and Figs. 10.2a and Fig. 10.2b indicate the effects of DIC and LLL treatments on grain and Stover water use efficiencies (WUEg and WUEs, kg/m^3). We can conclude that the changes in WUEg and WUEs took the same trend as of the vegetative growth parameters under investigation, that is, leaf area, plant height, leaf length and number of leaves per plant (Chapter 9 of this volume). This may be due to the positive effects of DIC and LLL treatments on the vegetative growth parameters of corn

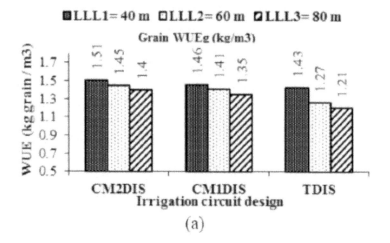

(a)

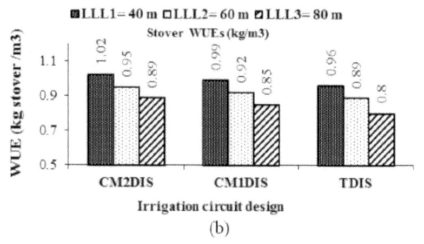

(b)

FIGURE 10.2   Effects of three closed circuit drip irrigation designs and three lateral line lengths on WUEg and WUEs.

mentioned here. According to WUEg and WUEs values, DIC and LLL can be ranked in the orders: CM2DIS > CM1DIS > TDIS and LLL1 > LLL2 > LLL3. Differences in WUEg only among DIC were significant at 1% level.

Differences in WUEg among LLL treatments were significant at 1% level, except that between LLL2 and LLL3. On the other hand, difference in WUEs was significant at 1% level only between LLL1 and LLL3. The effects of the interaction DIC × LLL on WUEg were significant at 1% level, except those among the interactions: CM2DIS × LLL3, CM1DIS x LLL2 and CM2DIS × LLL2, CM1DI × LLL1. The effects of interaction DIC × LLL on WUEs were significant at 1% level, except among the inter-action: CM2DIS × LLL3, TDIS × LLL2. The highest values of WUEg and WUEs (1.51 and 1.02 kg/m³) and the lowest one (1.2 and 0.80 kg/m³) were obtained in the interactions: CM2DIS × LLL1 and TDIS × LLL3, respectively.

10.3.3 FERTILIZERS USE EFFICIENCY (FUE)

Table 10.6 and Fig. 10.3 show the effects of DIC and LLL treatments on (N, P_2O_5 and K_2O) fertilizers use efficiency (FUE_N, FUE_{P2O5} and FUE_{K2O}) of corn. According to the FUE values of the three fertilizers used, the DIC and LLL treatments can be ranked in the orders: CM2DIS > CM1DIS > TDIS and LLL1 > LLL2 > LLL3. Differences in FUE among DIC between any two DIC treatments and /or LLL ones were significant at 1% level, except that between (CM2DIS, CM1DIS; TDIS) and (LLL2, LLL2; LLL3) in the case of (FUEN). Under the effects of LLL, there were significant differences at 1% level in FUE_N. While dif-ferences in FUE_{P2O5} and FUE_{K2O} were significant except that between interactions CM1DIS × LLL2, CM2DIS × LLL3 and CM2DIS x LLL2,

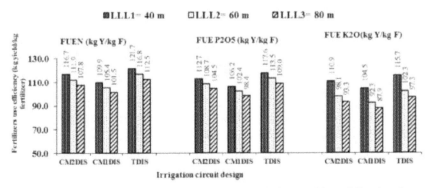

FIGURE 10.3 Effects of different irrigation circuit designs and lateral line lengths on fertilizer use efficiency (FUE).

CM1DIS x LLL2. The effects of the interactions DIC $\times$ LLL treatments on FUE were significant at 1% level among some interactions. The highest values of FUE_N, FUE_{P2O5} and FUE_{K2O} (116.7, 109.9 and 121.7 kg of yield per kg of fertilizer) and the lowest ones (93.3, 87.9 and 97.3 kg of yield per kg fertilizer) were obtained in the interactions: CM2DIS $\times$ LLL1 and TDIS $\times$ LLL3, respectively. These data are supported by Baligar and Bennett [3].

The results conclude that FUE took the same trend as of vegetative growth parameters (Chapter 9 of this volume), yield and WUE. This conclusion may be attributed to the direct relationship between WUE and FUE, found by Tayel et al. [20].

10.4   CONCLUSIONS

We can conclude the followings:
1. The highest values of WUEg and WUEs under DIC and LLL treatments LLL were (1.51 and 1.02 kg per m^3), whereas the lowest ones were (1.21 and 0.80 kg per m^3), in the interactions CM2DIS $\times$ LLL1 and TDIS $\times$ LLL3, respectively.
2. WUEg in (LLL1, LLL2, LLL3) treatments under CM2DIS and CM1DIS compared to TDIS were increased by (5, 12, 13%) and (2, 10, 11%), respectively. On the other hand WUEs in (LLL1, LLL2; LLL3) treatments under CM2DIS and CM1DIS compared to TDIS were increased by (5, 6, 9%) and (3, 4, 6%), respectively.
3. The highest values of FUEN, FUEP2O5 and FUEK2O (116.7, 109.9, 121.7 kg of yield per kg of fertilizer); and the lowest ones were (93.3, 87.9; 97.3 kg yield/kg fertilizer) were achieved in the interactions: CM2DIS $\times$ LLL1 and TDIS $\times$ LLL3, respectively.
4. FUEN, FUEP2O5 and FUEK2O in (LLL1, LLL2; LLL3) treatments under CM2DIS and CM1DIS compared to TDIS were increased by (5, 12, 13%) and (1.5, 9.8, 10.6%), respectively.
5. The efficiency parameters (WUEg, WUEs, FUEN, FUEP2O5 and FUEK2O) had positive effects on the orders: CM2DIS > CM1DIS > TDIS and LLL1 > LLL2 > LLL3. The highest and lowest values were in interactions: CM2DIS $\times$ LLL1 and TDIS $\times$ LLL3, respectively.

10.5 SUMMARY

Field experiments were carried out at Experimental Farm of Irrigation and Drainage Authority Project, Al-Hassa Governorate, Al-Hassa City, Hufof State of KSA in a sandy loam soil. This chapter discusses research studies to evaluate effects on water and fertilizer use efficiencies of corn (*Zea mays-L*, cv. Giza-155) crop by three closed circuit drip irrigation designs (CM2DIS, CM1DIS, TDIS) and three lateral line lengths (LLL1, LLL2, LLL3) in Kingdom of Saudi Arabia. Effects of these treatments on water use efficiency (WUE) and fertilizer use efficiency (FUE) were evaluated..

Based on WUE and FUE (Kg/m3), DIC and LLL can be ranked in the orders: TDIS < CM1DIS < CM2DIS and LLL3 < LLL2 < LLL1, respectively. The effects of interactions DIC × LLL on WUE and FUE mentioned above were significant at 1% level. The highest values of WUEg and WUEs (Kg/m^3) were 1.51 and 1.02. Whereas the lowest ones (1.21 and 0.80 Kg/m^3) were observed in the interactions: CM2DIS × LLL1; TDIS × LLL3, respectively. The highest values of FUEN, FUEP2O5 and FUEK2O (116.7, 109.9, 121.7 kg yield per kg of fertilizer) and the lowest ones (93.3, 87.9, 97.3 kg yield per kg fertilizer) were observed in the interactions: CM2DIS × LLL1 and TDIS × LLL3, respectively.

KEYWORDS

- closed circuit
- closed circuit irrigation design
- corn
- drip Irrigation
- fertigation
- fertilizer
- fertilizer use efficiency
- grain yield
- irrigation depth
- irrigation interval
- Kingdom of Saudi Arabia, KSA
- lateral line

- **Stover yield**
- **straw yield**
- **USA**
- **water use efficiency**

REFERENCES

1. Allen, R. G., L. S. Pereira, D. Raes and M. Smith, 1998. *Crop evapotranspiration-Guidelines for computing crop water requirements.* FAO Irrigation and Drainage paper 56. Rome

2. Bacon, M. A., 2004. *Water use efficiency in plant biology.* CRC Press.

3. Baligar, V. C. and O. L. Bennett, 1986. NPK – fertilizer efficiency – A situation analysis for the tropics. *Fert. Res.,* 10:147–164.

4. Bhatti, A. U. and M. Afzal, 2001. Plant nutrition management for sustained production. Deptt. Soil and Envir. Sci., NWFP Agri. Univ. Peshawar – Pakistan. Pp. 18–21.

5. Brady, N. C., 1974. Supply and availability of phosphorus and potassium. In: *The Nature and Properties of Soils.* (Ed.) R.S. Buckman. Macmillan Publishing Co., Inc., New York. Pp. 456–480.

6. Fulton, J. M., 1970. Relationship among soil moisture stress plant population, row spacing and yield of corn. *Can. J. Plant Sci.,* 50:31–38.

7. Gee, G. W. and J. W. Bauder, 1986. Particle – size analysis. Pages 383–412. In: Klute (ed.) *Methods of soil analysis. Part 1.* ASA and SSSA, Madison, WI.

8. Howell, T. A., A. Yazar, A. D. Schneider, D. A. Dusek and K. S. Copeland, 1995. Yield and water use efficiency of corn in response to LEPA irrigation. *Trans. ASAE.,* 38(6):1737–1747.

9. Keller, J. and D. Karmeli, 1975. *Trickle irrigation design.* 1st edition by Rain Bird Sprinkler Manufacturing Corporation, Glendora, California, 91740, USA.

10. Klute, A., 1986. Moisture retention. p. 635–662. In: A. Klute (ed.) *Methods of soil analysis. Part 1.* ASA and SSSA, Madison, WI.

11. Laboski, C. A. M., R. H. Dowdy, R. R. Allmaras and J. A. Lamb, 1998. Soil strength and water content influences on corn root distribution in a sandy soil. *Plant and Soil,* 203:239–247.

12. Mansour, H. A., 2006. *The response of grape fruits to application of water and fertilizers under different localized irrigation systems.* M.Sc. Thesis, Faculty of Agriculture, Agric., Ain Shams University, Egypt, pp. 78–81.

13. Mansour, H. A., 2012. *Design considerations for closed circuits of drip irrigation system.* Ph.D. Thesis, Faculty of Agriculture, Agric., Ain Shams University, Egypt.

14. Memon, K. S., 1996. Soil and Fertilizer Phosphorus. In: *Soil Science.* (Ed.) Elena Bashir and Robin Bantel. National Book Fund. Islamabad., pp. 308- 311.

15. Modarres, A. M., R. I. Hamilton, M. Dijak, L. M. Dwyer, D. W. Stewart, D. E. Mather and D. L. Smith, 1998. Plant population density effects on maize inbred lines grown in short-season environment. *Crop Sci.,* 38:104- 108.

16. Ogola, J. B. O., T. R., Wheeler and P. M. Harris, 2002. The water use efficiency of maize was increased by application of fertilizer N. *Field Crops Research*, 78(2–3):105–117.

17. Rebecca, B., 2004. Soil Survey *Laboratory Methods Manual*. USDA Soil Survey Laboratory Investigations Report No. 42, Room 152, 100 Centennial Mall North, Lincoln, NE 68508–3866. (402) 437–5006.

18. Safi, B., M. R. Neyshabouri, A. H. Nazemi, S. Masiha and S. M. Mirlatifi, 2007. Subsurface irrigation capability and effective parameters on onion yield and water use efficiency. *Journal of Scientific Agricultural*, 1:41–53.

19. Steel, R. G. D and J. H. Torrie, 1980. *Principles and Procedures of Statistics*. A biometrical approach. 2nd Ed., McGraw Hill Inter. Book Co. Tokyo, Japan.

20. Tayel, M. Y., I. Ebtisam, Eldardiry and M. Abd El-Hady, 2006. Water and fertilizer use efficiency as affected by irrigation methods. *American-Eurasian J. Agric. & Environ. Sci.*, 1(3):294–300.

CHAPTER 11

PERFORMANCE OF DRIP IRRIGATED SOYBEAN

M. Y. TAYEL, H. A. MANSOUR, and S. KH. PIBARS

CONTENTS

11.1   INTRODUCTION

Soybean is one of the most important crops for oil and protein contents in the world [4]. The present world production is about 6.2 million tons of soy-

Modified and printed from *Mohamed Yousif Tayel, Hani Abdel-Ghani Mansour, Sabreen Khalil Pibars, 2013. Effect of closed circuits drip irrigation system and lateral lines length on growth, yield, quality and water use efficiency of soybean crop. Agricultural Sciences, 4(2):85–90. Open access source* at http://www.scirp.org/journal/as/.

In this chapter: 1 feddan = 0.42 hectares = 4200 m² = 1.038 acres = 24 kirat. A feddan (Arabic) is a unit of area. It is used in Egypt, Sudan, and Syria. The feddan is not an SI unit and in Classical Arabic, the word means 'a yoke of oxen': implying the area of ground that can be tilled in a certain time. In Egypt the feddan is the only nonmetric unit, which remained in use following the switch to the metric system. A feddan is divided into 24 Kirats (175 m²). In Syria, the feddan ranges from 2295 square meters (m²) to 3443 square meters (m²).

bean seed that is cultivated on 45 million ha. The growth periods most sensitive to water deficit of soybean are last part of the flowering stage and the early part of the pod formation [4]. When water supply is limited, water can be saved by reducing the irrigation supply during the vegetative period and near crop maturity [7]. Irrigation is an important and an increasingly common practice in Georgia and other South-eastern states of USA for soybean production [8], as shown by an expansion of irrigated acreage from almost 9000 ha in 2000 to > 40,000 in 2008 Soybean yield had been reported as low as 807 kg/ha in 1980 and as high as 2220 kg/ha in 2003 [18]. This large difference is mainly due to droughts, evidencing the need for supplemental irrigation in Georgia, despite humid climate. Several studies conducted for a wide range of environments have demonstrated that soybean yield increases with irrigation [2, 3, 12, 18].

Closed drip irrigation circuits have used in attempts to overcome the drop in pressure at the end of the lateral line of drip irrigation system [10]. Mansour [10] carried out laboratory and field experiments to study the effects of closed drip irrigation circuit with one manifold for lateral lines (CM1DIS), with two manifolds for lateral lines (CM2DIS), traditional drip irrigation system as a control (TDIS), and lateral line lengths (LLL) on some hydraulic characteristics of the drip irrigation system, corn yield, water and fertilizer use efficiencies. Research results of this study can be summarized as: (i) Relative to TDIS, both CM2DIS and CM1DIS improved the studied hydraulic characteristics (pressure head, friction loss, flow velocity, lateral discharge, uniformity coefficient, coefficient of variation), corn yield, water and fertilizer use efficiencies; (ii) The mean effects of both DIC and LLL treatments on the studied parameters were significant at 1%; (iii) The effects of DIC × LLL on the parameters under investigation were significant at 1% level; and (iv) Based on the observed data treatments can be stated in the ascending orders: TDIS < CM1DIS < CM2DIS and LLL3 < LLL2 < LLL1.

11.2   MATERIALS AND METHODS

A split-plot randomized complete block design with three replications was used in this study. Laboratory tests were conducted to evaluate three irrigation lateral lines (LLL1 = 40 m, LLL2 = 60 m, and LLL3 = 80 m) and three drip irrigation circuits (DIC): closed circuit with one manifold of drip irrigation system (CM1DIS), closed circuits with two manifolds for lateral lines (CM2DIS), and traditional drip irrigation system (TDIS) as a control. Tables 11.1–11.3 indicate some (physical and chemical) characteristics of soil and irrigation water, respectively. Irrigation network was similar to that is mentioned in Chapter 4.

TABLE 11.1   Some Physical Properties of the Soil*

Sample depth, cm	Particle Size Distribution, %				Texture class	F.C.	W.P.	A.W.
	C. sand	F. sand	Silt	Clay		θ% (w/w)		
0–15	3.4	29.6	39.5	27.5	CL	32.35	17.81	14.44
15–30	3.6	29.7	39.3	27.4	CL	33.51	18.53	14.98
30–45	3.5	28.5	38.8	28.2	CL	32.52	17.96	14.56
45–60	3.8	28.7	39.6	27.9	CL	32.28	18.61	13.67

*Particle size distribution after [5] and moisture retention after [9];

C.L.: Clay Loam; F.C.: Field Capacity; W.P.: Wilting Point; A.W.: Available Water.

TABLE 11.2   Some Chemical Properties of the Soil (Saturated Extracted)*

Sample depth cm	pH 1:2.5 –	EC dS/m	Soluble cations				Soluble anions			
			Ca^{2+}	Mg^{2+}	Na^+	K^+	$(CO_3)^{-2}$	$(HCO_3)^{-1}$	$(SO_4)^{-2}$	Cl^{-1}
							meq/L			
0–15	7.3	0.35	1.5	0.39	1.52	0.12	0	0.31	1.52	1.67
15–30	7.2	0.36	1.51	0.44	1.48	0.14	0	0.41	1.56	1.63
30–45	7.3	0.34	1.46	0.41	1.4	0.13	0	0.39	1.41	1.63
45–60	7.4	0.73	2.67	1.46	3.04	0.12	0	0.67	2.86	3.82

*Chemical properties after [11].

TABLE 11.3   Some Chemical Properties of Irrigation Water

pH 1:2.5 –	SAR	EC dS/m	Soluble cations				Soluble anions			
			Ca^{2+}	Mg^{2+}	Na^+	K^+	$(CO_3)^{-2}$	$(HCO_3)^{-1}$	$(SO_4)^{-2}$	Cl^{-1}
							meq/L			
7.3	1.14	0.37	0.76	0.24	2.6	0.13	0	0.9	0.32	2.51

11.2.1 IRRIGATION SCHEDULING

Irrigation intervals (I, days) were calculated as follows:

$$I = d/ETc \qquad (1)$$

where: d = net water depth applied in each irrigation (mm), and ETc = crop evapotranspiration (mm/day). Net irrigation depth (d, mm) is estimated as follows:

$$d = AMD \times ASW \times Rd \times P \qquad (2)$$

where: AMD = allowable soil moisture depletion (%), ASW = available soil water, (mm water/m of soil depth), Rd = effective root zone depth (m) or irrigation depth (m), and p = percentage of soil wetted area (%).

$$AW \ (v/v, \ \%) = ASW \ (w \ /w, \ \%) \times B.D. \qquad (3)$$

where: AW = available water, B.D. = soil bulk density (gm·cm^{-3}). Irrigation intervals in this study were 4 days in all three irrigation systems.

11.2.2 MEASURING SEASONAL EVAPOTRANSPIRATION (ETC)

The ETc was computed using Class A Pan evaporation method for estimating (ETo) on daily basis. The climatic data were from the nearest meteorological station and the calculations are summarized as in Table 11.4. The modified pan evaporation equation was used:

$$ETo = Kp \ x \ Ep \qquad (4)$$

where: ETo = reference evapotranspiration (mm/day), Kp = pan coefficient of 0.76 for Class A pan placed in short green cropped and medium wind area. Ep = daily pan evaporation (mm/day) = seasonal average of 7.5 mm/day, [1].

The reference evapotranspiration (ETo) is then multiplied by a crop coefficient Kc for particular growth stage to determine crop consumptive

TABLE 11.4. Water Requirements For Soybean Grown At the Experimental Site

Item	Month					
	Apr.	May	Jun.	Jul.	Aug.	Sep.
Epan (mm/day)	6.34	6.92	7.97	9.59	9.32	7.17
Kp				0.76		
Kc	0.72	0.82	0.93	1.18	1.2	1.23
Kr	0.25	0.63	0.95	1	1	1
ETo (mm/day)	4.75	5.26	6.06	7.29	7.09	5.45
ETc (mm/day)	0.85	2.72	5.35	8.6	8.51	6.7
Ks				100% (1.00)		
Eu				90% (1.11)		
Lr				10%		

Growth stage, total days = 152				
Item	Planting establishment	Rapid vegetative growth	Flowering and seed fill	Maturation and harvesting
Length	15 Apr.–14 May	14 May–13 Jun	13 Jun.–12 Aug.	12 Aug.–11 Sep.
Number of days for irrigation	30	30	61	31
IRn (mm/month)	15.0	92.8	176.6 293.3	290.2 81.1
IRg (mm/month)	49.3	158.8	198.6 264.5	268.2 27.3

use at that particular stage of soybean crop. The reduction factor (Kr) was calculated using Eq. (6).

$$ETc = ETo \times Kc \tag{5}$$

$$Kr = GC + 1/[2(1 - GC)] \tag{6}$$

$$Ea = Ks \times Eu \tag{7}$$

$$IWRg = IWRn \times Ea + Lr \tag{8}$$

where: GC = ground cover percentage, Ea = irrigation efficiency (%), Eu = emission uniformity (%), Ks = reduction factor of soil wetted, IWRg = gross irrigation water requirements (mm of water depth), IWRn = net irrigation water requirements (mm) and Lr = extra amount of water needed for leaching. Irrigation efficiency (Ea) was calculated by Eq. (7). IWRg was calculated with Eq. (8).

11.2.2 CROP ESTABLISHMENT

Soybean seeds (*Glycine max-L, Rils-75*) were planted on April 15th of 2012. The row-to-row spacing was 0.7 m and plant to plant spacing was 0.15 m down the row, giving a plant density of 55,500 plants per fed according to (ISU). Each row was irrigated by a single straight lateral drip line in the closed circuits and traditional drip irrigation plots. Figure 11.1 shows that the total experimental area was 4536 m². Under each of the tested drip irrigation circuits, plot areas of were 168, 252 and 336 m² for each LLL1, LLL2 and LLL3, respectively. Soybean was harvested on September 11 of 2012. Irrigation of soybean was ended 10 days before harvest. Fertilization program was based according to the recommended doses throughout the growing season using fertigation method. These amounts of fertilizers (NPK: 20–20–10) were 74.6 kg/fed of N, 33.0 kg/fed of K_2O, and 60.5 kg/fed of P_2O_5. For all plots, weed and pest control measures were according to the recommendations for soybean crop.

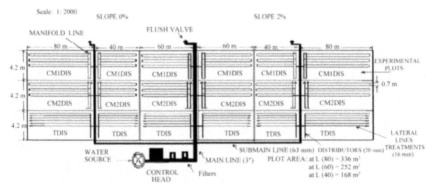

FIGURE 11.1   Field Layout of experimental plots: DIC, (CM2DIS, CM1DIS and TDIS); and (LLL1 = 40 m; LLL2 = 60 m and LLL3 = 80 m) treatments.

11.2.3 MEASUREMENTS OF SOYBEAN PLANT GROWTH, YIELD AND QUALITY

Plant measurements and observations were started 21 days after planting, and were terminated on the harvest date. Evaluation parameters included: leaf area (cm²) by plano-meter, plant height (cm) by ruler, total grain and straw yield (Kg/fed) by weighing balance, oil and protein contents (g/kg). Oil and protein contents were determined in Grain Quality Laboratory using near-infrared analysis. All plant samples were dried at 65°C until constant weight was achieved. Grain yield was determined by hand harvesting the 8 m sections of three adjacent center rows in each plot and was adjusted to 15.5% water content. In all treatments plots, the grain yields of individual rows were determined in order to evaluate the yield uniformity among the rows.

MSTATC program (Michigan State University) was used to carry out statistical analysis. Treatments mean were compared using the technique of analysis of variance (ANOVA) and the least significant difference (L.S.D.) between treatments at 1% [13].

11.3 RESULTS AND DISCUSSION

11.3.1 LEAF AREA AND PLANT HEIGHT

Table 11.5 indicates the effects of DIC and LLL treatments on leaf area (cm²) and plant height (cm). The treatments can be ranked in descending orders: CM2DIS ≥ CM1DIS < TDIS and LLL1 < LLL2 < LLL3. Differences in leaf area and plant height values between means of the two factors were significant at 1% level except that between CM2DIS and CM1DIS for both leaf area and plant height and between LLL1 and LLL2 for plant height. The effects of DIC × LLL on plant height and leaf area were significant at 1% level. The superiority of the studied growth parameters under (CM2DIS and CM1DIS relative to TDIS) and (LLL1, LLL2 relative to LLL3) can be noticed. This superiority was due to improving the water and fertilizer distribution uniformities [10, 15].

11.3.2   GRAIN AND STRAW YIELD

Table 11.5 shows the effects of DIC and LLL treatments on grain and straw yield (kg/fed). These treatments can be ranked in descending order: CM2DIS > CM1DIS > TDIS and LLL1 > LLL2 > LLL3. Differences in grain and straw yields between means of any two treatments were significant at 1% level except that between CM1DIS and CM2DIS in straw yield. The effects of the DIC × LLL on grain and straw yields were significant at 1% level. The highest and lowest values of both grain and straw yield were recorded in the interactions: CM2DIS × LLL1 and TDIS × LLL3, respectively. This superiority was due to improving both vegetative growth, water and fertilizers distribution uniformities [5, 10].

11.3.3   OIL AND PROTEIN CONTENTS

Table 11.5 indicates the effects of DIC and LLL treatments on soybean oil and protein contents (g/kg). These can be ranked in the ascending orders: TDIS < CM1DIS < CM2DIS and LLL3 < LLL2 < LLL1, respectively. According to the main effects of DIC on oil and protein contents, the differences in oil and protein were significant among all DIC and LLL treatments at 1% level except that between CM1DIS and CM2DIS for oil. The highest and lowest oil and protein values were obtained in the interactions: CM2DIS × LLL1 and TDIS × LLL3, respectively.

11.3.4   WATER USE EFFICIENCY FOR GRAIN AND STRAW YIELD

Table 11.5 shows the effects of DIC and LLL treatments on water use efficiency for grain and straw (WUEg and WUEs, kg/m^3). It can be observed that the changes in WUEg and WUEs followed the trends similar to those for plant growth, grain, and straw parameters and thus took the trend of grain quality (oil and protein contents).

Based on the positive effects of DIC and LLL treatments on WUEg and WUEs, these can be ranked in descending orders: CM2DIS > CM1DIS > TDIS and LLL1 > LLL2 > LLL3, respectively. Differences in WUEg

TABLE 11.5 Effects of Irrigation Circuits Designs and Lateral Lines Lengths on Soybean Plant Growth, Yield and Quality, At an Operating Pressure = 1 atm and slope = 0%

DIC	LLL	Leaf area	Plant height	Yield		Quality		Water use efficiency	
				Grain	Straw	Oil	Protein	WUEg	WUEs
	m	cm²	cm	kg/fed.		g/kg		kg/m³	
CM2DIS	40	7.91a	94.38a	657.5a	588.6a	183.5a	366.2a	0.150a	0.134a
	60	7.62cb	92.26dc	648.3cb	557.5d	181.2d	364.1d	0.148b	0.127d
	80	6.56f	91.15h	641.6db	531.8h	178.1 g	361.8 g	0.146d	0.121h
CM1DIS	40	7.64b	94.29ba	642.2b	583.7b	183.3b	365.3b	0.146cd	0.133b
	60	7.43dc	92.35c	628.3e	546.4e	180.6e	363.6e	0.143e	0.125e
	80	6.22h	91.52f	597.7 g	537.2 g	177.6h	361.3h	0.136 g	0.123 g
TDIS	40	6.85e	92.11e	605.3f	574.3c	182.4e	365.1c	0.138f	0.131c
	60	6.51 g	91.18 g	593.4hg	542.8fe	179.7f	362.2f	0.135h	0.124f
	80	5.92i	90.23ih	586.2i	519.6i	176.3i	360.4i	0.134ih	0.119i
(1) × (2)	LSD0.01	0.23	0.12	8.2	4.8	0.14	1.3	0.002	0.001
(1) Means	CM2DIS	7.36a	92.59a	649.1a	559.3a	180.9a	364.0a	0.148a	0.128a
	CM1DIS	7.09ba	92.72ba	622.7b	555.8ba	180.5ba	363.4ba	0.142b	0.127ba
	TDIS	6.43c	91.17c	595.0c	545.6c	179.5c	362.6c	0.136c	0.124c
	LSD0.01	0.48	0.51	15.8	7.2	0.6	0.7	0.004	0.002

Growth, yield and quality characteristics (average values)

TABLE 11.5 (Continued)

Growth, yield and quality characteristics (average values)

DIC	LLL	Leaf area	Plant height	Yield		Oil	Quality	Water use efficiency	
				Grain	Straw		Protein	WUEg	WUEs
	m	cm²	cm	kg/fed.			g/kg	kg/m³	
(2) Means	40	7.47a	93.60a	635.0a	582.2a	183.1a	365.5a	0.145a	0.133a
	60	7.19ba	91.93b	623.3b	548.9b	180.5ba	363.3b	0.142b	0.125b
	80	6.23c	90.97c	608.5c	529.5c	177.3c	361.2c	0.139c	0.121c
	LSD0.01	0.72	0.71	12.4	26.4	3.1	1.9	0.003	0.005

DIC: Irrigation circuit design; L.L.L.: Lateral line length; CM2DIS: Closed circuits with tow manifolds separately; CM1DIS: Closed circuits with one manifold; TDIS: Traditional drip irrigation system.

and WUEs between means of any two treatments were significant at 1% level except that between CM1DIS and CM2DIS for WUEs. The effects of the DIC × LLL on WUEg and WUEs were significant at 1% level. The highest and lowest values of WUEs were obtained in CM2DIS × LLL1 and TDIS × LLL3, respectively. We can notice that the soybean WUEg, WUEs, oil and protein contents took the similar trend as vegetative growth and yield parameters. These findings can be attributed to the close correlation between vegetative growth, grain yield from side and quality of oil and protein contents from the other one [10, 14–17].

11.4 CONCLUSIONS

Based on the mean values of performance parameters of soybean (leaf area, plant height, yield, oil and protein contents, and water use efficiency), the treatments can be ranked in the ascending orders: TDIS < CM1DIS < CM2DIS and LLL3 < LLL2 < LLL1. Differences in the means of the parameters among treatments were significant at 1% level. The effects of the DIC × LLL on the data were significant at 1% level. The highest values of the data and the lowest ones were achieved in the interactions: CM2DIS × LLL1 and TDIS × LLL3, respectively.

11.5 SUMMARY

During the growing season of 2012, soybean crop was established in clay loam soil at the Experimental Farm of Faculty of Agriculture, Southern Illinois University at Carbondale (SIUC), USA. This chapter discusses the effect of three drip irrigation systems (closed circuit drip irrigation system with one and two manifolds for lateral lines, CM1DIS and CM2DIS, and traditional drip irrigation system, TDIS) and three lateral lines lengths (LLL1 = 40 m, LLL2 = 60 m, and LLL3 = 80 m) on performance of soybean (plant growth, yield, oil and protein content, and water use efficiency). Plants were drip irrigated every 4 days. The fertilizers (N, K_2O and P_2O_5) were fertigated at recommended dosages. Based on the mean values of soybean crop growth parameters (leaf area, plant height, yield, oil and protein contents, and water use efficiency), the effects of treatments can be ranked in the ascending orders: TDIS < CM1DIS < CM2DIS and LLL3 < LLL2 < LLL1. Differences in the means of the values of the parameters among treatments were significant at 1% level. The effects of the DIC × LLL on the parameters were significant at 1% level. The highest and lowest values of these parameters were observed in the interactions: CM2DIS × LLL1 and TDIS × LLL3, respectively.

KEYWORDS

- **closed circuit irrigation design**
- **drip irrigation**
- **evapotranspiration**
- **fertigation**
- **fertilizer**
- **grain yield**
- **irrigation depth**
- **irrigation interval**
- **lateral line**
- **oil content**
- **plant height**
- **protein content**
- **soybean**
- **straw yield**
- **water use efficiency**

REFERENCES

1. Allen, R.G., Pereira, L.A., Raes, D. and Smith, M., 1998. *Crop evapotranspiration.* FAO Irrigation and Drainage Paper, 56, pages 293.
2. Bajaj, S., Chen, P., Longer, D.E., Shi, A., Hou, A., Ishibashi, T. and Brye, K.R., 2008. Irrigation and planting date effects on seed yield and agronomic traits of early maturing soybean. *Journal of Crop Improvement*, 22:47- 65.
3. Dogan, E., Kirnak, H. and Copur, O., 2007. Deficit irrigations during soybean reproductive stages and CROP- GRO-soybean simulations under semiarid climatic conditions. *Field Crops Research*, 103:154–159.

4. FAO, 1979. Yield response to water. FAO Irrigation and Drainage Paper 33, pages 193.

5. Gee, G.W. and Bauder, J.W., 1986. Particle size analysis. In: Klute, I.I. (ed.), *Methods of Soil Analysis*. Soil Science Society of America, Madison, pages 383–412.

6. Gercek, S., Boydak, E., Okant, M. and Dikilitas, M., 2009. Water pillow irrigation compared to furrow irrigation for soybean production in a semiarid area. *Agricultural Water Management*, 96:87–92.

7. Megh R. Goyal, 2012. *Management of Drip/Trickle or Micro Irrigation*. Apple Academic Press Inc., Oakville – ON – Canada.

8. Harrison, K.A., 2009. *Irrigation survey*. Cooperative Extension Service, College of Agricultural and Environmental Sciences, The University of Georgia, Tifton.

9. Klute, A., 1986. Moisture retention. In: Klute, A. (ed.), *Methods of Soil Analysis*, Soil Science Society of America, Madison, 635–662.

10. Mansour, H.A., 2012. Design considerations for closed circuits of drip irrigation system. Ph.D. Thesis, Ain Shams University, Cairo.

11. Rebecca, B., 2004. *Soil survey laboratory methods manual*. Soil Survey Laboratory Investigations Report, No. 42.

12. Sincik, M., Candogan, B.N., Demirtas, C., Büyükacangaz, H., Yazgan, S. and Goksoy, A.T., 2008. Deficit irrigation of soybean (*Glycine max L. Merr.*) in a subhumid cli- mate. *Journal of Agronomy and Crop Science*, 194:200–205.

13. Steel, R.G.D. and Torrie, J.H., 1980. *Principles and procedures of statistics, A biometrical approach*. 2nd Edition, McGraw Hill Book Company, Tokyo.

14. Tayel, M.Y., Mansour, H.A. and El-Gindy, A.M., 2012. Effect of different closed circuits and lateral line lengths on dripper and lateral discharge. *Journal of Applied Sciences Research*, 8:2725–2731.

15. Tayel, M.Y., Mansour, H.A. and El-Gindy, A.M., 2012. Effect of different closed circuits and lateral line lengths on uniformity coefficient and coefficient of variation. *Journal of Applied Sciences Research*, 8:2741–2748.

16. Tayel, M.Y., Mansour, H.A. and Lightfoot, D. A., 2012. Effect of different closed circuits and lateral line lengths on pressure head and friction loss. *Agriculture Sciences Journal*, 3:392–399.

17. Tayel, M.Y., Mansour, H.A. and Lightfoot, D. A., 2012. Effect of different closed circuits and lateral line lengths on flow velocity and velocity head. *Agriculture Sciences Journal*, 3:531–537.

18. United States Department of Agriculture, National Agriculture Statistics Service, 2009. http://www.nass.usda.gov/Quick_Stats/

CHAPTER 12

DRIP IRRIGATION IN RICE

R. K. SIVANAPPAN

CONTENTS

12.1 INTRODUCTION

More than 40% of the total 280 million-ha of irrigated land in the world are planted in rice. More than 90% of the world rice is produced in Asia, where many major rivers are now tapped out during the dry season and where competition for water for urban and farm use is escalating. Finding ways of irrigating rice with less water is critical to sustaining the harvest of this crop.

The Rice plant is unique in that its root does not need to take in oxygen from air pockets in the soil so that the plant can thrive in water logged conditions. Flooding rice field can result in substantial evaporation losses until the crop cover is established. In North-east Sri Lanka, researchers found that evaporation accounted for 29% of total dry season water consumptive use for rice. By using the system of rice intensification (SRI) method, the water saving is about 40 to 45% and yield increase of about 30–40% compared to conventional method of flood irrigation. The average yield of paddy in Egypt (Nile Delta) is about 8 to 9 tons/ha. Some farmers in Tamil Nadu have taken 10 to 15 tons/ha. Farmers, who shifted to the direct seeding method, were able to use less water during both the preplanting and growing period and were able to get 9% increase in yield. Overall water productivity was increased by 25%.

In contrast to wheat and corn, none of the world's rice is irrigated by sprinkler irrigation. Sprinkler irrigation gives higher water application uniformity than that in gravity (surface) irrigation methods. Also sprinkler irrigation allows farmers to irrigate the crops adequately with less water.

In the USA where rice is seeded directly, researchers found that sprinkler irrigation may substitute flood irrigation in case of some rice varieties. In Arkansas state of USA, they found that sprinkler irrigated rice was able to save 50% of water and gave yields compatible to flood irrigation. Results were more mixed in Louisiana and Texas States of USA with sprinkler irrigated rice yielding 10–25% less than that in the flooded rice. Since sprinkler irrigation represents an entirely new crop management regime, it needs further research especially with low cost sprinklers.

Rice cultivation in India and other countries involves transplanting of seedlings in the puddled fields and fields are kept continuously flooded with 5 to 10 cm of water throughout the growing season. Unproductive water losses due to seepage and percolation from flooded rice fields vary from 50 to 60% of the total water input in the field. Hence it results in low water use efficiency (WUE). We need 3000 to 5000 L or kg of water to produce 1 kg of unprocessed rice (paddy). It not only leads to wastage of water but also causes environmental degradation and reduces fertilizer use efficiency (FUE). Therefore, every drop of water received at the field needs to be used effectively and wisely.

Use of drip irrigation technology can reduce the rice consumptive use when the rice is grown like an irrigated dry crop. Research on drip-irrigated rice is limited and recent that is being practiced in many countries. Under these practices, rice is grown in nonpuddled nonsaturated aerobic soil with optimally balanced soil water air conditions using supplementary irrigation and fertilizer and aiming at high yields per drop of water used. Irrigation scheduling is based on evapotranspiration (ET) with water-applied equivalent to replenish the soil water content in the root zone back to the field capacity on daily or alternate days. Thus, the soil water matches the crop ET requirements (plus any application efficiency losses). Field experiments indicate that the water requirement was 1670 mm for low land rice field and 772 mm for drip irrigated with an yield potential of about 5 to 7 tons/ha. Additional benefits of fertigation through drip irrigation include: savings in energy/labor/fertilizers, no leaching; higher water productivity and nutrient use efficiency (NUE), etc. It also reduces the emission of methane gas in rice ecosystem, which is a major environment issue today. It is expected that fertigation via drip system can reduce nitrate pollution in community water bodies.

Drip irrigation with fertigation for paddy crop is being experimented in many countries including India during the last few seasons in many universities and Research stations in the world.

This chapter will focus on the research on drip-irrigated rice at Central Institute of Agricultural Engineering (CIAE), Bhopal India. This research study compares 'SRI method under drip irrigation with fertigation' with conventional rice cultivation as practiced by farmers with transplantation of young rice seedlings, to attain highest water productivity.

12.2   MATERIALS AND METHODS

The soil at the experimental site is clayey in texture (Vertisol) with an average annual rainfall of about 1100 mm during kharif season (June through September: *kharif* is an Indian name). The practice of transplanting of young seedlings that is being practiced in SRI method was adopted with drip irrigation to compare with conventional transplanting aged seedlings with check basins irrigation system. A drip tape having emitting point spacing of 30 cm × 30 cm^2 with 2 lph (liters per hour) discharge was used in the study to evaluate the drip irrigated paddy with conventional irrigated paddy. Recommended doses of fertilizer were applied in both the irrigation methods. Manual weeding as well as loosing of soil for aeration was carried out thrice at 20, 35, 50 days after transplanting in the conventional method. In drip irrigated paddy (DIP), soil was loosened one more time at 75 days after transplanting.

Different growth parameters of rice (plant height at maturity, number of tillers/plant, SPAD Values at flowering, number of panicles/plant, length of panicles) were monitored. Crop was planted during *kharif* season spread over 120 days of crop duration. The total rainfall was about 696.3 mm during crop period. Supplemental irrigation was provided either through drip irrigation or by flooding as may be the case.

12.3 RESULTS AND DISCUSSION

The yield, water used and other details in drip method and conventional method are given in Table 12.1 [4]. The number of productive tillers per plant was as high as 22 in case of drip-irrigated paddy. The results indicated that values of all growth parameters were higher in drip-irrigated paddy compared to values for conventional irrigation method.

This study revealed that water saving was 47% in the drip irrigation over the conventional irrigation method. The WUE in drip-irrigated paddy was 0.66 kg/m^3 compared to 0.37 kg/m^3 in conventional method. The result also indicated 21% saving in fertilizer application in the drip fertigation method. The cost of the drip fertigation was about Rs.87,000/ha (Rs 61.00 = 1.00 US$: Rs. is an abbreviation for Indian rupee). It was calculated that the return per unit of water (Rs./m^3) was 0.84 for conventional method and 4.85 in drip irrigation method, respectively.

TABLE 12.1   Performance Parameters For Paddy Under Two Irrigation Methods

Particulars	Units	Drip irrigation	Flood irrigation (conventional method)
Supplemental water used.	mm	291.40	553.70
Rainfall (during the crop period)	mm	696.30	696.30
Rice yield	tons/ha	6.57	4.64
Total electricity consumption	kWH	133.7	229.5
Water required to produce 1 kg of paddy.	liters	1500	2700

It has been suggested by the investigators in India that the SRI method can save 40% of water and increase in rice yield by about 30–35%. In the sprinkler method, the water saving was 50% without affecting the yield (based on results in USA). The cost of sprinkler irrigation is about Rs. 20,000 to 25,000/ha compared to Rs. 80,000 to 90,000/ha for drip with fertigation method.

Further research were carried out by the author in the paddy breading station at Tamil Nadu Agricultural University, Coimbatore, India during 1970s to 1980s. The author has indicated 40% saving in water without affecting the yield in water saving method of irrigation [6]. It was estimated that percolation losses for different varieties of paddy varied from 23.8 to 53.8% in conventional irrigation method. The water evaporation of the standing water in the field was about 1/3 of the total water used. Authors also estimated ET and crop coefficient (k), which can be used to give daily requirements of water for drip-irrigated paddy.

12.4 SUMMARY

This chapter compares the research results on drip-irrigated paddy with the traditional irrigated paddy. In this chapter, author used results of studies on different irrigation methods (surface, sprinkler and drip irrigation) at three different places. To take irrigation planning and management decisions, it is suggested to evaluate rice water requirements and cost – benefit ratios for different irrigation methods in rice cultivation. To decide which irrigation method is economical, it is suggested that research studies should be conducted using system of rice intensification (SRI) method of cultivation, sprinkler irrigation and drip irrigation with fertigation.

- **consumptive use**
- **conventional irrigation**
- **cost–benefit ratio**
- **crop coefficient**
- **drip irrigation**
- **evapotranspiration**

- fertigation
- fertilizer saving
- fertilizer use efficiency
- flooding
- growth parameters
- irrigation planning
- paddy
- percolation
- rice
- sprinkler irrigation
- supplemental irrigation
- system of rice intensification
- Tamil Nadu Agricultural University
- water productivity
- water saving
- water use efficiency

KEYWORDS

REFERENCES

1. Bouman, B. A. M., 2002. *Water wise rice production*. International Rice Research Institute, Los Baños.
2. Pollak, Paul and Sivanappan R. K., 1998. The potential contribution of low cost drip irrigation to the improvement of irrigation productivity in India. IDE, USA.
3. Postel, Sandra, 1999. *Pillar of sand can the irrigation miracle last*. New York: W. W. Norton Company.
4. Ramana Rao, K.V., et al., 2013. Drip irrigation in paddy crop. *Kisan World*, 40(10):21–23.
5. Sivanappan, R. K., 2013. Drip irrigation in rice. *Journal Plant Horti. Tech.*, Bangalore, 2013(Oct – Nov):6–8.
6. Sivanappan, R. K. and E. S. A., Saifudin, 1977. Water saving method of irrigation for high yielding rice crop. *Madras Agricultural Journal*, 64(11):745 –747.
7. Sivanappan, R. K., 1978. Annual reports: 1978–1980. College of Agric. Eng., Tamil Nadu Agric. Univ., Coimbatore, India.

8. Sivanappan, R. K. and K.V. Ramana Rao, 1980. Estimation of ET and crop coefficient in high yielding variety of paddy for Coimbatore region. *Revista ILRISO*, Anno XXIX 29(N2):Gingno.

9. Swaminathan, K. R. and R. K. Sivanappan, 1976. A mathematical model for ET in paddy. *Journal of Agricultural Engineering*, New Delhi, 23(3):1–3.

10. Tamil Nadu Agricultural University (TNAU), 2013. Proceedings of International Conference in Drip Fertigation in Rice, Coimbatore, India.

APPENDIX I: PHOTOS OF DRIP IRRIGATED RICE

CHAPTER 13

EVALUATION OF EMITTER CLOGGING FOR DRIP IRRIGATED SNAP BEANS

KH. P. SABREEN, H. A. MANSOUR, M. ABD EL-HADY, and E. I. ELDARDIRY

CONTENTS

13.1 INTRODUCTION

The agricultural sector in Egypt consumes 81% from the total available water and about 1.25 million tons of fertilizer annually [6]. Increasing

---

In this chapter: One *feddan* (Egyptian unit of area) = 0.42 ha.

Modified and printed from *Sabreen, Kh. Pibars, Mansour, H. A., M. Abd El-Hady, and Ebtisam I. Eldardiry, 2014. Maximize utilization from fertigation management for snap bean (Phaseolus Vulgaris l.) under sandy soil. Journal of Agriculture and Veterinary Science, 7(7):25–30. Open access article at: www.iosrjournals.org.*

fertilizer use leads not only to soil pollution but also contaminates the products. This problem forces the scientists to find out new techniques to solve such problems. One of these techniques is using the fertigation system to increase the efficiency of fertilization and irrigation systems. Drip irrigation system has the potential advantage of higher efficiency in supplying water and nutrients to plants [13, 18]. In addition, water and nutrients can be applied directly near the root zone, which increase yield and the irrigation performance under field conditions, and saves water saving. Charles [4] has reported some advantages of fertigation, such as: easy application, use under adverse conditions, low chemical hazard, conservation of proper soil structure, possible control of pests and weeds, and decreasing the adverse effect of salinity. However, the disadvantages of this system include extra capital expenditure, incidence of clogging of emitters, incidence of salinity build-up, and need for technical know-how.

Efficient use of water in any irrigation system is becoming important particularly in arid and semiarid regions, where water is a scarce commodity. There are specific problems in the management of sandy, namely: the excessive permeability, low water and nutrient holding capacities. Therefore, the proper management is helpful not only in the use of irrigation water but also in sandy soil amelioration efforts. Fertilizers suitable for fertigation are: technical grade salts (e.g., potassium sulfate), acids (e.g., nitric acid), bases (e.g., potassium hydroxide), polymers (e.g., polyphosphate) or chelates (e.g., iron EDTA). They are injected into the irrigation water already in solution (i.e., predissolved in water). Hochmuth [8] reported that maximum fertigation efficiency requires knowledge of crop nutrient requirements during different growth periods, soil nutrient supply, chemigation technology, irrigation scheduling, crop and soil monitoring techniques. If properly managed, fertigation through drip irrigation can reduce overall fertilizer application rates and minimize adverse environmental impacts [14]. Locascio and Smajstrala [12] stated that fertigation increased crop yield compared to the fertilizer applied just before planting.

Snapbean (*Phaseolus vulgaris L.*) is one of the important pulse crops in Egypt, and is cultivated during the winter season. Higher productivity, nutritive status, less water requirement, greater remunerative value and constant market demand make this crop more popular among the farmers.

Incorporating this crop in the cropping sequence can minimize the irrigation demand on ground water and also improve the soil productivity [17]

The interaction between water and nutrient is another important aspect of irrigation management to enhance the water use efficiency (WUE) of any crop. Among three major nutrients (nitrogen, phosphorus and potassium), the least information is available on interaction between irrigation water and phosphorus interaction. Pre-sowing irrigation, combined with phosphorus, enhances root proliferation rate as well as prolongs root growth period, but this effect is not prominent under water scarce situations [11].

This chapter discusses utilization of fertigation management in drip-irrigated snap bean (*phaseolus vulgaris* l.) under sandy soils of Egypt. Authors discuss the effects of fertigation/ irrigation time ratio, type of fertilizers on the water distribution uniformity, crop yield, WUE and FUE of the snap bean.

13.2   MATERIAL AND METHODS

During the summer seasons of 2012–2013, the experiment was conducted at Research and Production Station of National Research Centre, Nubaria region, Behera Governorate, Egypt. The soil at the experimental site was sandy in texture, very poor in organic matter content (0.65%) with a moderately alkaline pH (8.2), soil EC (0.35 dS/m), and $CaCO3$ (7.02%) before the initiation of first year experiment. The field capacity, wilting point and available water values were 11.1, 4.7 and 6.4% on weight basis, respectively. The source of irrigation water at experimental site is well water with EC 0f 0.39 dS/m and pH of 7.63.

Snap bean (*Phaseolus vulgaris L.* cv. Bronco) seeds were sown on 10th of February. Seeds were sown in hills 5–7 cm apart on two sides ridge with a row spacing of 90 cm and dripper spacing of 30 cm down the lateral drip line. The experiment was laid out in a randomized complete block design having six treatments and triplicated in 5.4 m × 4.8 m plot. The second year experiment was superimposed on the experimental plots of first year. The crop lasted 118 days from transplanting to final harvest, during first and second years of study. Flowering stage of snap beans started at about 45 days from planting in both seasons. Harvesting of fresh beans was started 20 days after flowering at an interval of 5 days.

For irrigation scheduling and estimation of potential evapotranspiration (ETo), the climatic data was used from Metero station of NRC Farm. Table 13.1 presents the climatology parameters during two growing seasons. Soil moisture was kept at not less than 60% of water content at field capacity. The snap bean is sensitive to drought and water depletion more than 40% of FC can affect negatively on the flowering formation, rate of pod filling and crop yield.

Total water use efficiency is defined as the ratio of economic yield per feddan to seasonal water consumption. During the growing seasons, accumulated ETo, highest/lowest values of ETo were recorded.

Authors noticed that from recorded values that high temperature was associated with increase in relative humidity and wind speed in April. The high temperature contributes to increase evaporation and hence evapotranspiration (ET), which affects the crop irrigation water requirements.

Fertilizers were fertigated using an injection device (Fig. 13.1), except super phosphate (15% P2O5), which was applied 50% of 100 kg/fed at planting and the rest 50% was fertigated during the growing season (80% commercial grade phosphoric acid). Superphosphate was applied to the experimental site during land preparation and before planting, because phosphorus is highly immobile in soil. Usually transport and uptake of phosphorus are regulated by diffusion [10]. Two types of fertilizers were used: completely soluble (19:19:19 for N-P-K) and traditional (ammonium sulfate, 20%N and potassium sulfate, 48–50% $_{K2SO4}$). Recommended ratio (40N, 20P, and 30K) by the Agricultural Extension was used. Both nitric acid and potassium sulfate (0, 13, 43% of N, P and K) were used to modify the soluble fertilizer ratio based on recommendations by Boman et al. [1, 20].

Uniformity of water distribution was determined for each treatment by selecting 25 emitters at random from each treatment, before starting the experiment and at the end of the experiment. The discharge rates of the emitters were estimated and emission uniformity was calculated from the following equation [1, 2, 5].

$$EU = 100 \times \{[(Qn/Qa) + (Qa/Qx)]/2\} \tag{1}$$

TABLE 13.1 Metrological Data For the Growing Period, During 2012–2013

Period	2012						2013					
	Temp.		Relative humidity		Wind speed	ETo	Temp.		Relative humidity		Wind speed	ETo
	high	low	high	low			high	low	high	low		
days	°C		%		m/s	mm/day	°C		%		m/s	mm/day
Jan 5	14.5	7.4	92.1	57.1	15.9	0.9	13.8	7.8	89.3	52.5	14.2	0.9
Jan 11	14.5	8.6	87.6	50.0	15.1	1.3	13.8	9.1	85.0	46.0	13.4	1.3
Feb 10	16.5	6.7	86.5	43.2	12.4	2.0	15.7	7.0	83.9	39.7	11.0	2.0
Feb 10	16.2	9.3	96.9	53.6	12.5	1.9	15.4	9.8	94.0	49.3	11.1	1.9
Feb 8	18.9	10.8	106.3	58.4	14.8	2.1	17.9	11.3	103.1	53.8	13.2	2.1
Mar 10	17.6	9.7	98.7	52.3	18.2	2.4	16.7	10.2	95.7	48.1	16.2	2.4
Mar 10	19.1	10.9	81.3	41.1	19.9	2.7	18.1	11.4	78.9	37.8	17.7	2.6
Mar 11	20.2	10.7	94.5	45.5	13.8	2.9	19.2	11.3	91.6	41.9	12.3	2.8
April 10	26.6	13.2	96.4	35.5	16.2	3.4	25.3	13.9	93.5	32.7	14.4	3.3
10	24.3	12.9	81.9	35.7	17.3	3.5	23.1	13.5	79.4	32.8	15.4	3.5
10	25.4	12.9	92.8	40.8	13.7	3.9	24.1	13.5	90.0	37.5	12.2	3.8
8	23.7	14.4	83.6	38.3	15.0	3.4	22.5	15.2	81.0	35.3	13.4	3.3

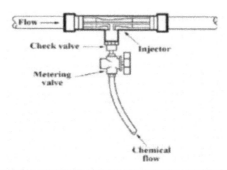

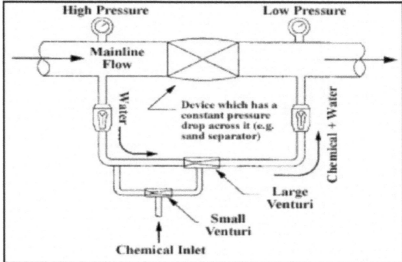

FIGURE 13.1 Chemical injectors based on venturi to create adequate pressure differentials for efficient chemigation.

where: EU = Field emission uniformity, %; Qn = The average of the lowest (1/4) of the emitters flow rate, lph; Qa = The average of the all emitters flow rate, lph; and Qx = The average of the highest (1/8) of the emitters flow rate, lph.

The cross section diameter of the long-path emitter was 0.7 mm. Emitter discharge was 4 lph with a lateral line length of 30 m. Emitter spacing was 30 cm down the lateral length. The emitter is considered laminar-flow-type (Re<2000). Nine emitters from each lateral were chosen for calculating the clogging ratio at the beginning and at the end of the growing season for both seasons. Three emitters at the beginning, three at middle and three at the end of the lateral were tested for flow rate. Clogging ratio was calculated using the following equations:

$$E = [qu/\ qn] \times 100 \tag{1}$$

$$CR = (1 - E) \times 100 \tag{2}$$

where: E = the emitter discharge efficiency (%); qu = emitter discharge, at the end of the growing season (lph); qn = emitter discharge, at the beginning of the growing season (lph); and CR = the emitter clogging ratio (%).

13.3   RESULTS AND DISCUSSION

Clogging data is shown in Table 13.2 and Fig. 13.2 for fertigation/irrigation time ratios of 2/3, ¾, and 1/2. Clogging values under traditional fertilizers can be arranged in ascending order: 2/3 < 3/4 <1/2. This can be attributed to the rest of irrigation water, which was used to flush the irriga-

TABLE 13.2   Effects of Fertilizer Type and Fertigation/Irrigation Time Ratio on Snap Bean Yield and Water Use Efficiency in Drip Irrigated Sandy Soil.

Fertilizer type	Fertigation/ irrigation time ratio	Total yield ton/ fed.	Irrigation water requirements (m³/season)		WUE, kg/m³	
			Consumed	Calculated	Actual	Calculated
Completely soluble	3/4	4.93			2.87	4.03
	2/3	4.40			2.56	3.59
	1/5	4.25			2.47	3.47
	Avg.	4.53	1720	1242	2.63	3.70
Traditional fertilizer	3/4	3.52			2.05	2.88
	2/3	3.75			2.18	3.06
	1/5	3.00			1.74	2.45
	Avg.	3.42			1.99	2.80
LSD 5%		0.32			0.24	1.11

tion system. Although ½ fertigation time gave the highest value and had 50% more irrigation water to flush the system, yet half time of fertigation was not enough to inject fertilizers without impurities. With regard to the completely soluble fertilizer treatment, clogging in fertigation were less than that in traditional method and can be arranged in descending order as: 3/4> 2/3> 1/2. It may be mainly due to quantity of rest of irrigation water that was not enough to flush the irrigation system [5, 7]. Results indicate that regardless of fertigation time, completely soluble fertilizer was superior and had a lowest value of clogging, which may be mainly due to decrease in the impurities.

Meanwhile, under drip irrigation system, the liquid fertilizers improved water distribution efficiency (WDE) from 80 to 84.1% as shown in Table 13.2 and Fig. 13.2. The decrease in WDE may be due to physical change that can occur when the short fertigation time (1/2) can increase density and viscosity of the fertilizer solution, which in turn can increase friction losses in laterals and decrease the emitter discharge. This is due to lowering of pH of irrigation water, and consequently preventing the precipitation of salts inside the laterals and emitters [15, 16].

Table 13.2 show that yields of fresh snap beans was increased under completely soluble fertilizers comparing to the traditional fertilizers.

The regression analysis between the main factors under investigation snap bean yield and fertigation time is shown in Fig. 13.3. We can observe

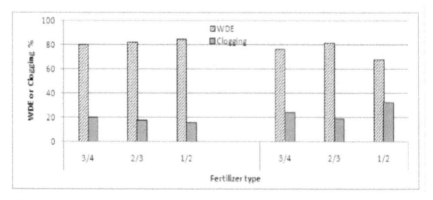

FIGURE 13.2   Effects of fertilizer type and fertigation/ irrigation time ratio on water distribution efficiency (WDE) and clogging.

that crop yield and time ratio were negatively correlated with a coefficient of correlation of 0.956 that was significant at 1% level. The coefficient of determination was 0.92 and the regression coefficients were significant at 1% level. We can conclude that in order to maximize utilization of fertilizers, fertigation/irrigation time ratio must be taken into consideration. Also, maximum yield was obtained with highly soluble fertilizer at time ratio of ¾. Traditional fertilizers took more time to mix in the solution solubility before fertigation process began and also used more water in the fertilizer tank [3].

Snap bean production is an important determinant of the economic yield. The total yield at final harvest was significantly 32.5% higher under completely soluble fertilizer (4.52 ton/fed.) compared to that under traditional one (3.42 ton/fed.) as shown in Table 13.2. The differences in yield for significant between the types of fertilizers. Furthermore, significantly higher yield was observed when the fertigation time ratio was increased. Increasing the fertigation time by 10 min increased the yield by 12% (from 2/3 to ¾ of time ratio) under completely soluble fertilizers. Whereas under traditional fertilizers, there was a reduction of 6.5% in same sequence. It was also observed that injection fertilizer period of ¾ was superior under completely soluble fertilizers and 2/3 time ratio was the best under traditional ones. Similar results of improved yield have been reported by Ibrahim [9].

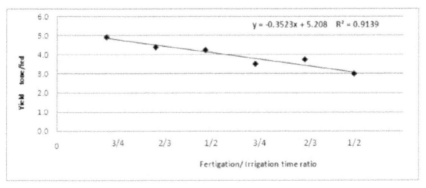

FIGURE 13.3   Linear regression analysis between yield (tons/feddan) and fertigation time ratio regardless of the type of fertilizer.

Water requirement of snap beans per season were calculated using the climatic data and ET equations and were also actually measured in the field. Of course, the actual irrigation requirement (1720 m^3/season) was 42% higher than the calculated value (1242 m^3/season), as shown in Table 13.2. Data show that there was a negative correlation between time ratio and WUE values especially under completely soluble fertilizers. But under traditional fertilizers, the highest value of WUE was obtained with fertigation time ratio of 2/3 and the lowest one was recorded in fertigation at ½ irrigation time.

Regardless of fertigation time, completely soluble fertilizers gave 32% higher value of WUE (2.63 kg yield/m^3 irrigation water) than that under traditional fertilizer (1.99 kg yield/m^3 irrigation water).

Fertilizers use efficiency for N-P-K of snap bean is shown in Table 13.3 under two types of fertilizers and for three time ratios. Data indicate that decrease in fertigation time was associated with decrease in FUE for

TABLE 13.3   Effects of Types of Fertilizers and Fertigation/Irrigation Time Ratios on the Fertilizer Use Efficiency (FUE), For Drip Irrigated Snap Beans

Fertilizer type	Fertigation/ Irrigation time ratio	FUE		
		N	P	K
		kg of yield/kg of fertilizer		
Completely soluble	3/4	123.25	0.25	4.11
	2/3	110.00	0.22	3.67
	1/2	106.25	0.21	3.54
	Avg.	113.17	0.23	3.77
Traditional fertilizer	3/4	88.00	0.18	2.93
	2/3	93.75	0.19	3.13
	1/2	75.00	0.15	2.50
	Avg.	85.58	0.17	2.85
LSD 5%		3.66	0.06	0.09

the studied macronutrients, except under traditional fertilizer, where fertigation time ratio of 2/3 gave the best values followed by the time ratio of
3/4. Also, FUE values under completely soluble fertilizer are higher than
the traditional one. Similarly in fertigation, applied fertilizer through the
drip system is placed to the active plant root zone and improves fertilizer
use efficiency according other investigators.

Regardless of the fertigation time, completely soluble fertilizer gave
higher values of FUE than traditional one and the percentage increase
were 32, 35 and 32% of FUE for N, P and K, respectively. And fertigation
time ratio of ¾ has a superior effect on FUE followed by 2/3 and ½ of time
ratios. There is no significant differences between FUE for P at fertigation
time of ¾ and 2/3.

Under completely soluble fertilizer, the percentage increase in FUE for
N was 12, 16% (comparing ¾ with 2/3 and ½) and 3.5% comparing 2/3 with
½ fertigation time for N. Whereas, the percentage increase in FUE was 13.6,
19.0, 4.8% for P; and 12.0, 16.0 and 3.7% for K, under three time ratios
of 3/4, 2/3, and 1/2, respectively. However, a different trend was observed
under traditional fertilizers: The 2/3 fertigation time gave highest value of
FUE followed by ¾ then ½. The increase in FUE for 2/3 fertigation time
above ¾ and ½ were 6.5, 25; 5.6, 26.7; and 6.8 and 25.2%, respectively.

Regardless of the fertilizers type, FUE values of fertigation time can
be arranged in descending order: ¾ > 2/3 > ½ for N, P and K fertilizers.
The percentage increase in FUE, under fertigation time of 3/4 comparing
with 2/3 and ½, was 3.6, 16.6; 5.0, 16.7; and 3.5, 16.6% for N, P and K,
respectively.

Agricultural grade fertilizers are generally not suitable for use in fertigation through drip irrigation because of high impurities, which may be
insoluble and can lead to clogging of drippers. For this reason technical
grade fertilizers are normally required in fertigation of fewer impurities
and proportionally higher levels of desired mineral nutrients [5, 9].

13.4 CONCLUSIONS

The use of fertigation is gaining popularity because of its efficiencies in
nutrient management, time and labor and potentially a greater control on

crop performance. Fertigation potentially offers many advantages over conventional methods to manage fertilizer needs of a crop. Although fertigation is an exciting and potentially profitable technology to horticultural production systems, yet it also requires significant investment in equipment, advanced management skills, constant monitoring and an understanding of the specific nutrient needs. The use of acidic fertilizers temporarily unclogs emitters. The irrigation and chemical injection systems should be thoroughly washed and flushed with fresh water, especially after the injection of acids into the system. Rule of thumb is to fertigate during middle third of irrigation cycle.

13.5   SUMMARY

Field experiment was conducted during the summer seasons of 2012 and 2013 at the Research and Production Station, National Research Centre, Nubaria region, Behera Governorate, Egypt to study the effects of three fertigation/irrigation time ratios and two types of fertilizers on the water distribution uniformity, water use efficiency and fertilizer use efficiency of the some macronutrients (N, P and K) of snap bean (*Phaseolus vulgaris L. cv.* Bronco).

Completely soluble fertilizers produced 21% higher yield of fresh snap beans compared with traditional fertilizers under different modern chemigation systems.

Clogging values under types of fertilizers can be arranged in ascending order: 2/3< 3/4< ½. It was observed that highly soluble fertilizer at a time ratio of ¾ gave the maximum yield. Significantly higher yield was observed with increasing the fertigation time. Increasing fertigation time by only 10 min increased the yield by 12% (from 2/3 to ¾ fertigation time) under completely soluble fertilizers. Under traditional fertilizers, the reduction in yield was 6.5% in same sequence. Also, time ratio of ¾ was superior under completely soluble fertilizers while 2/3 fertigation time was the best under traditional ones. Regardless of fertilizer type, FUE values can be arranged in the descending order: ¾ > 2/3 > ½ for N, P and K nutrients. The percentage increase in FUE under time ratio of 3/4 comparing with 2/3 and ½ was 3.6,16.6; 5.0,16.7; and 3.5,16.6% for N, P and

K, respectively. Decreasing fertigation time was associated with decrease in FUE for the studied macronutrients, except under traditional fertilizer. Fertigation during 2/3 irrigation time was the best followed by the highest value of fertigation time (3/4 from irrigation time).

KEYWORDS

- **Clogging**
- **drip irrigation**
- **emitter**
- **fertigation**
- **fertilizer use efficiency**
- **fertilizers**
- **irrigation scheduling**
- **macro nutrients**
- **sandy soil**
- **snap bean**
- **soluble fertilizer**
- **time ratio**
- **water use efficiency**

REFERENCES

1. Boman, B., and T. Obreza, 2002. Fertigation nutrient sources and application considerations for citrus, University of Florida, IFAS Circular 1410.
2. Boman, B., S. Shukla, and D. Haman, 2004. Chemigation equipment and techniques for citrus. University of Florida, IFAS Circular 1403.
3. Burt, C., K. O'Connor, and T. Ruehr, 1998. Fertigation. California Polytechnic State University San Luis Obispo, Irrigation Training and Research Center.
4. Charles, M., 2007. Fertigation of chemicals. *FERTIGATION*.
5. El-Gindy, A. M, M. Y. Tayel, K. F. El-Bagoury, and Kh. A. Sabreen, 2009. Effect of injector types, irrigation and nitrogen treatments on emitters clogging. *Misr J.Ag. Eng.*, 26(3):1263–1275.
6. FAO, 2005. *Fertilizer use by crop in Egypt*. First edition published by FAO, Rome. 57 pages.

7. Fares, A., and F. Abbas, 2009. Irrigation systems and nutrient sources for irrigation. University of Hawaii at Mānoa, College of Tropical Agriculture and Human Resources, publication SCM-25.

8. Hochmuth, G. J., 1992. Fertilizer management for drip-irrigated vegetables in Florida. *Hort. Technol.*, 2:27–32.

9. Ibrahim, A., 1992. Fertilization and irrigation management for tomato production under arid conditions. *Egyptian J. Soil Sci.*, 32(1):81–96.

10. Kargbo, D., J. Skopp and D. Knudsen, 1991. Control of nutrient soils, sediments, residuals, and waters. Southern Cooperative Series. *Agron. J.,* 83:1023–1028.

11. Li, F. M., Q. H. Song, H. S. Liu, F. R. Li and X. L. Liu, 2001. Effects of presowing irrigation and phosphorus application on water use and yield of spring wheat under semiarid conditions. *Agric. Water Manage.*, 49:173–183.

12. Locascio, S.J., and A.G. Smajstrla, 1995. Fertilizer timing and pan evaporation scheduling for drip irrigated tomato. In: Lamm F. R. (ed.) *Microirrigation for a Changing World: Conserving Resources/Preserving the Environment.* ASAE Publ. 4–95. Pages 175–180.

13. Ould, Ahmed B.A., T. Yamamoto, M. Inoue M. and H. Anyoji, 2006. Drip irrigation schedules with saline water for sorghum under greenhouse condition. *Trans JSIDRE,* 244:133–141.

14. Raun, W. R., J. B. Solie, G. V. Johnson, M. L. Stone, R. W. Mullen, K. W. Freeman, W. E. Thomason, and E.V. Lukina. 1999. Improving nitrogen use efficiency in cereal grain production with optical sensing and variable rate application. *Agron. J.,* 94:815–820.

15. Sabreen, Kh. A. P., K. F. El-Bagoury, M. Y. Tayel, and A. M. El-Gindy, 2009. Hydraulic performance of fertigation applicators. *Journal Biol. Chem. Environ. Sci.,* 4(1):1049–1065.

16. Sagi, G., 1990. Water quality and clogging of irrigation systems in Israel. In: *1989 Water and Irrigation Bulletin (in Hebrew)*, 280:57–61.

17. Sarkar, S., S. R. Singh and Y. Singh, 2000. Effective use of harvested water in relation to productivity and water use pattern of Rajmash (Indian word for French beans) as an intercrops. *J. Indian Soc. Soil Sci.,* 48:824–826.

19. Tayel M. Y., Sabreen, Kh. P., and H. A. Mansour, 2013. Effect of drip irrigation method, nitrogen source, and flushing schedule on emitter clogging. *Agric. Sci.,* 4(3):131–137.

CHAPTER 14

EVALUATION OF EMITTER CLOGGING

M. Y. TAYEL, KH. PIBARS SABREEN, and M. A. MANSOUR

CONTENTS

14.1 INTRODUCTION

Due to water shortages in many parts of the world today, drip irrigation is becoming very popular [13, 23]. Fertigation is the application of chemicals through irrigation systems [13] and it has increased dramatically during the past 15 years, particularly for sprinkler and drip irrigation systems [24]. Fertigation technology will continue to grow since it results in savings of fertilizers and labor, and better uniformity of fertilizer distribution.

Modified and printed from *Tayel, M.Y., Sabreen, Kh. Pibars and Mansour, H. A., 2013. Effect of drip irrigation method, nitrogen source, and flushing schedule on emitter clogging. Agricultural Sciences, 4(2):131–137. Open Access at: http://www.scirp.org/journal/as/.*

Although drip irrigation has numerous advantages, yet it has some limitations also like emitter clogging. Emitter clogging will increase the maintenance cost of drip irrigation systems and reduce the working life and water use efficiency [3, 13, 18, 29]. Therefore, emitter clogging can determine whether drip irrigation system can succeed [30]. Complete or partial blocking of drippers reduce the application uniformity of both water and fertilizers and negatively affects plant growth [7]. Effective fertigation program requires knowledge of [5]:

1. Plant characteristics: optimum daily nutrient consumption rate and root distribution in the soil.
2. Nutrient characteristics: solubility and mobility.
3. Irrigation water quality: pH, total soluble ions of Ca^{++}, Mg^{++}, Na^+, P, Fe^{++}, B, CO_3^-, HCO_3^-, SO_4^- and suspended solids.
4. No interaction should occur among fertilizers that are injected in the system and/or among fertilizers and irrigation water resulting in formation of precipitates.
5. Fertigation is carried out properly and according to the recommendations.
6. Good irrigation scheduling is used according to soil water holding capacity, profile depth, climate and crop type.
7. Distribution uniformity of irrigation water is critical for uniform fertilizer application.
8. Over irrigation during fertigation not only wastes water but can leach fertilizers below the root zone and can pollute ground water.
9. Select appropriate chemical injector.
10. Under saline conditions, salinity problem can be intensified by fertigation and improper irrigation management.
11. Irrigation systems should be monitored more closely during fertigation process.
12. Be aware of balance of "cations – anions."
13. Selection of correct form of nitrogen form that must be compatible.
14. Appropriate flushing schedule of irrigation system components after fertigation.
15. Fertigate during middle third of irrigation duration.
16. Always add chemicals to water to avoid accidents.

17. Successful fertigation requires precise calculation of injection rate [10].
18. Knowledge regarding solubility of different fertilizers; and basic know-how of fertigation equipment's.

Emitter clogging is physical, chemical and biological agents [1, 11, 12, 13, 15, 20]. Goyal [13] has described these agents in detail. Two or more of these clogging causes may occur at the same time [2, 8]. Emitter clogging can also be due to extreme small passages of water and low flow rate through the emitters [7]. More Clogging of emitters is more serious at the end of the drip laterals than at the beginning probably due pressure head loss [21]. Normal fertilizers also generally tend to clog the emitters [14]. Ozekici [19] carried a study on the effects of different fertigation practices on clogging of inline emitters using Samandag region well water in Turkey. He showed that different fertilizer treatments had significant effect on emitter clogging. Fertilizers containing both Ca^{++} and SO_4^- caused higher clogging compared to the others.

Chang [4] found that as water flow in drip irrigation system (DIS) slows down and/ or the chemical composition of the water changes, then chemicals precipitate and/or microbial flocks and slimes begin to form and grow, thus promoting emitter clogging. The effects on emitter clogging has been evaluated due to injector types (by-bass pressurized tank (J_1), venture injector (J_2), positive displacement injector pump (J_3)) irrigation treatments (50, 75 and 100% ETc: I_1, I_2, I_3) and nitrogen treatments (60, 90 and 120 kg of N/feddan: N_1, N_2, N_3) and their interaction on emitter clogging [22, 26]. According to the values of percentage of emitter clogging, the treatments under investigation were written in the descending orders: $J_3 < J_2 < J_1$, $I_3 < I_2 < I_1$ and $N_1 < N_2 < N_3$. They added that the interactions, J× I, J×N, I × N and J× I × N, had significant effects at 5% level on emitter clogging.

This chapter presents the effects of nitrogen source, surface drip irrigation and subsurface drip irrigation systems and flushing scheduling on emitter clogging.

14.2   MATERIALS AND METHODS

This research study was carried out in a sandy soil at the Experimental Farm of the National Research Center in Nubaria, Behura Governorate of Egypt.

TABLE 14.1   Chemical Properties of the Irrigation Water at Nubaria

pH	EC	Soluble ions meq/l							SAR
		Cations				Anions			
1:2.5	dS/m	Ca++	Mg++	Na+	K+	HCO3–	SO4–	Cl–	
7.63	0.39	1.02	0.51	2.43	0.22	0.13	1.34	2.71	2.8

14.2.1 IRRIGATION WATER CHARACTERISTICS

The source of irrigation water at experimental site was well water (the total depth of well: 45 m; water depth from ground surface: 4–5 m; and diameter of well pipe: 15 cm). Screen filter (2"/2" inlet and outlet diameters; 35 m^3/h of discharge rate and filtration unit: 120 mesh). Water samples were taken analyzed for chemical analysis (Table 14.1).

14.2.2 FERTILIZER INJECTOR (VENTURI TYPE)

A venturi injector is a tapered constriction, which operates on the principle that a pressure drop accompanies the increase in velocity of the water as it passes through the constriction [13]. It was installed on a by-bass arrangement placed on an open container having the fertilizer solution. The injector is constructed of a PE tube 1.5" diameter. The venturi was provided with a ball valve, which creates a differential pressure, thus allowing the injector to produce a vacuum. N and K_2O fertilizers were injected in two doses. The irrigation and injection processes lasted 2 h and 3 h depending on the chemigation duration, respectively.

14.2.3 EXPERIMENT LAYOUT AND TREATMENTS

During the growing seasons of 2010 and 2011, field experiments were conducted using split-split- plot complete randomized design with three replications. Super phosphate (15.5% P_2O_5) was broadcasted @ 200 kg/ fed., using traditional method of fertilization application. This amount was divided into two doses (1stduring soil preparation and 2nd after month from planting date). Peanut seeds (*Arachishypogaea* L. cv. Giza10) were planted in first week of May.

- Two irrigation methods were used in the main plots: surface drip irrigation and subsurface drip irrigation (SDIS and SSDIS).
- N fertilizer treatments were in the sub subplots. Three sources of N-fertilizer were used namely: NS1: NH4NO3, NS2: (NH4)2SO4 and NS3: Ca(NO3)2
- Three frequencies of flushing were used: no flushing, one flushing, and a monthly flushing (FL1, FL2; FL3) during the irrigation period

The doses of nitrogen and potassium fertilizers (N and K_2O) were 110 and 150 kg/fed., respectively as recommended by Ministry of Agriculture and Land Reclamation.. Peanut crop was harvested in the second week of September (i.e., growing season lasted 130 days).

14.2.4 EMITTERS CLOGGING

The flow cross-section diameter of the long-path emitter was 0.7 mm with discharge rate of 4 lph. The emitter spanning down the lateral drip line was 50 cm. Subsurface drip irrigation laterals were laid at a depth of 20 cm. The emitter type was laminar flow (Re < 2000) [16]. To estimate the emitter flow rate, cans and a stopwatch were used. Nine emitters from each lateral were chosen to evaluate the clogging ratio at the beginning and at the end of the growing season for two seasons. Three emitters at the beginning, three at middle and three at the end of the lateral were tested for flow rate. Clogging ratio (CR) was calculated as follows [9]:

$$E = [qu/\ qn] \times 100 \qquad (1)$$

$$CR = (1 - E) \times 100 \qquad (2)$$

where: E = the emitter discharge efficiency, (%); qu = emitter discharge, at the end of the growing season (lph); qn = emitter discharge, at the beginning of the growing season (lph); and CR = the emitter clogging ratio, (%).

14.2.5 STATISTICAL ANALYSIS

All data were statistically analyzed using a split-split plot design with three replications with analysis of variance (ANOVA) to evaluate effects of main and interaction [25]. Means among treatments were compared using least significant difference (LSD) at $P = 0.05$.

14.3 RESULTS AND DISCUSSION

Table 14.2 shows the effect of following treatments:
1. Irrigation systems: surface drip irrigation and subsurface drip irrigation (SDI and SSDI).
2. Nitrogen source: NH_4NO_3, $(NH_4)_2SO_4$ and $Ca(NO_3)_2 = Ns_1$, Ns_2 and Ns_3. and
3. Number of flushing's per season: 0, 1 and 4 (FL_1, FL_2, and FL_3).

According to the percentage of emitter clogging values, the treatments can be written the ascending orders: SDI < SSDI, $FL_3 < FL_2$ and $< FL_1$, $NS_1 < NS_2 < NS_3$. Differences in the values of emitter clogging between any two treatments were significant at 5% level.

Table 14.3 and Figs. 14.1 and 14.2 indicate the effects of the interactions among treatments on emitter clogging.

TABLE 14.2 Effects of Irrigation Systems, Flushing Frequency and Nitrogen Source on Percentage Emitter Clogging

Treatments	Clogging percentage	
SDI	7.706	a
SSDI	16.133	b
FL_1	14.543	a
FL_2	11.799	b
FL_3	9.417	c
NS_1	9.681	c
NS_2	12.097	b
NS_3	13.982	a

14.3.1 EFFECTS OF SUBSURFACE DRIP IRRIGATION (SSDI)

The effects of the SSDI × FL on emitter clogging percentage are given in Table 14.3 and Fig. 14.1. The ascending orders, SSDI×FL_3 < SSDI × FL_2 < SSDI×FL_1, illustrate the effects on emitter clogging regardless of NS used. The ascending orders, SSDI×NS_1 < SSDI×NS_2 < SSDI × NS_3, illustrate the effects on emitter clogging regardless of FL (Table 14.3 and Fig. 14.2).

TABLE 14.3 Effects of Interactions Among Irrigation Systems, Flushing Frequency and Nitrogen Source on Emitter Clogging Percentage

Treatments	Clogging ratio (%)		Treatment	Clogging ratio (%)	
SDIxFL_1	11.130	d	SDIxFL_1xNS_1	8.17	e
SDIxFL_2	9.810	e	SDIxFL_1xNS_2	10.84	f
SDIxFL_3	8.570	f	SDIxFL_1xNS_3	13.61	h
SSDIxFL_1	15.380	a	SDIxFL_2xNS_1	4.87	g
SSDIxFL_2	14.060	b	SDIxFL_2xNS_2	6.98	a
SSDIxFL_3	12.820	c	SDIxFL_2xNS_2	8.01	b
SDIxNS_3	8.690	f	SDIxFL_3xNS_1	3.90	d
SDIxNS_1	9.900	e	SDIxFL_3xNS_2	5.21	c
SDIxNS_2	10.900	d	SDIxFL_3xNS_3	7.76	f
SSDIxNS_1	12.950	c	SSDIxFL_1xNS_1	15.64	e
SSDIxNS_2	14.150	b	SSDIxFL_1xNS_2	18.21	d
SSDIxNS_3	15.150	a	SSDIxFL_1xNS_3	20.81	c
FL_1xNS_1	12.120	d	SSDIxFL_2xNS_1	14.33	b
FL_1xNS_2	13.320	b	SSDIxFL_2xNS_2	17.53	a
FL_1xNS_2	14.320	a	SSDIxFL_2xNS_2	19.74	i
FL_2xNS_1	10.790	g	SSDIxFL_3xNS_1	11.20	e
FL_2xNS_2	12.000	e	SSDIxFL_3xNS_2	13.81	a
FL_2xNS_3	13.000	c	SSDIxFL_3xNS_3	14.62	j
FL_3xNS_1	9.550	h			
FL_3xNS_2	10.760	g			
FL_3xNS_3	11.760	f			

Means with different letters within each column are significant at 5% level.

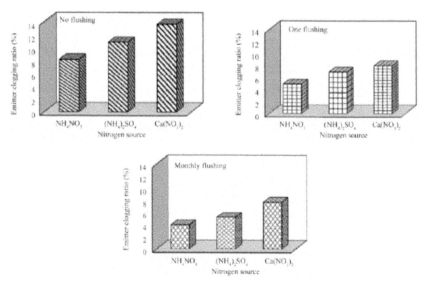

FIGURE 14.1 Effects of flushing frequency and nitrogen source on emitter clogging ratio (%) under surface drip irrigation.

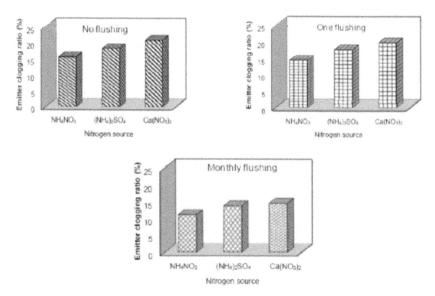

FIGURE 14.2 Effects of flushing frequency and nitrogen source on emitter clogging ratio (%) under subsurface drip irrigation.

14.3.2 EFFECTS OF SURFACE DRIP IRRIGATION (SDI)

Based on values of emitter clogging (Table 14.3) and the interaction SDI × FL can be stated in the ascending orders: $SDI \times FL_3 < SDI \times FL_2 < SDI \times FL_1$, regardless of NS used. The effects of the interaction, SDI×NS, on emitter clogging percentage regardless of FL can be written in the ascending orders: $SDI \times NS_1 < SDI \times NS_2 < SDI \times NS_3$. Regardless of the irrigation methods, the effects of the interactions FL ×NS can be arranged in ascending orders:

$$FL_1 \times NS_1 < FL_1 \times NS_2 < FL_1 \times NS_3; FL_2 \times NS_1 < FL_2 \times NS_2 < FL_2 \times NS_3; \text{ and}$$
$$FL_3 \times NS_1 < FL_3 \times NS_2 < FL_3 \times NS_3.$$

Differences in emitter clogging percentage, between any two interactions from those mentioned in Table 14.3, were significant at 5% level. This can be attributed to the presence of the ions: Ca^{++}, Mg^{++}, HCO_3^-, CO_3^- either in irrigation water or in injected fertilizer or in both that can cause formation of precipitates in the different irrigation system components in the following sequence: $CaCO_3$, $CaSO_4$ and $MgSO_4$. Also increasing flushing frequency increases the removal of precipitates physically out of the irrigation system.

It is obvious that emitter clogging percentage under SSDI exceeds that under the SDI (Figs. 14.1 and 14.2). This may be due to one or more of the following causes:

1. Flushing processes are active under SDI relative to SSDI,
2. The ions Ca^{++}, Mg^{++}, HCO_3^- CO_3^- and SO_4^- in soil take part in formation of precipitates on emitter outlets,
3. The overburden pressure of soil layer on the lateral line under SSDI may cause partial deformation in the shape of the cross sections area thus increasing friction loss and subsequently decrease both pressure and water flow velocity [27, 28].
4. Emitter discharge under SSDI can also decrease as a result of positive pressure in the soil water matrix creating a back-pressure at the emitter orifice,
5. Biofilm formation is a primary problem in emitter clogging under SDI. This problem is exacerbated under SSDI because external

soil particles are sucked to the biofilm at the emitter outlets thus increasing clogging. It is expected that these soil particles would be siphoned back to the emitter inside when the irrigation system is shut down for different reasons since the emitters used are nonpressure compensating ones [27].

6. Flushing process is less effective in biofilm removal without acidification and/or chlorination due to its low specific gravity and high adhesive characteristics [21, 27].

7. Root intrusion of the crop and some weeds [6],

8. Water dissipation by emitter is negatively affected to some extent by soil resistance under SSDI,

9. More salts can be accumulated on the emitter outlets under SSDI due to both water loss via evaporation and water uptake by plant roots,

10. The micro changes in lateral slopes (up or down) decrease water flow in the laterals under SSDI, and

11. Many fine soil particles may find their way to inside the lateral drip lines due to digging, laying and covering processes.

14.3.3 EFFECTS OF SECOND INTERACTION ON EMITTER CLOGGING

Tables 14.3 indicates the effects of second interaction $I \times NS \times FL$ on emitter clogging. Differences in clogging percentage between any two interactions were significant at 5% level. According to the interactions under study, emitter-clogging percentage varied from 3.9 to 20.18. The maximum value of emitter clogging was 20.18% and the minimum was 3.9%, that were observed in the interactions: $SSDI \times FL_1 \times NS_3$ and $SDI \times FL_3 \times NS_1$, respectively.

14.4 CONCLUSIONS

Field experiments were conducted on peanut (var. Giza 5) crop grown in sandy soil at the National Research Center farm, Nubaria, Behura Governorate, Egypt, during two successive growing seasons of 2010–2011 to study the effects nitrogen sources, methods of drip irrigation and number of flushing's/season on emitter clogging percent. Treatments used were: 110 kgN/fed. in the form of NH_4NO_3, $(NH_4)_2SO_4$ and $Ca(NO_3)_2 = NS_1$, NS_2 and NS_3, two drip irrigation methods: surface drip irrigation and sub-surface drip irrigation (SDI; SSDI) and number of flushing's per growing season 0, 1, and 4 (Fl_1, Fl_2; Fl_3).

Concerning the main effects of the treatments on emitter clogging percent, the ascending orders were: SDI<SSDI, $Fl_3<Fl_2<Fl_1$ and $NS_1<NS_2<NS_3$. The differences in emitter clogging percent between any two treatments within the same order were significant at 5% level. The effects of 1st and 2nd interactions on emitter clogging between and/or among treatments were significant at 5% level. Regardless of the source of the ions (Ca^{++}, Mg^{++}, HCO_3^-, CO_3^- and SO_4^- in irrigation water, soil, and fertilizers), we must avoid the reactions to form the insoluble salts altogether to prevent formation of participates formation, especially under arid and semiarid climate. The maximum value of emitter clogging (20.19%) and the minimum one (3.9%) were obtained in the following interactions: $SSDI \times Fl_1 \times NS_3$ and $SDI \times Fl_3 \times NS_1$, respectively. Analysis of soil/irrigation water/fertilizers is essential to understand the mechanics of clogging. Since emitter clogging under SSDI surpassed that under SDI, the SSDI must be equipped with pressure-compensating emitters and vacuum relieve valve.

14.5 SUMMARY

Field experiments were carried out at the National Research Center farm, Nubaria area, Behura Governorate, Egypt to study the effects of nitrogen source, flushing schedule and irrigation methods on emitter clogging. Peanut seeds (*Arachishypogaea* L. cv. Giza 5) were planted in sandy soil during two successive growing seasons (2010–2011) in the 1st week of May and harvested after 130 days. Treatments used are: two irrigation methods (surface drip irrigation and subsurface drip irrigation: SDI, SSDI); nitrogen source (NS: NH_4NO_3, $(NH_4)_2SO_4$ and $Ca(NO_3)_2 = Ns_1$, Ns_2 and Ns_3) and flushing frequency (FL: 0, 1 and 4 = FL_1, FL_2 and FL_3).

The experimental design was split-split plot and three replications. The main effects of treatments on clogging percentage can be written in the ascending orders: $SDI < SSDI$, $FL_3 < FL_2 < FL_1$, $NS_1 < NS_2 < NS_3$. Concerning the 1st interaction, the ascending orders were: $SDI \times Fl_3 < SDI \times Fl_2 < SDI \times FL_1$, $SDI \times NS_1 < SDI \times NS_2 < SDI \times NS_3$, $SSDI \times FL_3 < SSDI \times FL_2 < SSDI \times FL_1$, $SSDI \times NS_1 < SSDI \times NS_2 < SSDI \times NS_3$, $FL_1 \times NS_1 < FL_1 \times NS_2 < FL_1 \times NS_3$, $FL_2 \times NS_1 < FL_2 \times NS_2 < FL_2 \times NS_3$ and $FL_3 \times NS_1 < FL_3 \times NS_2 < FL_3 \times NS_3$. The differences between any two treatments and/ or any two interactions in clogging percent were significant at 5% level. The effect of the 2nd interaction on clogging percent was significant at 5% level. The maximum value of clogging (20.18%) and the lowest one (3.9%) were archived in the interactions: $SSDI \times FL_1 \times NS_3$ and $SDI \times FL_3 \times NS_1$, respectively.

KEYWORDS

- clogging
- clogging agents
- EC
- Egypt
- emitter clogging
- flushing frequency
- flushing number
- irrigation method
- Nitrogen source
- precipitation
- SAR
- subsurface drip irrigation
- surface drip irrigation
- water analysis
- water quality

REFERENCES

1. Adin, A., 1987. Clogging in irrigation system reusing pond effluent and its prevention. *Water Sci. Technol.*, 19–23.

2. Bucks, D. A., and F. S., Nakayama, 1984. Problems to avoid clogging with drip/trickle irrigation systems. *Proc. Amer. Soc. Civil Eng. Specially Conf. Irri. and Drainage Div.*, 24–84.

3. Bucks, D. A., F. S., Nakayama and R. G. Gilbert, 1977. Clogging research on drip irrigation. Proceedings 4th Annual International Drip Irrigation Association Meeting, pp. 25–31.

4. Chang, C. A., 2008. Drip lines and emitters: chlorination for disinfection and prevention of clogging. http://www.informaworld.com/smpp/content~content=a792065664 ~db=all~jumptype

5. Charles, M. and C. E. Burt, 1997. Fertigation chemicals. Irrigation Training and Research Center (ITRC), California Polytechnic State University, Cal Poly San Luis Obispo, CA 93407, www.itrc.org pp. 97–101.

6. Choi, C. Y., Suarez-Rey, E. M., 2004. Subsurface drip irrigation for Bermuda grass with reclaimed water. *Trans. ASAE*, 47(6):1943–1951.

7. Dasberg, S., and E., Bresler, 1986, *Drip irrigation manual*. International Irrigation Information Center, Publ. No. 9, Voleani Center, Bet Dagan, Israel, 61 pp.

8. De Troch, F., 1988. *Irrigation and drainage*. Gent univ., Belgium.

9. El-Berry, A. M., G. A. Bakeer, and A. M. Al-Weshali, 2003. The effect of water quality and aperture size on clogging of emitters. <http://afeid.montpellier.cemagref.fr/Mpl2003/AtelierTechno/AtelierTechno/Papier%20Etier/N%C2%BO48%20-%20EGYPTE_BM.pdf>

10. Fares, A. and F. Abbas, 2009. Injection rates and components of a fertigation system. College of Tropical Agriculture and Human Resources, University of Hawai at Manoa, Engineer's Notebook. pp. 1–4.

11. Ford, H. W., 1984. The problem of emitter clogging and methods for control. *Citrus Ind.*, 46–52.

12. Gilbert, R. G., and H. W. Ford, 1986. Operational principles emitters clogging. In *Trickle irrigation for crop production*, Eds. F.S. Nakayama and D.A. Bucks. New York: pages 42–163.

13. Megh R. Goyal, 2012. *Management of Drip/Trickle or Micro Irrigation*. Apple Academic Press Inc.

14. Hebbar, S. S., B. K. Ramachandrappa, H. V. Nanjappa and M. Prabhakar, 2004. Studies on NPK drip fertigation in field grown tomato (Lycopersicon esulentum Mill.). *Eur. J. Agron.*, 21:117–127.

15. Hills, D. J., F. M. Nawar, and P. M., Waller, 1989. Effect of chemical clogging on drip-tape irrigation uniformity. *Trans. ASAE*, 32(4):1202–1206.

16. James, L. G., 1988. *Principles of farm irrigation system design*. John Wiley and Sons, pp. 264- 268.

17. Lamm, F. R., and Camp, R. C., 2007. Subsurface drip irrigation. In: Lamm, F.R., Ayars, J.E., Nakayama, F.S. (Eds.), *Microirrigation for Crop Production. Design, Operation, and Management*. Elsevier, Amsterdam, pp. 473–551.

18. Oron, G., J. Demalach, Z. Hoffman, and R. Cibotaru, 1991. Subsurface micro irrigation with effluent. *J. Irrig. Drain. Eng. ASCE*, 117(1):25–36.

19. Ozekici, B., 1998. Clogging factors in drip irrigation systems. Final Report. University of Cukurova, Faculty of Agriculture, Department Research Project and C.U. Rectorate Research Fon. BAP-TYS-95–06, Adana, Turkey (Turkish).

20. Pillsbury, A. F and A. Degan, 1975. *Sprinkler irrigation*. FAO Agricultural Development Paper No. 88, Rome.
21. Ravina, I., Paz, E., Sofer, Z., Marcu, A., Shisha, A., Sagi, G., Yechialy, Z., and Lev, Y., 1997. Control of clogging in drip irrigation with stored treated municipal sewage effluent. *Agric. Water Manage.*, 33(2–3):127–137.
22. Sabreen, Kh. P., 2009. *Fertigation technologies for improving the productivity of some vegetable crops*. Ph.D. Thesis, Faculty of Agriculture, Ain Shams University, Egypt.
23. Sahin, U., O. Anapal, M. F. Donmez and F. Sahin, 2005. Biological treatment of clogged emitters in a drip irrigation system. *J. Environ. Manage.* 76:338–341.
24. Segars, B., 2010. *Efficient fertilizer use manual-fertigation*. pp. 1–8.
25. Snedcor, G. W. and W. G. Cochran, 1982. *Statistical Methods*. 7th Ed. The Iowa State Univ. Press, Iowa, USA.
26. Tayel, M. Y., A. M. El-Gindy, K.F. El-Bagoury, and Kh. A. Sabreen, 2009. Effects of injector types, irrigation and nitrogen treatments on emitters clogging. *Misr. Journal of Agricultural Engineering MJAE*, 23(3):1263–1276.
27. Tayel, M. Y., H. A. Mansour and David A. Lightfoot, 2012. Effect of different closed circuits and lateral lines length on: I- pressure head and friction loss. *Agric. Sciences*, 3(3):392–399.
28. Tayel, M. Y., H. A. Mansour and David A. Lightfoot, 2012. Effect of different closed circuits and lateral lines length on: II-flow velocity and velocity head. *Agri. Sciences*, 3(4):531–537.
29. Zhai, G. L., M. C. Lv, H. Wang and H.A. Xiang, 1999. Plugging of micro irrigation system and its prevention. *Trans. Chin. Soc. Agric. Eng.*, 15(1):144–147.
30. Zhu, L. Y., and Cui, C. L., 2005. Discussion on problem of block-up in drip irrigation underground and its treatment. *Res. Soil Water Conserv.*, 12(2):111–112.

CHAPTER 15

EVAPOTRANSPIRATION FOR CYPRESS AND PINE FORESTS: FLORIDA, USA[1]

DAVID M. SUMNER

CONTENTS

15.1   INTRODUCTION

The importance of evapotranspiration (ET) in the hydrologic cycle has long been recognized; in central Florida, evapotranspiration is second

---

[1]Reprinted with permission from David M. Sumner, Chapter 8: Evapotranspiration for cypress and pine forests, in Evapotranspiration: Principles and Applications for Water Management. 2014. Apple Academic Press Inc.

This chapter is an edited version of *Sumner, D.M., 2001. Evapotranspiration from a Cypress and Pine Forest Subjected to Natural Fires, Volusia County, Florida, 1998–1999: U.S. Geological Survey Water-Resources Investigations Report 01–4245, 56 p. Open access available at: http://fl.water.usgs. gov/Abstracts/wri01_4245_sumner.html.*

only to precipitation in magnitude. Of the approximately 1,320 millimeters (mm) of mean annual rainfall in central Florida, 680 to 1,220 mm have been estimated to return to the atmosphere as evapotranspiration [44, 49]. Despite the importance of evapotranspiration in the hydrologic cycle, the magnitude, seasonal and diurnal distributions, and relation to environmental variables of evapotranspiration remain relatively unknown. Uncertainty in evapotranspiration from nonagricultural vegetation is particularly apparent. The mixed cypress wetland and pine flat wood forest cover examined in the present investigation is common in central Florida, as are the fires that burned much of the forest during the study. Accurate estimates of evapotranspiration from commonly occurring land covers are fundamental to the quantitative understanding necessary for prudent management of Florida's water resources.

The eddy correlation method (or eddy covariance method) has been used successfully to directly measure evapotranspiration in Florida by Bidlake and others [3]; Knowles [22]; and Sumner [44]. This micrometeorological method offers several advantages to alternative water-budget approaches (lysimeter or regional water budget) by providing more areal integration and less site disruption than lysimeters, by eliminating the need to estimate other terms of a water budget (precipitation, deep percolation, runoff, and storage), and by allowing relatively fine temporal resolution (less than 1 h).

Evapotranspiration can be estimated by using evapotranspiration models. These models also provide insight into the relative importance of individual environmental variables in the evapotranspiration process. The Priestley-Taylor model [34] for evaporation from a wet surface (potential evapotranspiration) was modified to allow for nonpotential conditions [13], and has successfully simulated evapotranspiration in the Florida environment [15, 22, 44].

The U.S. Geological Survey (USGS), in cooperation with the St. Johns River Water Management District and the County of Volusia, began a 4-year study in 1996 to estimate the temporal pattern of evapotranspiration in the Tiger Bay watershed, Volusia County, Fla., a forested watershed, and to develop a quantitative description of the effect of environmental variability on evapotranspiration from forested areas in Florida. This analysis can provide guidance in the estimation of evapotranspiration and the description of the relation between the environment and evapotranspiration in other areas with similar environmental characteristics. During the study period, the watershed experienced a severe drought and natural fires, which provided the opportunity to study the effects of such extreme events on the evapotranspiration process.

This chapter presents daily estimates of evapotranspiration during a 2-year period from a forested watershed (Tiger Bay, Volusia County, Fla.), which was subjected to natural fires, and provides evaluations of the causal relations between the environment and evapotranspiration. Measurements were made on a nearly continuous basis from January 1998 through December 1999 at an evapotranspiration station just outside the watershed, using eddy correlation and meteorological instrumentation. An evapotranspiration model based on the Priestley-Taylor equation was used to estimate evapotranspiration for burned and unburned areas and to quantify the relation between evapotranspiration and the environment. A water budget of the watershed was constructed to assess the validity of the eddy correlation-measured evapotranspiration totals for the 2-year period.

15.1.1 DESCRIPTION OF THE STUDY AREA

The study area is the approximately 7,500-hectare Tiger Bay watershed within Volusia County, Fla. (Fig. 15.1). The watershed was almost completely forested in January 1998, but was subjected to extensive burning and logging during the study period. The watershed characteristics are typical of many areas within the lower coastal plain of the south-eastern United States—nearly flat, slowly draining land with a vegetative cover consisting primarily of pine flat wood uplands interspersed within cypress wetlands. The northern part of the watershed mostly is within the 9,500-hectare Tiger Bay State Forest; the southern part of the watershed primarily is privately owned land used for timber production. The watershed is within the relatively flat Talbott Terrace physiographic area [37]. More than 90% of the watershed is at an altitude of 11 to 13 meters (m). Small variations in local topography result in areal variations in hydroperiod. A low-lying wetland can be inundated much of the year, whereas an adjacent upland, less than a few tens of centimeters (cm) elevated above

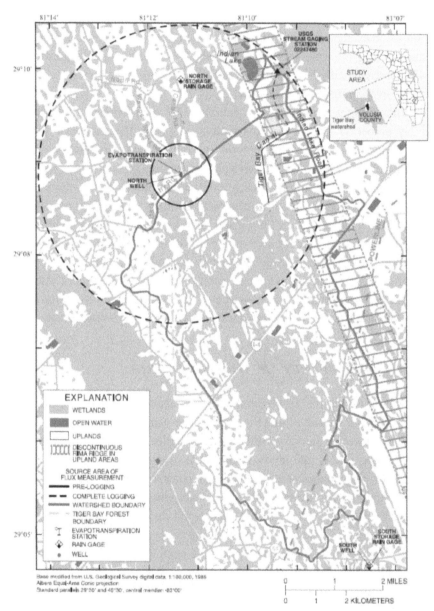

FIGURE 15.1   Location of Tiger Bay watershed.

the wetland, may only occasionally or never exhibit standing water. Most of the surface runoff from the watershed is through interconnected wetlands [36].

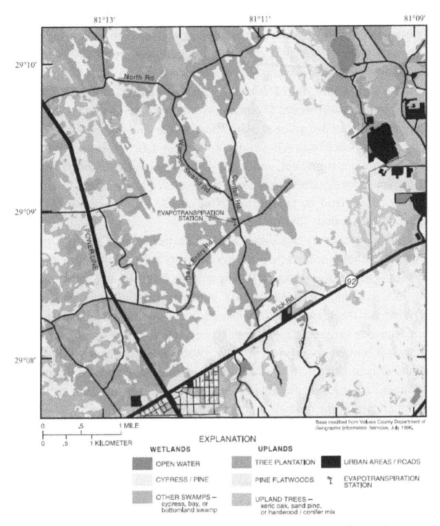

FIGURE 15.2   Distribution of vegetation in vicinity of evapotranspiration station.

More than 95% of the watershed is forested. Two tree species dominate the forest cover in the watershed: slash pine (evergreen) and pond cypress (deciduous; leaves drop in November-December with regrowth in March–April). The distribution of vegetation in the vicinity of the evapotranspiration station is shown in Fig. 15.2.

Vegetation in the watershed reflects the variation in hydroperiod [40]. Wetlands are dominated by pond cypress (*Taxodium ascendens*), with lesser amounts of other wetland tree species including blackgum (*Nyssa biflora*), loblolly bay (*Gordonia lasianthus*), and red maple (*Acer rubrum*). The understory of wetlands consists of a wide variety of plants including leather fern (*Acrostichum danaeifolium*), marsh fern (*Thelypteris palustris*), cinnamon fern (*Osmunda cinnamomea*), swamp lily (*Crinum americanum*), maidencane (*Panicum hemitomon*), red root (*Lachnanthes caroliniana*), hooded pitcher plant (*Sarracenia minor*), St. John's Wort (*Hypericum fasciculatum*), yellow colic root (*Aletris lutea*), pipewort (*Eriocaulon decangulare*), and white-topped sedge (*Rhynchospora colorata*). Water level varies from about 0.3 m above land surface to as much as 1 m below land surface in low-lying areas, although these areas are inundated more than 50% of the time [40].

Uplands generally are either slash pine tree (*Pinus elliottii*) plantations or naturally seeded pine flatwoods (primarily slash pine with some longleaf pine (*Pinus palustris*)). These areas have an understory including saw palmetto (*Serenoa repens*), gallberry (*Ilex glabra*), wax myrtle (*Myrica cerifera*), red root (*Lachnanthes caroliniana*), and broomsedge (*Andropogon virginicus*). Understory vegetation in the pine plantations is control-burned about every 3 years. Water level varies from about 0.1 m above land surface to as much as 2 m below land surface in uplands; however, water levels are always greater than 2 m below land surface in the small part of the uplands within the Rima Ridge (Fig. 15.1). The Rima Ridge consists of discontinuous remnants of terrace deposits parallel to the present-day coastline [37]. Vegetation on the ridge areas includes sand live oak (*Quercus geminata*) and sand pine (*Pinus clausa*). Most of the limited urbanization within the Tiger Bay watershed is on the Rima Ridge.

Brush fires burned extensively throughout peninsular Florida during spring 1998 as a result of a severe drought. A high-pressure system remained stationary over the State, blocking the normal pattern of convective thunderstorms [48]. During the 3-month period, April-June, National Oceanic and Atmospheric Administration (NOAA) stations at Daytona Beach and DeLand recorded about 10 and 30% of long-term, average precipitation, respectively. Brush fires, ignited by lightning strikes, began in Volusia County on June 19, 1998, and continued until rainfall resumed in late June and early July, burning about 55,000 hectares (one-fifth of the County) and about 40% of the watershed (Fig. 15.3). Although areas

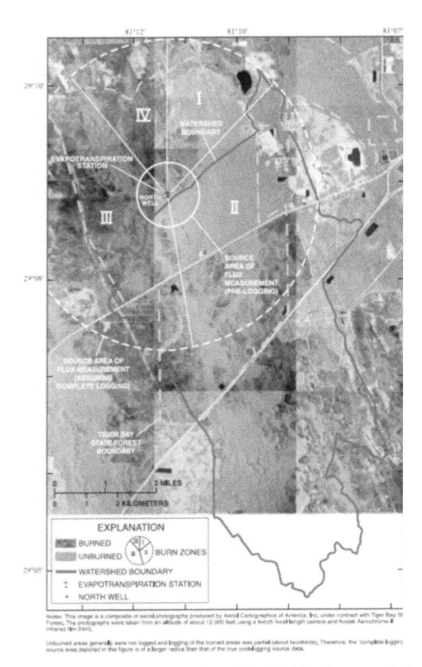

FIGURE 15.3   Infrared photograph (July 7, 1998) of vicinity of evapotranspiration station showing area: burned during fires of June 1998.

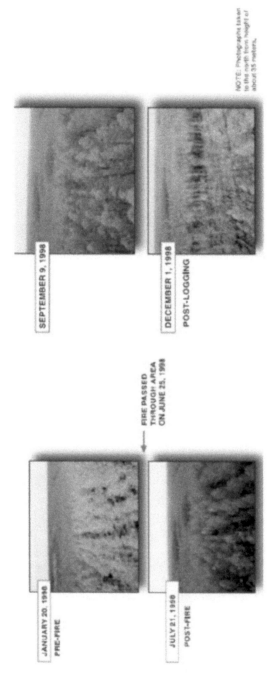

FIGURE 15.4 Photographic times series of vegetation in vicinity of evapotranspiration station.

of both wetlands and uplands were burned during the June-July fires, a comparison of Figs 15.1 and 15.3 reveals that upland areas were burned more extensively than wetland areas. Re-growth of understory vegetation occurred rapidly after the fires ceased and the rains began. Emergent growth of red root (*Lachnanthes caroliniana*) in burned areas was particularly evident. Some trees were killed by the fire, whereas other burned trees were merely damaged and exhibited leaf regrowth soon after the fire (Fig. 15.4). Large-scale harvesting of insect-infested, firedamaged trees (both living and dead trees) occurred during the months following the fires. Of the approximately 4,800 hectares that burned within the 9,500-hectare Tiger Bay State Forest, about 3,200 hectares were logged (Catherine Lowenstein, Tiger Bay State Forest, oral communication, 2000).

Fires moved from west-to-east through the area of the evapotranspiration station on June 25, 1998. Damaged trees in the vicinity of the evapotranspiration station were logged during November–December 1998.

The two dominant soil groups of the watershed also reflect the areal variation in hydroperiod and vegetation [2]. Wetlands tend to be underlain by organic soils (hyperthermic family of Terric Medisaprists) of the Samsula-Terra Ceia-Tomoka group that are very poorly drained. The uplands

FIGURE 15.5   Krypton hygrometer (foreground) and sonic anemometer (background) mounted at top of tower at evapotranspiration station.

FIGURE 15.6   Evapotranspiration station being serviced by hydrologic technician.

tend to be underlain by poorly drained soils (sandy, siliceous, hyperthermic family of Ultic Haplaquods) of the Pomona-Wauchula group that have a dark, organic-stained subsoil underlain by loamy material.

The climate of central Florida is humid subtropical and is characterized by a warm, wet season (June–September) and a mild, relatively dry season (October–May). During the dry season, precipitation commonly is associated with frontal systems. Rainfall averages about 1,350 mm/yr in Volusia County [37]. More than 50% of the annual rainfall generally occurs during the wet season when diurnal thunderstorm activity is common. Mean air temperature in the study area is about 21 °C, ranging from occasional winter temperatures below 0 °C to summer temperatures approaching 35 °C. Diurnal temperature variations average about 12 °C.

Rainfall to the watershed leaves the basin as runoff, evapotranspiration, or deep leakage from the surficial aquifer system to the underlying Upper Floridan aquifer [21, 33]. Intermittent runoff gaged at Tiger Bay canal along the northern edge of the watershed (Fig. 15.1) averaged 0.47 cubic meters per second (m^3/s) or about 200 millimeters per year (mm/yr) from 1978 to 1999 [53]. Evapotranspiration has been estimated to average about 990 mm/yr over Volusia County [37] and about 890 mm/yr in the Tiger Bay watershed [7]. Previous researchers have documented relatively small differences in the annual evapotranspiration rates from the two primary land covers. Bidlake and others [3] estimated annual cypress evapotranspiration (970 mm) to be only 8.5% less than that from pine flatwoods (1,060 mm), based on studies conducted in Sarasota and Pasco Counties, Fla. Liu [24] estimated average annual evapotranspiration from both covers to be 1,080 mm, based on a study conducted in Alachua County, Fla.

The hydraulic head in the surficial aquifer system within the watershed generally is above that of the underlying Upper Floridan aquifer. Consequently, water leaks downward from the surficial aquifer system, through the intermediate confining unit, to the Upper Floridan aquifer. Deep leakage was estimated (based on ground-water flow simulations) to have been about 56 mm/yr prior to ground-water development, but in 1995, the rate was estimated to have doubled to 112 mm/yr, as a result of lowering the hydraulic head in the Upper Floridan aquifer by pumping (Stan Williams, St. Johns River Water Management District, oral communication, 2000).

15.2 METHODS AND MATERIALS: MEASUREMENT AND SIMULATION OF EVAPOTRANSPIRATION

Evapotranspiration was measured at a site just outside the study area (Fig. 15.1) using the eddy correlation method in a manner similar to that described by Sumner [44]. The site chosen for the evapotranspiration station was within an 18.3-m-tall, 30-year-old pine plantation (Fig. 15.2). Eddy correlation instrumentation was mounted on a 36.5-m-tall Rohn 45G communications- type tower at the site (Figs. 15.5 and 15.6), and data were collected for a 2-year period from January 1, 1998, to December 31, 1999. Other meteorological instrumentation also was deployed on or around the tower to collect data for evapotranspiration modeling and to provide ancillary data for the eddy correlation analysis. Instrumentation used in the study is described in Table 15.1. Measured daily values of evapotranspiration were used to calibrate evapotranspiration models (modified Priestley-Taylor). Evapotranspiration was estimated for burned and unburned areas

TABLE 15.1   Study Instrumentation

Parameter	Instrument	Height(s) above land surface, meters
Evapotranspiration	CSI eddy correlation system including model CSAT3 3-D sonic anemometer and model KH$_2$O krypton hygrometer.	36.5
Air temperature/ Relative humidity	CSI model HMP35C temperature and relative humidity probe.	1.5, 9.1, 18.3 and 35
Net radiation	REBS model Q-7.1 net radiometer.	35
Wind speed & direction	RMY model 05305–5 wind monitor – AQ	35
Photosynthetically active radiation (PAR)	LI-COR, Inc., Model LI-190SB quantum sensor.	35
Soil moisture	CSI model CS615 water content reflectometer.	0.0 to −0.3
Precipitation	TE model 525 tipping bucket rain gage and NovaLynx model 260–2520 forester's (storage) rain gages (two)	18.3 (tipping bucket) and 1.0 (storage)
Water level in well	Druck, Inc., Model PDCR950 pressure transducer.	−2.0
Datalogging	CSI model 21X and model 10X data loggers; 12 volt deep-cycle batteries (two); 20 watt solar panels (two)	0 to 1.0

Note: Negative height is depth below land surface.
CSI, Campbell Scientific, Inc.; REBS, Radiation and Energy Balance Systems, Inc.; RMY, R. M. Young, Inc.; TE, Texas Electronics, Inc.

using the calibrated evapotranspiration models. A water budget for the watershed over the study period was constructed based on measured or estimated values of precipitation, evapotranspiration, runoff, leakage, and storage.

15.2.1 MEASUREMENT OF EVAPOTRANSPIRATION

15.2.1.1 Eddy-Correlation Method

The eddy correlation method [9, 45] was used to measure two components of the energy budget of the plant canopy: latent and sensible heat fluxes. Latent heat flux (λE) is the energy removed from the canopy in the liquid-to- vapor phase change of water, and is the product of the heat of vaporization of water (λ) and the evapotranspiration rate (E). Sensible heat (H) is the heat energy removed from the canopy as a result of a temperature gradient between the canopy and the air.

Both latent and sensible heat fluxes are transported by turbulent eddies in the air. Turbulence is generated by a combination of frictional and convective forces. The energy available to generate turbulent fluxes of vapor and heat is equal to the net radiation (R_n) minus the sum of the heat flux into the soil surface (G) and the change in storage (S) of energy in the biomass and air. The energy involved in fixation of carbon dioxide usually is negligible [5]. Net radiation is the difference between incoming radiation (shortwave solar radiation and long wave atmospheric radiation) and outgoing radiation (reflected shortwave and long- wave radiation; and emitted long wave canopy radiation). Energy is transported to and from the base of the canopy by conduction through the soil.

Assuming that net horizontal advection of energy is negligible, the energy-budget equation, for a control volume extending from land surface to a height z_s at which the turbulent fluxes are measured, is given in Eq. (1):

$$R_n - G_S = H + \lambda E \tag{1}$$

$$E = \overline{w\rho_v} = \overline{(\overline{w} + w')(\overline{\rho_v} + \rho_v')} \tag{2}$$

$$E = (\overline{\overline{w}\overline{\rho_v}} + \overline{\overline{w}\rho_v'} + \overline{w'\overline{\rho_v}} + \overline{w'\rho_v'}) \tag{3}$$

$$\overline{w'\rho'_v} = covariance(w, \rho_v) \tag{4}$$

In the energy-budget Eq. (1): the left side of represents the available energy and the right side represents the turbulent flux of energy; R_n is net radiation to or from plant canopy, in watts per square meter; G is soil heat flux at land surface, in watts per square meter; S is change in storage of energy in the biomass and air, in watts per square meter; H is sensible heat flux at height z_s above land surface, in watts per square meter; and ΔE is latent heat flux at height z_s above land surface, in watts per square meter. The sign convention is such that R_n and G are positive downwards; and H and ΔE are positive upwards.

The eddy correlation method is a conceptually simple, one-dimensional approach for measuring the turbulent fluxes of vapor and heat above a surface. For the case of vapor transport above a flat, level land- scape, the time-averaged product of measured values of vertical wind speed (w) and vapor density (ρ_v) is the estimated vapor flux (evapotranspiration rate) during the averaging period, assuming that the net lateral advection of vapor is negligible. Because of the insufficient accuracy of instrumentation available for measurement of actual values of wind speed and vapor density, this procedure generally is performed by monitoring the fluctuations of wind speed and vapor density about their means, rather than monitoring their actual values.

This formulation is represented in Eqs. (2)–(4), where: E is evapotranspiration rate, in grams per square meter per second; w is vertical wind speed, in meters per second; ρ_v is vapor density, in grams per cubic meter; and over-bars and primes indicate means over the averaging period and deviations from means, respectively.

The first term of the right side of Eq. (3) is approximately zero because mass-balance considerations dictate that mean vertical wind speed perpendicular to the surface is zero; this conclusion is based on an assumption of constant air density (correction for temperature-induced air-density fluctuations is discussed later in this chapter). The second and third terms are zero based on the definition that the mean fluctuation of a variable is zero. Therefore, it is apparent from Eq. (4) that vertical wind speed and vapor density must be correlated in order for the value of vapor flux to be nonzero. The turbulent eddies that transport water vapor (and sensible heat) produce fluctuations in both the direction and magnitude of vertical wind speed. The ascending eddies must on average be more moist than the descending eddies for evapotranspiration to occur, that is, upward air movement must be positively correlated with vapor density and down-ward air movement must be negatively correlated with vapor density.

15.2.2   Source Area of Measurements

The source area for a turbulent flux measurement defines the area (upwind of measurement location) contributing to the measurement. The source area can consist of a single vegetative cover if that cover is ade-quately extensive. This condition is met if the given cover extends suffi-ciently upwind such that the atmospheric boundary layer has equilibrated with the cover from ground surface to at least the height of the instru-mentation. If this condition is not met, the flux measurement is a compos-ite of fluxes from two or more covers within the source area. The source area is defined in this report as the area contributing to 90% of the sensor measurement. Schuepp and others [39] provide an estimate of the source area, and the relative contributions within the source area, based on an analytical solution of a one-dimensional (upwind) diffusion equation for a uniform surface cover. In this approach, source area varies with instrument height (z_s), zero displacement height (d), roughness length for momentum (z_m), and atmospheric stability. The instrument height in this study was 36.5 m. Campbell and Norman [8] proposed empirical relations based on canopy height (h) for zero displacement height ($d \sim 0.65h$) and roughness length for momentum ($z_m \sim 0.10h$.). Uniform canopy heights of 18.3 m (prelogging) and 0.3 m (assuming complete logging) were assumed in this analysis. The source area estimates were made assuming mildly unstable conditions; the Obukhov stability length [6] was set equal to -10 m. The source area increases as the height of the instrument above the vegetative canopy increases and as the roughness length for momentum decreases; therefore, the extensive logging that occurred following the fires enlarged the source area. The source area for the turbulent flux measurements (Fig. 15.7) was estimated to be within an upwind distance of about 1,000 m (prelogging) or 4,800 m (assuming complete logging). As stated earlier,

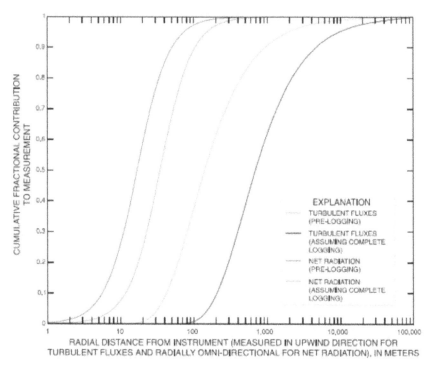

FIGURE 15.7   Radial extent of source areas of turbulent flux and net radiation measurements.

unburned areas generally were not logged and logging of the burned areas was partial (about two-thirds). Therefore, the "complete logging" source area depicted in Fig. 15.7 is of a larger radius than that of the true postlogging source area.

The site of the evapotranspiration station was chosen such that the source area of the turbulent flux measurements would be representative of the relative mix of wetlands and uplands in the prefire watershed (Fig. 15.1). Before the fire and associated logging, the source area of the turbulent flux measurement (Fig. 15.1) consisted of: 43.7% upland, 56.1% wetland, and 0.2% lake. These relative fractions of wetland and upland were very close to those of the entire Tiger Bay watershed (43.8% upland, 55.5% wetland, and 0.7% lake) before the fires. Also, areas of wetland and upland within the pre fire source area were interspersed, indicating that turbulent flux measurements approximated a representative value of the composite mix of wetlands and uplands, regardless of the wind direction.

Fires within the watershed during spring 1998 changed the primary components of source area heterogeneity from wetland/upland to burned/unburned (Fig. 15.3) and complicated interpretation of the turbulent flux measurements. Burned and unburned areas were not well-interspersed, resulting in measurements that reflected varying fractions of burned and unburned areas, depending on the wind direction. Following the fires, turbulent fluxes representative of burned areas were measured, both pre- and postlogging, when the wind was from the north-west (zone IV in Fig. 15.3).

Turbulent fluxes representative of unburned areas were measured when the wind was from the east (zone II) throughout the study period. The absence of near-station burning in zone II, and therefore, a lack of subsequent near-station logging in this zone, resulted in a consistently small (radius of 1,000 m), and unburned, source area throughout the study period when the wind was from zone II. Turbulent fluxes representative of burned areas were measured following the fires and prior to logging when the wind was from the north-east (zone I). With the expansion of the source area associated with logging, however, the postlogging turbulent flux measurements were representative of a composite of burned and unburned areas when the wind was from zone I. Examination of the estimated [39] cumulative fractional contribution to the turbulent flux measurement as a function of upwind distance from the measurement (Fig. 15.7) provided information to approximate the relative degree of burned/unburned area compositing. Based on this approach, an estimate was made that postlogging turbulent flux measurements made when the wind was from zone I reflected a surface cover that was 75% burned and 25% unburned. Burned and unburned areas within zone III were relatively well interspersed and in approximately equal relative amounts following the fires. Therefore, postfire turbulent fluxes measured when the wind was from zone III were assumed to reflect a surface cover that was 50% burned and 50% unburned. Estimates of the relative contribution (as a function of wind direction and status of the surface cover) of burned vegetation to the measured turbulent flux signal are summarized in Table 15.2. These estimates were used to develop weighting coefficients

TABLE 15.2   Relative Fraction of Burned Vegetation Sensed by Eddy Correlation Instrumentation*

Burn zone i	Sector	g_i = Fractional contribution		
		Pre-fire	Post-fire/prelogging	Post-logging
I	0 to 45	0.0	1.0	0.75
II	45 to170	0.0	0.0	0.00
III	170 to 320	0.0	0.5	0.50
IV	320 to 360	0.0	1.0	1.00

*The sector is in degrees measured clockwise from north (Fig. 15.3); g_i is the fractional contribution of burned area within burn zone i to the measured latent heat flux when wind direction is from burn zone i.

indicative of the fraction of the turbulent flux measurement for a given day that reflected burned vegetation, which is further discussed later in this chapter.

15.2.3   INSTRUMENTATION

Instrumentation capable of high-frequency resolution must be used in an application of the eddy correlation method because of the relatively high frequency of the turbulent eddies that transport water vapor. Instrumentation included a three-axis sonic anemometer and a krypton hygrometer to measure variations in wind speed and vapor density, respectively (Fig. 15.5). The sonic anemometer relies on three pairs of sonic transducers to detect wind-induced changes in the transit time of emitted sound waves and to infer fluctuations in wind speed in three orthogonal directions. The measurement path length between transducer pairs is 10.0 cm (vertical) and 5.8 cm (horizontal); the transducer path angle from the horizontal is 60 degrees. In contrast to some sonic anemometers used previously [44], the transducers of this improved anemometer are not permanently destroyed by expo- sure to moisture, and thus are suitable for long-term deployment. Operation of the anemometer used in this study ceases when moisture on the transducers disrupts the sonic signal, but recommences upon drying of the transducers.

The hygrometer relies on the attenuation of ultra-violet radiation, emitted from a source tube, by water vapor in the air along the 1-cm path to the detector tube. The instrument path line was laterally displaced 10 cm from the midpoint of the sonic transducer path lines. Hygrometer voltage output is proportional to the attenuated radiation signal, and fluctuations in this signal can be related to fluctuations in vapor density by Beer's Law [59]. Similar to the anemometer, the hygrometer ceases data collection when moisture obscures the windows on the source or detector tubes. Also, the tube windows become "scaled" with exposure to the atmosphere, resulting in a loss of signal strength. The hygrometer is designed such that vapor density fluctuations are accurately measured in spite of variable signal strength; however, if signal strength declines to near-zero values, the fluctuations cannot be discerned. Periodic cleaning of the windows (performed monthly in this study) with a cotton swab and distilled water restored the signal strength. Eddy correlation instrument-sampling frequency was 8 Hertz with 30-minute averaging periods. The eddy correlation instrumentation was placed about 18.2 m above the tree canopy (Fig. 15.6). Data were processed and stored in a data logger near ground-level.

   To be representative of the surface cover, flux measurements must be made in the inertial sublayer, where vertical flux is constant with height and lateral variations in vertical flux are negligible [28]. Measurements made in the underlying roughness sublayer can reflect individual roughness elements (e.g., individual trees or gaps between trees), rather than the composite surface cover. Garrat [14] defines the lower boundary of the inertial sublayer to be at a height such that the difference of this height and the zero displacement height (d) is much greater than the roughness length for momentum (z_m). Employing Campbell and Norman's [8] empirical relations and assuming that "much greater than" implies greater by a factor of ten [10], leads to an instrument height (z_s) requirement of $z_s > [1.65*h]$. A factor of about two was used in this chapter as a conservative measure. As a conservative measure, the instrument height (36.5 m) used in this chapter was about twice canopy height.

15.2.4 CALCULATION OF TURBULENT FLUXES

Latent heat flux [*see* Eq. (5)] was estimated based on a modified form of Eq. (4). In Eq. (5): λE is latent heat flux, in watts per m^2; λ is latent heat of vaporization of water, estimated as a function of temperature [43], in joules per gram; ρ is air density that is estimated as a function of air temperature, total air pressure, and vapor pressure [28], in grams per cubic meter; H is sensible heat flux, in watts per m^2; C_p is specific heat capacity of air, estimated as a function of temperature and relative humidity [43], in joules per gram per degree Celsius; T_a is air temperature, in °C; F is a factor that accounts for molecular weights of air and atmospheric abundance of oxygen, and is equal to 0.229 gram-degree Celsius per joule; K_o is an extinction coefficient of hygrometer for oxygen, estimated as 0.0045 cubic meters per gram per centimeter [46]; K_w is an extinction coefficient of hygrometer for water, equal to the manufacturer-calibrated value, in cubic meters per gram per centimeter; and overbars and primes indicate means over the averaging period and deviations from the means, respectively. The second and third terms of the right side of Eq. (5) account for temperature induced fluctuations in air density [58] and for the sensitivity of the hygrometer to oxygen [45], respectively.

Similarly to vapor transport, sensible heat can be estimated by using Eq. (6). The sonic anemometer is capable of measuring "sonic" temperature based on the dependence of the speed of sound on this variable [19, 20]. Schotanus and others [38] related the sonic sensible heat based on measurement of sonic temperature fluctuations to the true sensible heat given in Eq. (6). Those researchers included a correction, for the effect of wind blowing normal to the sonic acoustic path that has been incorporated directly into the anemometer measurement by the manufacturer (Swiatek, E., 1998. Campbell Scientific, Inc., written communication), leading to a simplified form of the Schotanus and others [38] formulation given in Eq. (7).

In Eq. (7): T_s is the sonic temperature, in °C; and q is specific humidity, in grams of water vapor per grams of moist air. Fleagle and Businger [12] defined the specific humidity [q] in Eq. (8), based on the relation between specific humidity and vapor density.

In Eq. (8): ρ_v is vapor density, in grams per cubic meter; R_d is the gas constant for dry air (= 0.28704 joules per degree Celsius per gram); and P_a is atmospheric pressure, in pascals (assumed to remain constant at 100.7 kilopascals at top of tower at about 48 meters above sea level). Eq. (7) can be expressed in terms of fluctuations in the hygrometer-measured water vapor density rather than fluctuations in specific humidity as shown in Eq. (9).

$$\lambda E = \lambda \left(\overline{w'\rho'}_v + \left[\frac{\rho_v H}{\rho C_p} \right] * \left[\frac{1}{T_a + 273.15} \right] + \frac{F K_o H}{K_W (T_a + 273.15)} \right) \tag{5}$$

$$H = \rho C_p \overline{w' T_a'} \tag{6}$$

$$\overline{w' T_a'} = \overline{w' T_s'} - 0.51 (T_a + 273.15) \overline{w' q'} \tag{7}$$

$$q \approx \frac{\rho_v R_d (T_a + 273.15)}{P_a} \tag{8}$$

$$\overline{w' T_a'} = \frac{(T_a + 273.15)}{(T_s + 273.15)} \left(\overline{w' T_s'} - 0.51 R_d (T_a + 273.15)^2 \, \overline{w' \rho'}_v / P_a \right) \tag{9}$$

$$\left(\overline{w' c'} \right)_r = \overline{w' c'} \cos\theta - \overline{u' c'} \sin\theta \cos\eta - \overline{v' c'} \sin\theta \sin\eta \tag{10}$$

$$\cos\theta = \sqrt{\frac{(u^2 + v^2)}{(u^2 + v^2 + w^2)}} \tag{11}$$

$$\sin\theta = \frac{w}{\sqrt{(u^2 + v^2 + w^2)}} \tag{12}$$

$$\cos\eta = \frac{u}{\sqrt{(u^2 + v^2)}} \tag{13}$$

$$\sin\eta = \frac{v}{\sqrt{(u^2 + v^2)}} \tag{14}$$

Estimation of turbulent fluxes (Eqs. (5) and (6)) relies on an accurate measurement of velocity fluctuations perpendicular to the lateral airstream. The study area is relatively flat and level, indicating that the air stream is approximately perpendicular to gravity and the sonic anemometer was oriented with respect to gravity with a bubble level. Measurement of wind speed in three orthogonal directions with the sonic anemometer allows for a more refined orientation of the collected data with the natural coordinate system through mathematical coordinate rotations. The magnitudes of the coordinate rotations are determined by the components of the wind vector in each 30-minute averaging period.

The wind vector is composed of three time-averaged components (u, v, w) in three initial coordinate directions (x, y, z). Using a bubble level, direction z initially was approximately oriented with respect to gravity, and the other two directions were arbitrary. Tanner and Thurtell [47] and Baldocchi and others [1] outline a procedure in which measurements made in the initial coordinate system are transformed into values consistent with the natural coordinate system. First, the coordinate system is rotated by an angle η about the z-axis to align u along a transformed x-direction on the x–y plane. Next, rotation by an angle θ is performed about the y-direction to align w along a transformed z-direction. These rotations result in a natural coordinate system with mean values of wind speed along the transformed y and z axes equal to zero and the mean airstream pointed directly along the transformed x axis.

The coordinate rotation-transformed covariances needed to compute turbulent fluxes are described in Eq. (10), where:

$(\overline{w'c'})_r$ = is the rotated covariance;

c' = is the fluctuation in either vapor density (ρ_V), or virtual temperature (T_S); and

$\overline{w'c'}$, $\overline{u'c'}$, and $\overline{v'c'}$ = are covariances measured in the original coordinate system.

The [$\cos \theta$], [$\sin \theta$], [$\sin \eta$], and [$\cos \eta$] are defined in Eqs. (11)–(14), respectively. The presence of the tower and the anemometer produced spurious turbulence, which possibly impacted measured velocity fluctuations, particularly when the wind was from the tower-side of the sensor. Turbulent flux data for which "the inferred mis-leveling angle θ greater than $10°$" were excluded based on the assumption that spurious turbulence was the cause of the excessive amount of coordinate rotation.

15.2.5 CONSISTENCY OF MEASUREMENTS WITH ENERGY BUDGET

Previous investigators [3, 15, 17, 23, 29, 44, 50] have described a recurring problem with the eddy correlation method: A common discrepancy of the measured latent and sensible heat fluxes with the energy-budget equations (Eq. (1)). The usual case is that measured turbulent fluxes $(H + \lambda E)$ are less than the measured available energy $(R_n - S)$. Bidlake et al. [3] accounted for only 49 and 80% of the measured available energy with measured turbulent fluxes $(H + \lambda E)$ at cypress swamp and pine flat- wood sites, respectively. Turbulent fluxes measured above a coniferous forest by Lee and Black [23] accounted for only 83% of available energy. Several researchers [15, 17, 29] have shown that the eddy correlation method performs best in windy conditions (relatively high friction velocity, u^*). Friction velocity is directly proportional to wind speed, but also incorporated rates the frictional effects of the plant canopy and land surface on the wind and the effects of atmospheric stability [8]. Friction velocity can be computed with three dimensional sonic anemometer measurements of velocity fluctuations as [43] shown in Eq. (15).

Goulden and others [17] concluded that eddy correlation-measured values of carbon flux from a forest were underestimated when u^* was less than 0.17 m/s. German [15] noted that at u^* greater than 0.3 m/s, little discrepancy existed between measured available energy and measured turbulent fluxes. Possible explanations for the observed discrepancy between the measured turbulent fluxes and the measured available energy include: a sensor frequency response that is insufficient to capture high-frequency eddies; an averaging period insufficient to capture low-frequency eddies, resulting in a nonzero mean wind speed perpendicular to the airstream; drift in the absolute values of anemometer and hygrometer measurements resulting in statistical nonstationarity within the averaging period; lateral advection of energy; and overestimation of available energy. Lateral advection of energy is not a likely explanation because most of the studies reporting underestimation of turbulent fluxes were conducted at sites with adequately extensive surface covers. Measurement of the soil heat flux and storage terms of the available energy can be problematic, given the difficulty in making representative measurements of these terms; however, the turbulent flux underestimation occurs even with a daily composite of fluxes (in which case these terms generally are negligible).

Likewise, overestimation of net radiation seems unlikely, given the relative simplicity and laboratory calibration of net radiometers. For these reasons, it was assumed in this study that the available energy was accurately measured and that any error in energy-budget closure was associated with errors in measurement of turbulent fluxes. Moore [29] also noticed an under-estimation of turbulent fluxes and suggested that this under-estimation would likely apply equally to each of the turbulent fluxes (sensible and latent heat flux), leading to the conclusion that the ratio of the fluxes can be measured adequately. This assumption seems reasonable, given that the same turbulent eddies transport both sensible and latent heat, and therefore, any eddies that are missed by the instrumentation because of anemometer response or averaging period would have a proportionally equal effect on both turbulent fluxes. German [15] provided empirical support for this assumption at a saw-grass site in south Florida where simultaneous measurement of the ratio of fluxes was based on two approaches: the eddy correlation method (using instrumentation identical to that used in the present study) and the measurement of temperature and vapor pressure differentials between vertically separated sensors [4]. These independent approaches for estimating the ratio of turbulent fluxes were in reasonable agreement during the daylight hours when evapotranspiration predominated. Assuming that the ratio of turbulent fluxes is adequately measured by the eddy correlation method, the energy budget equation (Eq. (1)), along with turbulent fluxes (H and λE) measured using the standard eddy correlation technique, can be used to produce corrected (H_{cor} and λE_{cor}) turbulent fluxes in an energy-budget variant of the eddy correlation method, as shown in Eq. (16). As shown below, we get Eq. (19) for H_{cor}, after introducing the Bowen ratio [Eq. (17)].

$$u^* = \sqrt{\sqrt{\overline{u'c'}^2 + \overline{v'w'}^2}} \tag{15}$$

$$R_n - G - S = H_{cor} + \lambda E_{cor} = \lambda E_{cor}(1 + B) \tag{16}$$

Bowen ratio is defined in Eq. (17):

$$B = \frac{H}{\lambda E} \tag{17}$$

Rearranging Eq. (16), we get:

$$\lambda E_{cor} = \frac{R_n - G - S}{1 - B} \tag{18}$$

Combining Eqs. (17) and (18):

$$H_{cor} = R_n - G - S - \lambda E_{cor} \qquad (19)$$

Instrumentation was installed at the evapotranspiration station to provide estimates of soil heat flux (G) and changes in stored energy (S) in the biomass and air. Soil heat flux at a depth of 8 cm was measured at two representative locations using soil heat-flux plates. An estimate of the soil heat flux at land surface was computed based on the estimated change in stored energy in the soil above the heat flux plates. The changes in stored energy in the soil above the heat flux plates were estimated based on thermocouple-measured changes in soil temperature and estimates of soil heat capacity. The estimates of soil heat capacity were based on mineralogy, soil bulk density, and soil moisture content. Soil moisture content was measured using time-domain reflectometry (TDR) probes placed within the upper 8 cm of soil. Thermocouples were installed at multiple locations within the trunks of representative trees to allow for estimation of changes in storage of energy within the biomass. Estimates of biomass density (based on tree surveys) and biomass heat capacity (available from previous studies) also are required for calculation of changes in biomass-stored energy. Changes in storage of energy in the air generally are small in comparison with soil heat flux and biomass heat storage, but were estimated based on measurement of the temperature and relative humidity profile below the turbulent flux sensors. With the exception of the temperature and relative humidity sensors, all of the instrumentation intended to provide data to estimate soil heat flux and changes in stored energy was destroyed by earth-moving equipment used to construct a fire break around the evapotranspiration station a few hours before a fire passed through the area of the station.

Energy generally enters the soil surface and is stored in the biomass and air during the day and released at night. It was assumed that soil heat flux and changes in energy storage in the biomass and air were negligible over a diurnal cycle. This facilitated the evaluation of Eqs. (18) and (19), using daily composites of terms in these equations. This approach allowed for neglect of those terms of the energy budget that were not measured as a result of firedamaged instrumentation.

During periods of rapid temperature changes (e.g., cold front passage), however, the net soil heat flux and the net change in energy stored in the biomass and air over a diurnal cycle may not be negligible. As mentioned previously, problems such as scaling of hygrometer windows, moisture on anemometer or hygrometer, or excessive coordinate rotation can result in missing 30-minute turbulent flux data. These data must be estimated prior to construction of daily composites of turbulent fluxes. In the present study, regression analysis of measured turbulent flux data and photosynthetically active radiation (PAR) was used to estimate unmeasured values of turbulent fluxes. These regression-estimated values of turbulent fluxes are not as reliable as measured values. Therefore, the fraction of daily composited turbulent flux data derived from regression estimates was limited to 25% (up to 6 h per day). The procedure outlined above for culling, estimating, and compositing 30-minute turbulent flux data still resulted in missing values for some days.

15.2.6 SIMULATION OF EVAPOTRANSPIRATION

An evapotranspiration model was developed for estimating daily values of evapotranspiration representative of both burned and unburned areas. Post-fire measurements of evapotranspiration generally reflected a composite of evapotranspiration from burned and unburned vegetation. A model was developed that reflected the mixture of source area characteristics and allowed calculation of the evapotranspiration from each source area.

15.2.7 EVAPOTRANSPIRATION MODELS

The eddy correlation instrumentation can have extended periods of inoperation, as discussed previously. However, more robust meteorological and hydrologic instrumentation (sensors for measurement of net radiation, air temperature, relative humidity, PAR, wind speed, soil moisture, and water-table depth) can provide nearly uninterrupted data collection. Evapotranspiration models, calibrated to measure turbulent flux data and based on continuous meteorological and hydrologic data, can provide continuous

estimates of evapotranspiration. Evapotranspiration models also can provide insight into the cause-and-effect relation between the environment and
evapotranspiration.

Physics-based evapotranspiration models generally rely on the work
of Penman [32], who developed an equation for evaporation from wet
surfaces based on energy budget and aerodynamic principles. Penman
equation has been used to estimate evapotranspiration from well-watered,
dense agricultural crops (reference or potential evapotranspiration). In
Penman's equation, the transport of latent and sensible heat fluxes from
a "big leaf" to the sensor height is subject to an aerodynamic resistance.
The big leaf assumption implies that the plant canopy can be conceptual-
ized as a single source of both latent and sensible heat at a given height
and temperature. Inherent in the Penman approach is the assumption of a
net one-dimensional, vertical transport of vapor and heat from the canopy.
The Penman equation is shown in Eq. (20).

$$\lambda E = \frac{\Delta(R_n - G - S) + \frac{\rho C_p(e_s - e)}{r_a}}{\Delta + \gamma} \tag{20}$$

$$\lambda E = \frac{\Delta(R_n - S)}{\Delta + \gamma} \quad \text{for } e_s = e \tag{21}$$

$$\lambda E = \alpha \frac{\Delta(R_n - S)}{\Delta + \gamma} \quad \text{with Priestley-Taylor coefficient, } \alpha \tag{22}$$

$$\lambda E = (1 - w_b)\lambda E_u + w_b \lambda E_b \tag{23}$$

In Eq. (20): λE is latent heat flux, in watts per square meter; Δ is slope
of the saturation vapor-pressure curve, in kilopascals per degree Celsius;
G is soil heat flux at land surface, in watts per square meter; S is change
in storage of energy in the biomass and air, in watts per square meter; C_p is
specific heat capacity of the air, in joules per gram per degree Celsius; e_s is
saturation vapor pressure, in kilopascals; e is vapor pressure, in kilopascals;
r_h is aerodynamic resistance, in seconds per meter; and γ is the psychrometric constant = approximately 0.067 kilopascals per degree Celsius, but
varying slightly with atmospheric pressure and temperature. The first term
is known as the energy term; the second term is known as the aerodynamic

term. The Eq. (21) is simplification of the Penman equation for the case of saturated atmosphere ($e = e_s$), for which the aerodynamic term is zero.

However, Priestley and Taylor [34] noted that empirical evidence suggests that evaporation from extensive wet surfaces is greater than this amount, presumably because the atmosphere generally does not attain saturation. Therefore, the Priestley-Taylor coefficient [, Eq. (22)] was introduced as an empirical correction to the theoretical expression [Eq. (21)]. This formulation assumes that the energy and aerodynamic terms of the Penman equation are proportional to each other. The value of has been estimated to be 1.26, which indicates that under potential evapotranspiration conditions, the aerodynamic term of the Penman equation is about 21% of the total latent heat flux. Eichinger and others [10] have shown that the empirical value of has a theoretical basis: A nearly constant value of is expected under the existing range of Earth-atmospheric conditions.

Previous studies [13, 41, 44] have applied a modified form of the Priestley-Taylor equation. The approach in these studies relaxes the Penman assumption of a free-water surface or a dense, well-watered canopy by allowing α to be less than 1.26 and to vary as a function of environmental factors. The Penman-Monteith equation [27] is a more theoretically rigorous generalization of the Penman equation that also accounts for a relaxation of the these Penman assumptions. However, Stannard [41, 42] noted that the modified Priestley-Taylor approach to simulation of observed evapotranspiration rates was superior to the Penman-Monteith approach for a sparsely vegetated site in the semiarid rangeland of Colorado. Similarly, Sumner [44] noted that the modified Priestley-Taylor approach performed better than did that of Penman-Monteith for a site of herbaceous, successional vegetation in central Florida. Therefore, the modified Priestley-Taylor approach was chosen for the present investigation.

15.2.8 PARTITIONING OF MEASURED EVAPOTRANSPIRATION

An evapotranspiration model (daily resolution) was developed to partition the measured evapotranspiration into two components characteristic of the primary types of surface cover (burned and unburned) of the watershed during the study period. As mentioned previously, upland areas were more likely to have been burned during the June-July 1998 fires than wetland areas. Therefore, to some extent, the model results also reflect the variation between upland and wetland evapotranspiration. The model was of the form, as shown in Eq. (23), where: λE is measured latent heat flux at the station, in watts per m²; w_b is the fraction of the measured latent heat flux originating from burned areas, dimensionless; λE_u is latent heat flux from unburned areas, in watts per m²; and λE_b is latent heat flux from burned areas, in watts per m².

$$f_i = \frac{\sum_{i=1}^{48} PAR_k \delta_i \psi_k}{\sum_{i=1}^{48} PAR_k} \qquad (24)$$

$$w_b = \sum_{i=I}^{IV} g_i f_i \qquad (25)$$

In Eq. (24), the weighting coefficient (w_b) for a given day must incorporate the spatial distribution of surface cover types near the point of flux measurement (Fig. 15.3 and Table 15.2), the changing (upwind) source area for the measurement associated with changes in wind direction, and the diurnal changes in evapotranspiration. If the relative fraction of burned surface cover in the upwind source area remained constant for a given day (i.e., the wind direction remained from a given zone of a relatively uniform mixture of surface cover types), w_b would be simply the fraction of burned surface cover within the zone. Also, if evapotranspiration from each surface cover type remained constant during a given day, w_b would be simply the time-weighted average of the fraction of burned surface cover within the upwind source areas. However, intraday changes in source area composition, associated with changes in wind direction, and the strong diurnal cycle in evapotranspiration had to be considered during computation of day-by-day values of w_b. For example, suppose that the wind were from the west during the night and from the east during the day. In this situation, the measured daily evapotranspiration would be much more representative of the surface cover to the east because day- time evapotranspiration generally is much higher than nighttime evapotranspiration. Strong diurnal biases in wind direction (Fig. 15.8) exist in the study area, which can lead to situations such as that described. Therefore, weighting coefficients must reflect these diurnal patterns in evapotranspiration.

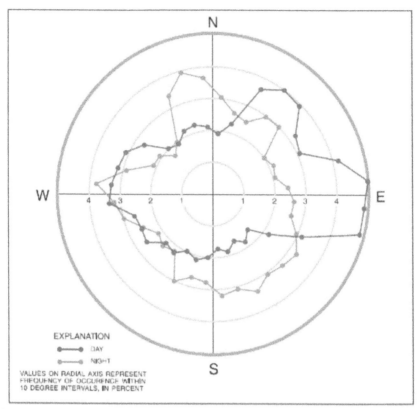

FIGURE 15.8   Wind direction frequency pattern at location of evapotranspiration station.

The diurnal pattern of evapotranspiration during a given day generally is strongly correlated with the diurnal pattern of incoming radiation, as can be inferred from the Priestley-Taylor equation [Eq. (22)] or seen empirically [44]. *PAR* was used as a surrogate for the factors that produce intraday variations in evapotranspiration for both surfaces cover types. Nighttime *PAR* is equal to zero, implying that only daytime winds from a given zone are assumed to contribute to the measured latent heat flux for a given day. Other factors (such as variations in air temperature) that contribute to the diurnal pattern of evapotranspiration were considered minor, compared to the effect of *PAR*, and were not considered in the determination of weights for use in the Eq. (23). The computation for the day-by-day values of w_b is derived in Appendix I at the end of this chapter and is shown in Eq. (24), where: g_i is the fractional contribution of burned area within burn zone i to the measured latent heat flux when wind direction is from burn zone i (Table 15.2); i is an index for the burn zones (Fig. 15.3); and f_i is the *PAR*-weighted fraction of the day that wind direction is from burn zone i and is computed using Eq. (25).

In Eq. (25): k is an index for the 48 measurements of 30-minute averages within a given day; $\delta_i(\Psi_\kappa)$ is a binary function equal to 1, if Ψ_κ is within burn zone i and otherwise equals 0; PAR_k is the measured *PAR* for time period k within a given day; and Ψ_κ is the wind direction for time period k within a given day.

In the evapotranspiration model [Eq. (23)]: Both λE_u and λE_b are simulated by the modified Priestley-Taylor equation [Eq. (22)] with individual Priestley-Taylor α functions. The α function for λE_u was assumed to remain unchanged throughout the 2-year study period; however, the α function for λE_b was divided into multiple time periods to reflect the radical change in surface cover of the burned areas following the fire, logging, and regrowth of vegetation. The measurements of average, daily evapotranspiration provided a standard with which to calibrate the Priestley-Taylor evapotranspiration model. Calibration of the Priestley- Taylor model involved quantification of the functional relations between the Priestley-Taylor's α and environmental variables. This quantification was achieved through identification of the form of the functional relation (trial-and-error approach) and estimation of the parameters of that relation (regression analysis) that produced optimal correspondence between measured and simulated values of latent heat flux.

The form of the calibrated model [Eq. (23)] allowed for evapotranspiration to be estimated for any mix of burned and unburned areas through appropriate specification of w_b. Daily values of evapotranspiration for burned and unburned areas were estimated with w_b equal to 1 and 0, respectively. Evapotranspiration from the watershed was estimated with w_b equal to 0 and 0.4 (burned fraction of watershed) prior to and following the fires, respectively. The potential evapotranspiration from the watershed was estimated with similar weighting, but with a Priestley-Taylor α equal to a constant value of 1.26.

15.2.9   MEASUREMENT OF ENVIRONMENTAL VARIABLES

Meteorological, hydrologic, and vegetative data were collected in the study area for several reasons: (i) as ancillary data required by the energy- budget variant of the eddy correlation method, (ii) as independent variables within the evapotranspiration model, and (iii) to construct a water budget for the Tiger Bay watershed. Meteorological variables monitored included net radiation, air temperature, relative humidity, wind speed, and *PAR*. These data were recorded by data loggers at 15-second intervals, using instrumentation summarized in Table 15.1, and the resulting 30-minute means were stored.

Two net radiometers, each deployed at a height of 35 m, provided redundant measurements of net radiation at the evapotranspiration station. Measured values of net radiation were corrected for wind-speed effects as suggested by the instrument manual for the Radiation and Energy Balance Systems, Inc., Model Q-7.1 net radiometer. In late 1999, missing net radiation data necessitated an estimate of net radiation based on a regression of *PAR* and net radiation. *PAR* consists of that part of incoming solar radiation that is used in plant photosynthesis and is highly correlated with incoming solar radiation. Based on data collected during 1993–1994 in Orange County – Florida, solar radiation (in watts per m^2) can be approximated (standard error of estimate = 11 watts per m^2) as 0.49 times *PAR* (in micromoles per second per m^2).

The source area of the net radiation measurement was estimated by using the approach of Reifsnyder [35] and Stannard [42]. The measurement of net radiation had a much smaller source area than the turbulent flux measurement (Fig. 15.7). About 90% of the source area for the net radiometers was within a radial distance of 55 m (prelogging) or 110 m (post logging). Therefore, the source area for the net radiometer in the near-vicinity of the evapotranspiration station was one of the following: (i) pine plantation (prelogging), (ii) burned pine plantation (postfire, but prelogging), or (iii) clear-cut, with understory regrowth (postlogging). Other covers also existed within the watershed, primarily wetlands and unburned pinelands. Lacking net radiation measurements over more than one cover, the assumption was made that net radiation measured at the unburned pine plantation was representative of all unburned surface covers. The period of record prior to the fire (the initial 175 days of 1998) was used to develop a regression-based predictor of net radiation as a function of *PAR*. This relation was used to estimate net radiation in unburned areas following the burning of the area around the evapotranspiration station. The net radiation measured at the evapotranspiration station following burning was assumed to be representative of all burned areas. Logging of the burned area near the evapotranspiration station occurred during a period of extensive logging through- out the watershed. Some error is introduced to the estimation of net radiation over burned areas because the logging was not simultaneous for all burned areas and because the logging over burned areas was not complete (as mentioned previously, two-thirds of the burned forest within Tiger Bay State Forest was logged). Estimates of daily net radiation for burned and unburned areas were composited as shown in Eq. (26) into a value consistent with the turbulent flux measurements [Eqs. (18) and (19)] using the weighting coefficient (w_b) previously defined [Eq. (24)].

$$R_N = (1 - w_b)\, R_{nu} + w_b\, R_{nb} \tag{26}$$

$$NDVI = \frac{NIR - Vis}{NIR + Vis} \tag{27}$$

In Eq. (26): R_n is composited net radiation, in watts per square meter; R_{nu} is net radiation for unburned areas, in watts per square meter; and R_{nb} is net radiation for burned areas, in watts per square meter.

A regression between postlogging, daily values of net radiation and *PAR* was used to estimate net radiation from burned and logged surfaces during the latter part of 1999 after net radiometer domes were damaged, perhaps by birds. Vegetation within the study area was mapped previously by Volusia County Department of Geographic Information Systems [56, 57] and Simonds and others [40]. Post-fire, infrared, aerial photographs were used to identify the areal distribution of burned vegetation in the watershed.

Temporal variations in vegetation were documented with monthly photographs taken from the tower at the evapotranspiration station and with normalized difference vegetation index (NDVI) data. NDVI data were provided by the USGS Earth Resources Observation Systems (EROS) Data Center through analysis of the Advanced Very High Resolution Radiometer (AVHRR) data [11, 52, 54] from operational National Oceanic and

Atmospheric Administration (NOAA) polar-orbiting satellites. NDVI is defined in Eq. (27), where: NIR is near-infrared reflectance measured in AVHRR band 2 (725–1100 nanometers); and Vis is visible reflectance measured in AVHRR band 1 (580–680 nanometers).

NDVI is highly correlated with the density of living, leafy vegetation. The physical basis for this correlation is the sharp contrast in the absorptivity of visible and near-infrared radiation by leaves, which absorb approximately 85% of incident visible radiation, but only 15% of near-infrared radiation [8]. Other ground covers (dead plant material, soil, and water) do not exhibit this extreme spectral differential in absorption. The AVHRR-computed NDVI data are provided at 2-week and 1-kilometer (km) by 1-km resolution. For the present study, NDVI data, within a 3-km by 3-km square and approximately centered on the location of the evapotranspiration station, were composited to quantify temporal trends in the density of living, leafy vegetation in the vicinity of the turbulent flux measurements during the study period.

Air temperature and relative humidity were monitored at the evapotranspiration station at heights of 1.5, 9.1, 18.3, and 35 m. The slope of the saturation vapor pressure curve (a function of air temperature) and vapor pressure deficit were computed in the manner of Lowe [25] using the average of air temperature and relative humidity values measured at these four heights. A propeller-type anemometer to monitor wind speed and direction and an upward-facing quantum sensor to measure incoming *PAR* were deployed at a height of 35 m at the evapotranspiration station.

Hydrologic variables that were monitored included precipitation, water-table depth, stream discharge, and soil moisture. Precipitation records were obtained from a tipping bucket rain gage mounted at a height of about 18.3 m at the evapotranspiration station and from two storage rain gages installed in forest clearings and monitored weekly (Fig. 15.9). Spatial variability in annual rainfall can be substantial within Volusia County, based on the long-term NOAA stations at DeLand and Daytona Beach (Fig. 15.9). The Daytona Beach area, on average, receives about 15% less annual rainfall than does the DeLand area [30, 31]. The uncertainty associated with the rainfall distribution between these two stations precluded the use of both stations for estimation of rainfall to the Tiger

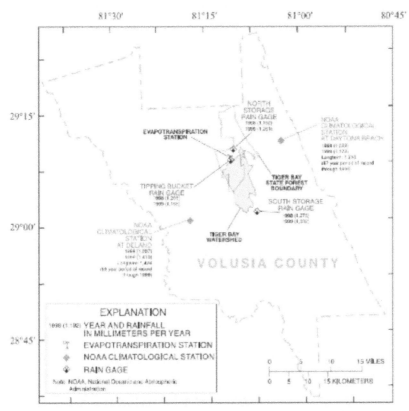

FIGURE 15.9   Location of rain gages in vicinity of Tiger Bay watershed.

Bay watershed during the study period. Rather, the rainfall totals from the two storage rain gages located near the watershed were averaged to provide estimates of rain- fall to the watershed. Tipping bucket rain gages can underestimate rainfall, particularly during high-intensity events; therefore, the tipping bucket gage monitored at the evapotranspiration station was used primarily to provide a high-resolution description of the temporal rainfall pattern, and the storage rain gages were used primarily to estimate cumulative rainfall.

Water-table depth was monitored at two surficial- aquifer system wells at opposite ends of the watershed. Water-level measurements were obtained at 30-minute intervals using a pressure transducer in the north well (USGS site identification number 2908130811111801), located at the evapotranspiration station. The south well (USGS site identification number 290119081074001), at the location of the south storage rain gage (Fig. 15.1), was measured weekly using an electric tape. Although the two wells monitored were located at opposite ends of the watershed (Fig. 15.1), both wells were within similar upland settings. Although the water-table depth in wet- land areas would be expected to be less than that measured in upland wells, water levels are expected to change at the same rate in the low relief environment of this watershed. Therefore, changes in the measured upland water-table depths can be regarded as indicators of changes in the representative water-table depth of the watershed.

Daily values of stream discharge for the only surface-water outflow from the Tiger Bay watershed, Tiger Bay canal near Daytona Beach (Fig. 15.1; USGS station number 02247480), were obtained from the USGS database [51, 53, 55]. Soil moisture at two representative locations at the evapotranspiration station was monitored using time-domain reflectometry (TDR) probes installed to provide an averaged volumetric soil moisture content within the upper 30 cm of the soil. Soil moisture measurements were made and recorded on the data logger every 30 min. The TDR probes were damaged by a fire in late June 1998, but were replaced in early August 1998. The soil moisture measurements made at the evapotranspiration station probably are indicative of only the uplands; wetlands commonly are inundated at times when shallow upland soils are not.

15.3   RESULTS AND DISCUSSION: EVAPOTRANSPIRATION MEASUREMENT AND SIMULATION

Most (73%) of the 30-minute resolution eddy correlation measurements made during the 2-year study period were acceptable and could be used to develop an evapotranspiration model to estimate missing data and to discern the effects of environmental variables on evapotranspiration. Unacceptable measurements resulted from failure of the krypton hygrometer or sonic anemometer, or because of excessive (more than 10 degrees) coordinate rotation in the postprocessing "leveling" of the anemometer data. Unacceptable data were most extensive in the evening and early morning hours (Fig. 15.10) because dew formation on the sensors during these times of day was common. This diurnal pattern of missing data was fortunate because turbulent fluxes are expected to be relatively small during the evening and early morning, when solar radiation is low. Missing data were estimated based on linear regression between the turbulent fluxes and *PAR* (Figs. 15.11 and 15.12). Because *PAR* is zero at night, this approach assigned constant values of latent and sensible heat flux to missing nighttime data. The assumed constant value of nighttime latent heat flux assigned to missing data was 9.04 watts per m^2 (Fig. 15.11). This value generally was small relative to daytime values of latent heat flux, and therefore, not significantly inconsistent with the assumption of negligible nighttime latent heat flux inherent in the development of weighting coefficients [Eqs. (23)–(25)]. Examples of measured and *PAR*-estimated turbulent fluxes are shown for a period in late February 1998 in Fig. 15.13.

Turbulent flux data exhibited pronounced diurnal patterns. The average diurnal pattern of turbulent fluxes and *PAR* (Fig. 15.14) indicates that the vast majority of evapotranspiration occurs in daytime, driven by incoming solar radiation. During average daytime conditions, both latent and sensible heat flux are upward, with most of the available energy partitioned to latent heat flux. At night, the land or canopy surface cools below air temperature, producing a reversal in the direction of sensible heat flux (Fig. 15.14). Although the average, nighttime latent heat flux is upward (Fig. 15.14), dew formation (downward latent heat flux) commonly occurs.

The relation between net radiation and *PAR* varied as a result of the fire, logging, and regrowth. Regressions between daily values of net radiation and *PAR* are shown in Fig. 15.15 for three periods: prefire, postfire/prelogging, and postlogging. The measured and estimated values of daily net radiation for burned and unburned areas are shown in Fig. 15.16. Measured values of *PAR*, a quantity highly correlated with incoming solar radiation, are shown in Fig. 15.17. The strong seasonality of net radiation evident in Fig. 15.16 was a consequence of the yearly solar cycle, which produces a sinusoidal input of solar radiation to the upper atmosphere. Deviations from the sinusoidal pattern (such as during September–October 1999) were largely the result of cloudy conditions that produced periods of low *PAR*. The cloudy and rainy period immediately after the fire resulted in relatively low values of *PAR* and low estimated values of net radiation in unburned areas. The measured (burned) net radiation, however, was relatively high,

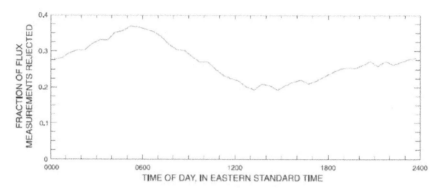

FIGURE 15.10   Diurnal pattern of rejected flux measurements.

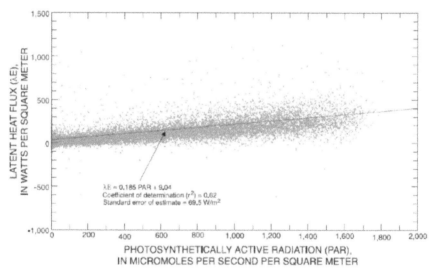

FIGURE 15.11   Relation between measured 30-minute averages of photosynthetically active radiation and latent heat flux (λE).

indicating that the surface reflectance of burned areas decreased markedly after the fire blackened much of the landscape. The measured net radiation for burned areas was about 20% higher than the estimated net radiation for unburned areas in the 6 months following the June 1998 fire. With the regrowth of vegetation, reflectance gradually increased to near prefire

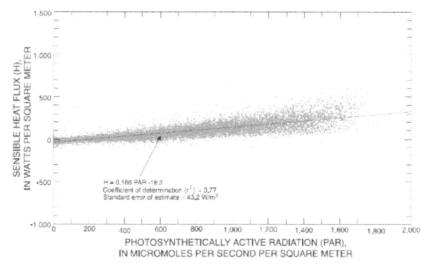

FIGURE 15.12   Relation between measured 30-minute averages of photosynthetically active radiation and sensible heat flux (H).

values in the postlogging period, and the differences between values of net radiation for burned and unburned areas were less distinct.

As described previously, daily composites of measured turbulent fluxes were constructed with the restriction that no more than 6 h of data for a given day could be missing and subject to estimation using the gross PAR-based relations (Figs. 15.11 and 15.12). This restriction limited the number of acceptable daily values of measured turbulent fluxes to 449 during the 2-year (730 days) study period. Only a small amount of the total turbulent flux (5.6 and 5.1% for latent and sensible heat flux, respectively) comprising the acceptable daily values was estimated by the PAR-based relation. As expected from previous studies, the available energy tended to be greater (measured turbulent fluxes accounted for only about 84.7% of estimated available energy) than the turbulent fluxes derived from the standard eddy correlation method (Fig. 15.18), and the energy-budget closure tended to improve with increasing friction velocity (Fig. 15.19). The measured turbulent fluxes generally accounted for estimated available energy at friction velocity values greater than about 0.6 m/s. The acceptable daily values of turbulent fluxes, computed by both the standard eddy correlation method [Eqs. (5) and (6)] and the energy-budget variant of the

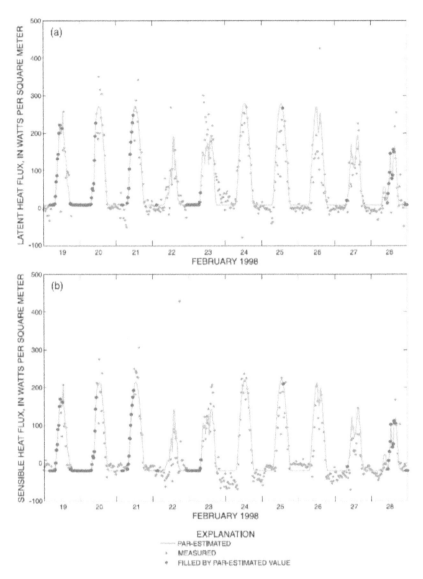

FIGURE 15.13 Measured and photosynthetically active radiation-estimated values of (a) latent heat flux and (b) sensible heat flux during 10-day period in late February 1998.

eddy correlation method [Eqs. (18) and (19)], are presented in Figs. 15.20 and 15.21.

These values represent the fluxes measured at the evapotranspiration station, and therefore, represent varying proportions of burned and

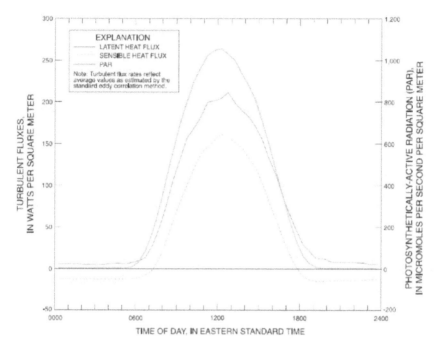

FIGURE 15.14   Average diurnal pattern of energy fluxes and photosynthetically active radiation.

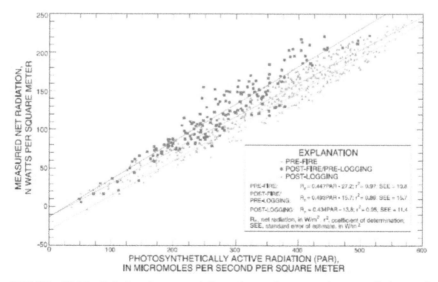

FIGURE 15.15   Relation between daily values of measured net radiation and photosynthetically active radiation.

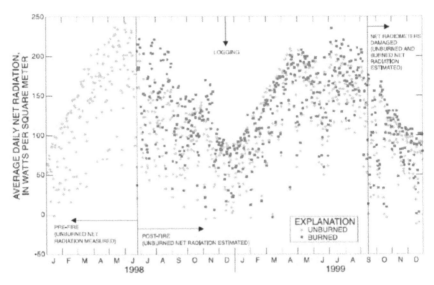

FIGURE 15.16   Average daily net radiation for burned and unburned areas.

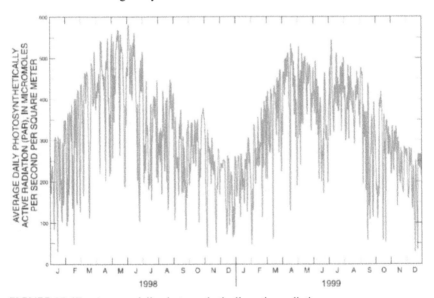

FIGURE 15.17   Average daily photosynthetically active radiation.

unburned source areas. The relative proportions varied widely following the fire (Fig. 15.22), with values ranging from those that were almost completely representative of unburned areas ($w_b = 0$) to those with 80% representative of burned areas ($w_b = 0.8$).

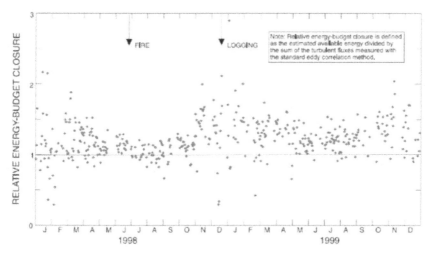

FIGURE 15.18   Temporal distribution of daily relative energy-budget closure.

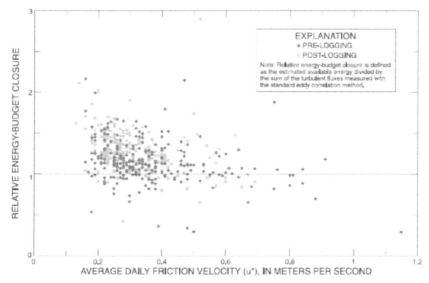

FIGURE 15.19   Relation between daily energy-budget closure and average daily friction velocity.

As a consequence of the previously mentioned discrepancy between available energy and measured turbulent fluxes, the standard eddy correlation method produced turbulent flux values that were, on average, only 84.7% of those produced by the energy-budget variant.

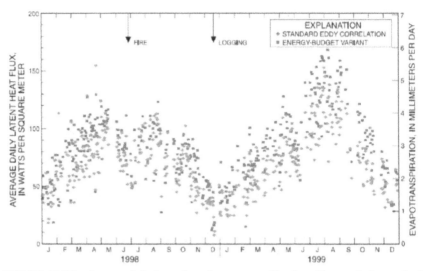

FIGURE 15.20 Average daily latent heat flux measured by the eddy correlation method and the energy-budget variant.

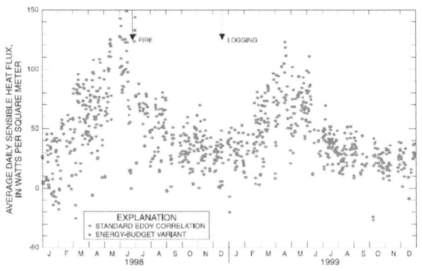

FIGURE 15.21 Average daily sensible heat flux measured by the eddy correlation method and the energy-budget variant.

15.3.1 CALIBRATION OF EVAPOTRANSPIRATION MODEL

Calibration of the evapotranspiration model was essentially a process of determining the best functional form of the modified Priestley-Taylor coefficient, α. The environmental variables considered as possible predictors of Priestley-Taylor's α (Eq. (22)) included: water-table depth, soil moisture, PAR, air temperature, vapor-pressure deficit, daily rainfall, NDVI, and wind speed. Of these variables, only water-table depth, soil moisture, and PAR were identified as significant determinants of Priestley-Taylor's α. Soil moisture was highly correlated with water-table depth (Fig. 15.23), and there fore, one of these variables can be excluded from the α function to avoid redundancy. To enhance the transfer value of this study, water-table depth was retained as a variable in the α function, and soil moisture was eliminated, because water-level data are more commonly available than soil moisture data. In other environmental settings, such as areas with a relatively deep water table or coarse-textured soils, the water table may be hydraulically de-coupled from the shallow soil moisture much of the time, and a

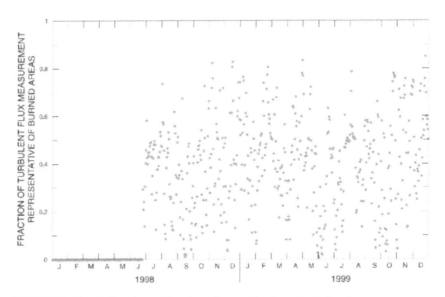

FIGURE 15.22 Daily values of fraction of burned fraction of turbulent flux measurement.

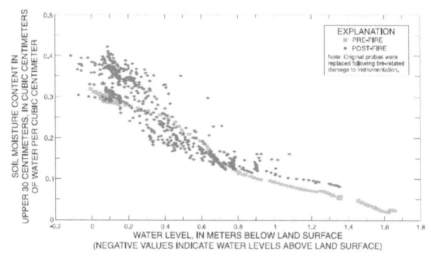

FIGURE 15.23   Relation between soil moisture content and water level.

different functional representation of α than was used in this study would be appropriate.

Priestley-Taylor's α was initially simulated with a three-part model incorporating the three different surface covers: (i) unburned areas; (ii) postfire/pre logging, burned areas (June 25 to December 16, 1998); and (iii) postlogging, burned areas (December 17, 1998 to December 31, 1999). The time divisions for the burned areas grossly approximated the observed variation in NDVI over the study period (Fig. 15.24). The effects of the fire and transient regrowth of vegetation (Fig. 15.4) on NDVI were evident (Fig. 15.24). In the almost 6 months prior to the fire (January 1–June 24, 1998), NDVI maintained a relatively constant value of about 0.5. NDVI sharply declined at the time of the fire, but recovered within 4 months to a value of about 0.4, which was maintained throughout the remainder of the study. As a simplification, the effect of the transient aspect of vegetative regrowth within the 4-month recovery period was not incorporated into the model for α. Instead, the function of α for this recovery period, as for all time periods, was a function solely of water-table depth and *PAR*.

Surprisingly, the annual pattern of leaf growth and drop for the deciduous cypress trees within the watershed was not apparent in values of NDVI,

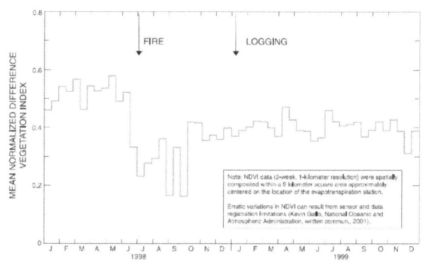

FIGURE 15.24   Temporal variability of normalized difference vegetation index (NDVI).

perhaps because of the exposure of understory vegetation following leaf drop. Simulations that attempted to use NDVI directly as an explanatory variable for variations in evapotranspiration were unsuccessful. This failure is perhaps related to erratic variations in NDVI (Fig. 15.24), which are a product of sensor and data registration limitations (Kevin Gallo, NOAA, written communication, 2001).

An analysis of error in the preliminary model showed a seasonal pattern in the residuals (difference of measured and simulated latent heat fluxes) within the postlogging period (Fig. 15.25). Measured evapotranspiration generally was overestimated in the early part of this period and underestimated in the late part of the period. The bias was apparently unrelated to changes in green leaf density, based on the relatively constant value of NDVI following logging (Fig. 15.24). Possible explanations for the model bias include factors not clearly identified by NDVI: phenological changes associated with maturation or seasonality of plants that emerged after the fire or successional changes in composition of the plant community within burned areas. To reflect the apparent change in system function during the postlogging period, this period was further subdivided into an early period (December 17, 1998 through April 22, 1999) and a late period (April 23 through December 31, 1999). This subdivision of the

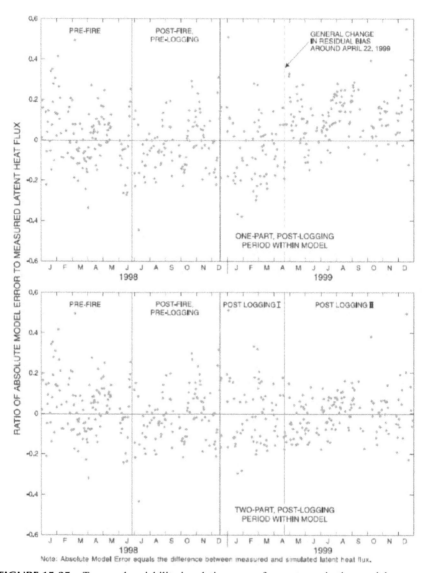

FIGURE 15.25 Temporal variability in relative error of evapotranspiration model.

postlogging period resulted in an improved model (standard error of estimate = 9.67 watts per m²), compared to the model with a single postlogging period (standard error of estimate = 10.82 watts per m²) and
reduced the seasonal bias in residuals (Fig. 15.25).

The general form of α was identical for all surface covers (Eq. (28)), although model parameter values varied with surface cover (Table 15.3):

$$\alpha_j = C_{1j} \, h_{wt} + C_{2j} PAR + C_{3j} \tag{28}$$

In Eq. (28): α_j is the Priestley-Taylor coefficient for the j-th surface cover; j is an index denoting the surface cover; $j = 1$ (unburned areas); $j = 2$ (burned areas during postfire/prelogging period; $j = 3$ (burned areas during initial postlogging period); and $j = 4$ (burned areas during final post logging period); C_{1j}, C_{2j}, and C_{3j} are empirical parameters that are estimated through regression, within the context of Eqs. (22)–(25); and h_{wt} is water table depth below a reference level placed at the highest water level measured (0.11 meters above land surface) at the evapotranspiration station (uplands environment) during the study period, in meters. h_{wt} is constrained to be greater than zero.

Regressions to estimate the model parameters within Eq. (28) were designed to minimize the sum of squares of error residuals between measured and simulated latent heat fluxes. Measured latent heat flux was used as the dependent variable of the regression; the right side of Eq. (22) contained the independent variables, as well as the unknown parameter (C_{1j}, C_{2j}, and C_{3j}; and $j = 1$–4). The values of λEu and λEb were estimated with Eq. (22), using the appropriate values of net radiation (R_{nu} and R_{nb} of Eq. (26) for λEu and λEb, respectively), and Eq. (28). The variable w_b was estimated with the Eqs. (24) and (25).

The form of α used in this study is similar to that used by German [15] for south Florida wetlands, where water level and incoming solar radiation were the sole determinants of α. In the study by German, however, the form of α involved both first and second order terms of incoming solar radiation. In this study, addition of the second-order PAR term added negligible improvement to simulation of evapotranspiration.

A comparison between simulated and measured values of latent heat flux is shown in Fig. 15.26 and regression statistics are shown in Table 15.3. The model exhibited little temporal bias (Fig. 15.25), even in the postfire/prelogging period when substantial transient changes (regrowth) in vegetative cover occurred in the burned areas. The lack of significant temporal bias supports the utilization of the particular discretization of time used in the model. More than 95% of the values of latent heat flux were within 25% of the measured values.

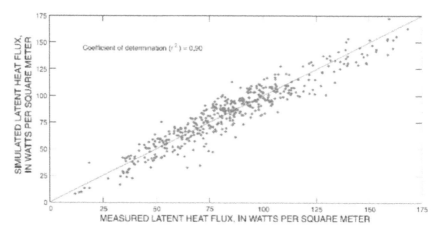

FIGURE 15.26 Comparison of simulated and measured values of daily latent heat flux.

15.3.2 APPLICATION OF EVAPOTRANSPIRATION MODEL

The calibrated evapotranspiration model [Eqs. (22) and (23)], with α values given by Eq. (28) and regression-derived parameters given in Table 15.3) described in the previous section was used to estimate average, daily values of evapotranspiration for both burned and unburned areas of the watershed during the 2-year study period. The model also provided a quantitative framework to examine the relation between evapotranspiration and the environment. The input variables for the model included daily values of net radiation (Fig. 15.16), PAR (Fig. 15.17), water-table depth at the evapotranspiration station (Fig. 15.27), and air temperature (Fig. 15.28).

Values of latent heat flux and evapotranspiration for January 1998 through December 1999 were estimated using the calibrated model (Fig. 15.29). Despite the relatively high net radiation in burned areas (Fig. 15.16), evapotranspiration from burned areas generally remained lower than that from unburned areas until spring 1999. This effect presumably was a result of destruction of transpiring vegetation by fire and then logging. Beginning in spring 1999 (postlogging II period for burned areas), evapotranspiration from burned areas increased sharply relative to unburned areas, sometimes exceeding evapotranspiration from unburned areas by almost 100%. From a simulation perspective, this change in evapotranspiration in spring 1999

TABLE 15.3 Summary of Parameters and Error Statistics For Daily Evapotranspiration Models

Parameters	Unburned area (j = 1)	Three-part model for burned area		
		Post-fire/pre-logging (j=2)	Post-logging I (j=3)	Post-logging II (j=4)
		Time period		
	01–1-1998 to 31–12–1999	25–06–1998 to 16–12–1998	17–12–1998 to 22–04–1999	23–04–1999 to 31–12–1999
C_{1j}	–0.175	–0.167	–0.312	–0.508
C_{2j}	–0.00102	–0.00147	–0.00031	0.00013
C_{3j}	+1.42	1.26	1.03	1.36

Error statistics: $r^2 = 0.90$; SEE = 9.67 and CV= 0.11

*Parameters C_1, C_2, and C_3 are defined by the equation: $\alpha_j = C_{1j} h_{wt} + C_{2j} PAR + C_{3j}$ where: j is an index denoting the surface cover; h_{wt} is water-table depth below a reference level placed at the highest water level measured (0.11 m above land surface) at the evapotranspiration station (uplands environment), in meters; and PAR is photosynthetically active radiation, in micromoles per m^2 per second. Error statistics: r^2, coefficient of determination of measured and simulated values of latent heat flux, dimensionless; SEE, standard error of estimate (in watts per m^2); CV, coefficient of variation, dimensionless, equal to SEE divided by the mean of the measured values of latent heat flux].

was clearly the result of the change in Priestley-Taylor α model parameters between the two postlogging periods. From a physics perspective, the possible explanation(s) for the change in evapotranspiration is identical to those described in the earlier discussion of the differentiation of the early and late post logging periods within the evapotranspiration model. Evapotranspiration from burned areas for the 10-month period after the fire (July 1998–April 1999) averaged about 17% less than that from unburned areas and, for the following 8-month period (May 1999–December 1999), averaged about 31% higher than from unburned areas. During the 554-day period after the fire, the average evapotranspiration for burned areas (1,043 mm/yr) averaged 8.6% higher than that for unburned areas (960 mm/yr).

Annual evapotranspiration from the watershed was 916 mm for 1998 and 1,070 mm for 1999, and averaged 993 mm. The extensive burning and logging that occurred during the study produced a landscape that was not typical of forested areas of Florida. The estimated evapotranspiration from unburned areas can be considered representative of more typical forest

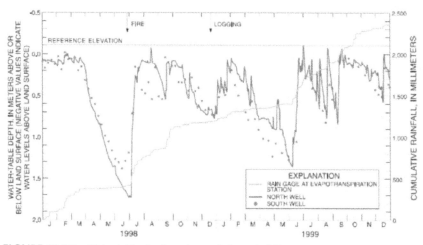

FIGURE 15.27   Water-table depth and cumulative rainfall.

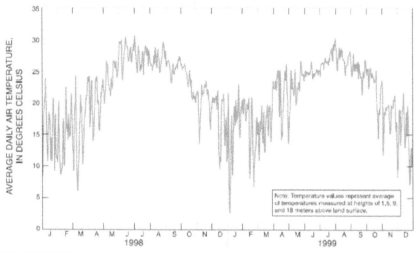

FIGURE 15.28   Average daily air temperature.

cover. Annual evapotranspiration from unburned areas was 937 and 999 mm for 1998 and 1999, respectively, and averaged 968 mm. Both actual and potential evapotranspiration showed strong seasonal patterns and day-to-day variability (Figs. 15.29 and 15.30). Actual evapotranspiration from the watershed averaged only 72% of potential evapotranspiration.

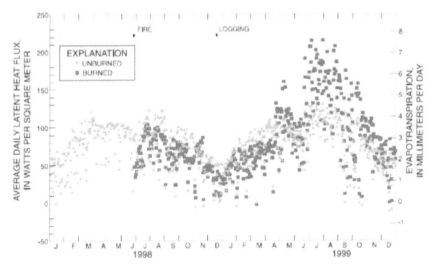

FIGURE 15.29   Average daily latent heat flux and evapotranspiration.

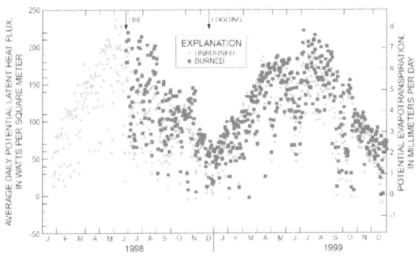

FIGURE 15.30   Average daily potential latent heat flux and potential evapotranspiration.

The effect of the extreme drought period in spring 1998 (Fig. 15.27) on turbulent fluxes was substantial (Figs. 15.29, 15.31, and 15.32). Turbulent fluxes usually emulate the general sinusoidal, seasonal pattern of solar radiation and air temperature [15, 22, 44]. The usual sinusoidal pattern of latent heat flux was truncated in spring 1998 (Fig. 15.29) because of

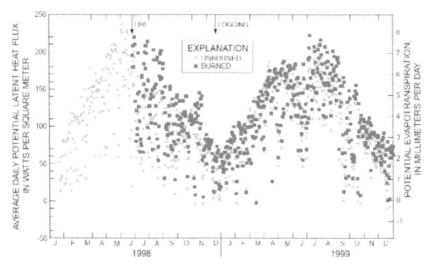

FIGURE 15.31   Average daily sensible heat flux.

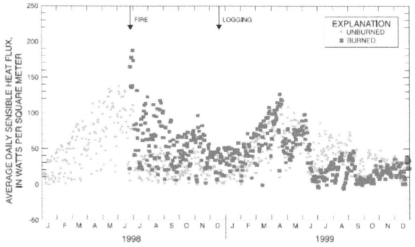

FIGURE 15.32   Average daily Bowen ratio.

a lack of available moisture (Figs. 15.27 and 15.33). The drought-induced reduction in latent heat flux was compensated by an increase in sensible heat flux (Fig. 15.31) with an associated increase in the Bowen ratio. Comparison of the Bowen ratio (Fig. 15.32) with the water-table and soil moisture records (Figs. 15.27 and 15.33) indicates that the moisture status of the watershed has a major role in the partitioning of the available energy. Relative evapotranspiration is a ratio of actual to potential evapotranspiration and was computed as α/1.26; and it decreased from about 1 in the early, wet part of 1998 to less than 0.50 during the drought (Fig. 15.34). After the drought ended in late-June and early-July 1998

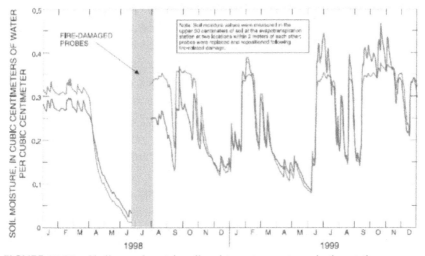

FIGURE 15.33   Shallow, volumetric soil moisture at evapotranspiration station.

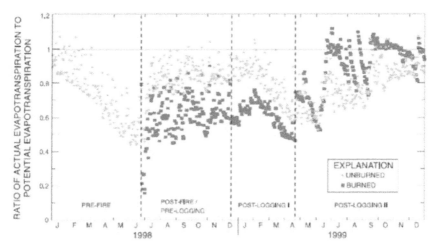

FIGURE 15.34   Temporal variability of daily values of relative evapotranspiration.

and water levels quickly returned to near land surface, evapotranspiration increased sharply. The evapotranspiration rate, however, averaged only about 60% of the potential rate in the burned areas, as compared to about 90% in the unburned areas. This discrepancy can be explained as a result of fire damage to vegetation.

Potential evapotranspiration rates for burned and unburned areas were similar (Fig. 15.30), although actual evapotranspiration rates for the two areas were quite distinct from each other (Fig. 15.29). The relation between actual and potential evapotranspiration was not a simple constant multiplier (e.g., a crop factor), but rather was time-varying as a function of water-table depth, PAR, and surface cover (Fig. 15.34). Relative evapotranspiration exceeded a value of 1 at times, probably as a result of experimental error, as well as the approximate and empirically derived nature of the assumed potential value of 1.26 for α. The potential evapotranspiration rates (Fig. 15.30) did not strongly reflect either the drought or surface burning and logging, as does the actual evapotranspiration.

Within the framework of the calibrated model, variations in the environmental variables contained in α (water-table depth and PAR) reduce actual evapotranspiration below potential evapotranspiration for a given surface cover. The evapotranspiration model indicated that relative evapotranspiration decreased as the depth to the water table increased (Fig. 15.35). The

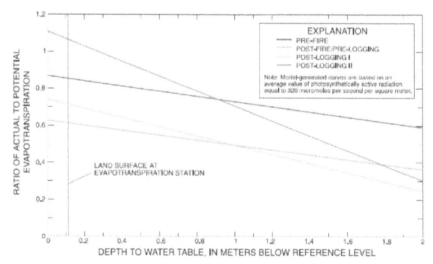

FIGURE 15.35   Relation between relative evapotranspiration and depth to water table.

range of water-table depths prevalent during the study period was slightly above land surface to about 1.75 m below land surface. Presumably, at some water-table depth greater than 1.75 m, relative evapotranspiration would reach an asymptotic constant value as vegetation becomes unable to access moisture below the water table. The rate of decline of relative evapotranspiration with water-table depth was greater for the postlogging period than for the prelogging period. This result is perhaps a manifestation of the replacement of many deep- rooted trees by shallow-rooted understory vegetation following the fires. Shallow-rooted plants would be less able to tap into deep soil moisture or the water table than would deep-rooted vegetation.

Water-table depth has been considered an important predictor of evapotranspiration in hydrologic analysis [49], but little empirical evidence has been available to define the relation between these two environmental variables. The USGS modular finite – difference ground-water flow model (MODFLOW) simulates relative evapotranspiration as a unique, piece-wise, linear function of water-table depth, where evapotranspiration declines from a potential rate when the water table is at or above land surface to zero at the "extinction depth" [26]. Contrary to the MODFLOW conceptualization of evapotranspiration, this study indicates that the variation in relative evapotranspiration is explained not only by water-table depth, but also by PAR. Relative evapotranspiration decreased with increasing *PAR* (Fig. 15.36), with the exception of the late postlogging period, which showed a slight increase in relative evapotranspiration with increasing *PAR*. This observation perhaps can be explained by assumptions within the Priestley-Taylor formulation that the energy and aerodynamic terms of the Penman equation are proportional to each other. Under nonpotential conditions, these two terms might deviate from the assumption of proportionality, but in such a manner that can be "corrected" through a functional relation between the multiplier α and a term (*PAR*) strongly correlated with the energy term.

Within the model developed in this study, net radiation and air temperature do not directly affect the Priestley-Taylor α and relative evapotranspiration, although net radiation has an indirect effect through the correlation of this variable with *PAR*. These variables, however, are important in the determination of evapotranspiration, as can be seen in Eq. (22). Evapotranspiration is directly proportional to $[\Delta/(\Delta + \gamma)]$, a term that is a function of temperature (Fig. 15.37).

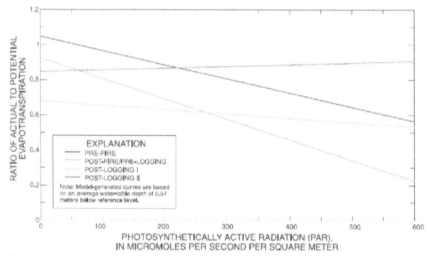

FIGURE 15.36 Relation between relative evapotranspiration and photosynthetically active radiation.

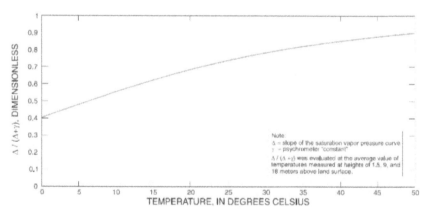

FIGURE 15.37 The Priestley-Taylor variable $\Delta/(\Delta + \gamma)$ as function of temperature.

For example, a change in air temperature from 20 to 30 °C will produce about a 14-percent increase in evapotranspiration, assuming the environment is otherwise unchanged. The relation of net radiation and evapotranspiration is one of direct proportionality. Net radiation displayed dramatic temporal variations, both day-to-day (as a result of variations in cloud cover) and seasonally (Fig. 15.16), making this variable the most important determinant of evapotranspiration. This conclusion is supported by a sensitivity analysis (Table 15.4) based on perturbing each environmental variable of the evapotranspiration model by an amount equal to the observed standard deviation of the daily values of that variable. All unperturbed variables were assumed equal to mean values. This analysis indicated that variations in net radiation explained the greatest amount of the variation in evapotranspiration. Variations in *PAR*, closely correlated with net radiation, explained a large amount of the variation in evapotranspiration prior to logging, but explained little of the variation after logging. Evapotranspiration was moderately sensitive to variations in air temperature. Variations in water-table depth explained a moderate amount of the variation in evapotranspiration prior to the fire; however, evapotranspiration became more sensitive to variations in water-table depth after logging.

The model developed in this study is subject to several qualifications. The form of the equation developed for α was empirical, rather than physics-based, and was simply designed to reproduce measured values of evapotranspiration as accurately as possible. The correlation between environmental variables complicates a unique determination of parameters.

The model was developed for a limited range of environmental conditions, and therefore, extrapolation of the model to conditions not encountered in this study should be done with caution. The measured (upland) water-table depth at the evapotranspiration station, used as an independent variable in the model, explained some of the variation in evapotranspiration from the mixed upland/wetland watershed. However, water-table depth is not uniform within the watershed and, in particular, water-table depth in wetland areas usually is less than in upland areas. Therefore, caution should be used in applying the model to estimate evapotranspiration based on water-table depth measurements made at other locations in the watershed. For these reasons, the evapotranspiration model described in this report should be viewed as a general guide, rather than as a definitive description of the relation of evapotranspiration to environmental variables. The fact that the model successfully ($r^2 = 0.90$) reproduced 449 daily measurements of site evapotranspiration over a wide range of seasonal and surface-cover values lends credence to the ability of the model to estimate evapotranspiration at the site.

TABLE 15.4 Sensitivity Analysis of Evapotranspiration Models to Environmental Variables

Environmental variables, X	Mean value	Standard deviation, σ
R_n for unburned	118.3	50.0
R_n for burned	127.6	49.6
PAR	320.0	118.3
T_a	21.7	5.4
h_{wt}	0.57	0.42

	ET1 (mean + σ)	ET2 (mean − σ)	% change, (+)	% change (−)
Unburned model				
R_n for unburned	4.15	1.69	42	-42
PAR	2.58	3.27	-12	12
T_a	3.16	2.64	8	-10
h_{wt}	2.71	3.14	-7	7
Post-fire/prelogging model				
R_n for burned	3.05	1.33	39	-39
PAR	1.64	2.74	-25	25
T_a	2.37	1.98	8	-10
h_{wt}	1.97	2.41	-10	10
Post-logging I model				
R_n for burned	3.32	1.45	39	-39
PAR	2.26	2.50	-5	5
T_a	2.58	2.15	8	-10
h_{wt}	1.97	2.80	-17	17
Post-logging II model				
R_n for burned	4.86	2.12	39	-39
PAR	3.53	3.44	1	-1
T_a	3.78	3.15	8	-10
h_{wt}	2.80	4.16	-19	19

Mean ET, mm/day = 2.92 for unburned; 2.19 for postfire/prelogging; 2.38 for post-logging I; and 3.49 for postlogging II models. R_n = Net radiation in watts per m^2; PAR = Photosynthetically active radiation in μmoles per m^2 per sec.; T_a = Air temperature in °C; h_{wt} = Water table depth below the reference level in meters.

Note: Values in table were computed using each of the four ET models defined in Table 15.3. Mean and standard deviation values are representative of daily values during the 2-year period of record and the R_n values are only representative of 1999.

15.3.3   WATER BUDGET

Construction of a water budget for the Tiger Bay watershed serves to provide a tool for watershed management and for assessing the integrity of the eddy correlation evapotranspiration measurements. The water budget for the watershed is given in Eq. (29), where: P is precipitation, in millimeters per year; ET is evapotranspiration, in millimeters per year; R is runoff, in millimeters per year; L is leakage to the Upper Floridan aquifer, in millimeters per year; and ΔS is rate of change in storage, in millimeters per year.

$$P - (ET + R + L - \Delta S) = 0 \qquad (29)$$

A water budget for the Tiger Bay watershed during the 1998–1999 study period is shown in Table 15.5 and Fig. 15.38. Precipitation (Figs. 15.9 and 15.27), evapotranspiration (Fig. 15.29), and runoff (Fig. 15.39) were measured or obtained as described previously in this chapter. The estimated value of deep leakage to the Upper Floridan aquifer (112 mm/yr) during 1995 (Stan Williams, St. Johns River Water Management District, oral communication, 2000) also was assumed to be appropriate for the study period (1998–1999). The rate of change in water- shed storage over the study period was not directly measured, but was estimated as the water-budget residual. The water budget (Tables 15.5 and 15.6; Fig. 15.38) indicated that about 76% of watershed rainfall was lost as evapotranspiration during the 2-year study. The ratio of evapotranspiration to rainfall was remarkably stable from year-to-year (74% in 1998 and 77% in 1999). This stability occurred despite the very different environmental

TABLE 15.5   Water Budget for Tiger Bay Watershed

Year	P	ET	R	L	ΔS
1998	1233	916	357	112	−152
1999	1396	1070	114	112	100
1998–1999	1315	993	236	112	−26

P, mm/year = Precipitation, (average of north & south rain gages: See Fig. 1); ET, mm/year = Evapotranspiration; R, mm/year = Runoff from watershed at Tiger Bay canal; L, mm/year = Estimated leakage to the upper Floridian aquifer; and ΔS, mm/year = Rate of change in watershed storage estimated as a water-budget residual.

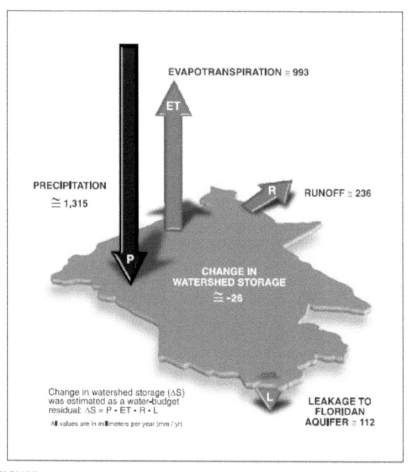

EVAPOTRANSPIRATION ≅ 993

ET

PRECIPITATION
≅ 1,315

R   RUNOFF ≅ 236

P

CHANGE IN
WATERSHED STORAGE
≅ -26

Change in watershed storage (ΔS)
was estimated as a water-budget
residual: ΔS = P - ET - R - L

All values are in millimeters per year (mm / yr)

L   LEAKAGE TO
FLORIDAN
AQUIFER ≅ 112

FIGURE 15.38   Water budget for Tiger Bay watershed during calendar years 1998–1999.

conditions prevailing during the study. Rainfall was a more consistent predictor of evapotranspiration than was potential evapotranspiration. The relative evapotranspiration varied rather greatly (67% in 1998 to 77% in 1999).

Runoff removed about 18% of the rainfall during the study period, but this percentage varied widely from year-to-year (29% in 1998 and 8% in 1999) as shown in Fig. 15.39. The runoff for 1998 was over three times that of 1999, despite the greater rainfall in 1999. This disparity can be explained

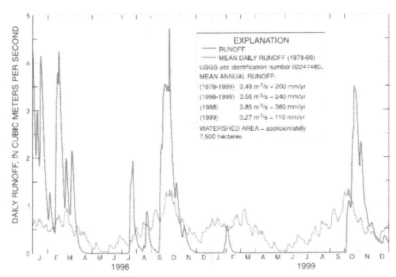

FIGURE 15.39   Runoff from Tiger Bay watershed.

TABLE 15.6   Potential Evapotranspiration [PET, mm/year] and Relative Rates of Annual Water-Budget Terms for Tiger Bay Watershed

Year	PET	Relative rates			
		ET/PET	**ET/P**	**R/P**	**L/P**
1998	1356	0.67	0.74	0.29	0.09
1999	1391	0.77	0.77	0.08	0.08
1998–1999	1374	0.72	0.76	0.18	0.09

ET = Evapotranspiration, mm/year; P = Precipitation, mm/year [average of North and South storage rain gages]; R, mm/year = Runoff from the watershed; L, mm/year = Estimated leakage to the upper Floridian aquifer for 1995.

largely by the antecedent water-table conditions for individual rain periods (Fig. 15.27).

A relatively large fraction of precipitation in 1998 occurred when the water- table depth was shallow, leading to relatively high rejection of infiltration and subsequent runoff. Additionally, the temporal distribution of precipitation affects the amount of watershed runoff. Runoff is maximized following short, but intense, rainfall during which the infiltration capacity of the soil is exceeded. This phenomenon may explain the disparate runoff responses in July 1998 (very intense rainfall and significant runoff) and June–July 1999 (less intense rainfall and no runoff). This disparity was noted despite similar total amounts of precipitation with similar antecedent water-table conditions for each of the two periods. An alternative explanation may be that the soils became hydrophobic as a result of the fire, contributing to relatively more runoff in July 1998. Also, seasonal or fire-related variations in evapotranspiration can result in variations in the amount of precipitation available as runoff. Deep leakage was a relatively small fraction of the rainfall (about 9%), although this water-budget term could increase (at the expense of runoff and evapotranspiration) if continued development of the Upper Floridan aquifer in the area increases the hydraulic gradient between the surficial aquifer system and the underlying Upper Floridan aquifer.

The consistency of the water-budget terms can be expressed by the absolute and relative water-budget closures as shown in Eqs. (31) and (32), where: C_a is absolute water-budget closure, in millimeters per year; C_r is relative water budget closure, in percent; and P, ET, R, L, and ΔS are the same as defined in Eq. (29).

$$C_a = P - (ET + R + L + \Delta S) \qquad (30)$$

$$Cr = (100C_a)/P \qquad (31)$$

Watershed storage (ΔS) was an unmeasured quantity within the water budget. Therefore, evaluation of water-budget closure was facilitated by the judicious choice of a time period when negligible change in storage occurred within the watershed. Based on the measured water levels in the watershed (Fig. 15.27), the time period from March 3, 1998, through September 23, 1999, was selected as an interval when change in watershed storage could be assumed to be zero.

The beginning and ending of this interval occurred at times when temporal changes in water level were relatively slight, implying that the water levels measured at the two monitor wells at the beginning and ending dates of the interval were probably representative of the watershed. The absolute value of the measured rate of change in water level was less than 6 mm/yr at both monitor wells over this 570-day interval.

Based on measured or estimated values of P (1,245 mm/yr), ET (1,048 mm/yr), R (132 mm/yr), and L (112 mm/yr), the absolute and relative water-budget closures were –47 mm/yr and 3.8%, respectively. The consistency of these independently measured water-budget terms provides support for, but not confirmation of, the reliability of the measured evapotranspiration. Compensating errors among water-budget terms or compensating errors within the temporal pattern of estimated evapotranspiration also could produce a consistent water budget.

Evapotranspiration was estimated during the present study using an energy-budget variant [Eq. (18)] of the eddy correlation method, rather than the standard eddy correlation method [Eq. (5)]. The water-budget analysis provided an independent means to evaluate the relative accuracies of the two eddy correlation methods. The standard method produced turbulent flux estimates that were, on average, about 84.7% of those produced by the energy-budget variant. Applying this fraction to the evapotranspiration total for the water budget period from March 3, 1998, to September 23, 1999, the absolute and the relative budget closures corresponding to the standard eddy correlation method are 113 mm/yr and 9.1 percent, respectively. These closure values are greater than the values reported for the energy-budget variant, consistent with the assumption that the energy-budget variant was more accurate than the standard eddy correlation method.

Additional support for the assumption that the energy-budget variant was preferable to the standard eddy correlation method could be discerned from a residual analysis that assumed that precipitation, leakage, runoff, and evapotranspiration were accurately measured and that a lack of water-budget closure can be explained solely by the residual-calculated storage term. The specific yield representative of the watershed was then computed as the rate of change of watershed storage divided by a representative rate of change in water level within the watershed. The specific yield, estimated in this manner, was evaluated for credibility as a means of identifying the preferred variant of the eddy correlation method. Specific yield is defined as the volume of water yielded per unit area per unit change in water level. Specific yield can range from near zero if the capillary fringe intersects land surface [16] to near unity for standing water. The specific yield of sandy soils (such as those in the uplands) ranges from 0.10 to 0.35 [18]. In this analysis, the representative rate of change in water-table depth for the watershed was assumed equal to the average rate of change in water-table depth at the two upland monitor wells (Table 15.7). As mentioned previously, upland and wetland water levels are expected to change at the same rate in the low relief environment of this watershed.

Results of the residual analysis, using evapotranspiration estimated by both approaches, are shown in Table 15.8. The energy-budget variant of the eddy correlation method produced specific yield estimates (0.24 in 1998, 0.27 in 1999, and 0.19 in 1998–1999) that were somewhat consistent between each of the three time periods and were within the range of possible values. The standard eddy correlation method produced estimates of specific yield that were inconsistent between each of the three time periods and were unreasonable (0.02 in 1998, 0.71 in 1999, and −0.94 in 1998–1999). The residual analysis of water budgets further supports the

TABLE 15.7   Average Rate of Change in Water Table Depth at Monitor Wells

Year	Δh_{North}	Δh_{South}	Δh_{avg}
1998	−660	−616	−638
1999	+432	+308	+370
1998–1999	−114	−154	−134

Δh_{North} = Rate of change in water-table depth at the ET North station in mm/year; Δh_{South} = Rate of change in water-table depth at the south rain gage in mm/year; and Δh_{avg} = average rate of change in water-table depth in mm/year = $[\Delta h_{North} + \Delta h_{South}]/2$.

TABLE 15.8   Comparison of Estimates of Specific Yield Based on ET Estimated With Energy Budget Variant and With the Standard Eddy Correlation Method

Year	Energy-budget variant		Standard eddy correlation method	
	ΔS	S_y	ΔS	S_y
1998	−152	0.24	−12	0.02
1999	100	0.27	268	0.71
1998–1999	−26	0.19	126	−0.94

Note: ΔS, mm/year = Rate of change in the watershed storage computed as a residual of the water balance method $[\Delta S = P − (ET + R + L)]$; P, mm/year = Average watershed precipitation; ET, mm/year = Evapotranspiration estimated by the energy-budget variant of the eddy correlation method is about 84.7% of the ET (Table 15.5) estimated by the energy-budget of the eddy correlation method; R, mm/year = Average watershed runoff; L, mm/year = Estimated leakage in 1995 to the Upper Floridian aquifer; S_y = Specific yield = $[\Delta S/\Delta h_{avg}]$, no units; Δh_{avg} = Estimated average rate of change in water-table depth (Table 15.7, mm/year).

assumption that the energy-budget variant of the eddy correlation method is more accurate than the standard method.

Based on this research, followings can be concluded: A 2-year (1998–1999) study was conducted to estimate evapotranspiration (ET) from a forested watershed (Tiger Bay, Volusia County, Florida), which was subjected to natural fires, and to evaluate the causal relations between the environment and ET. The watershed characteristics are typical of many areas within the lower coastal plain of the south-eastern United States – nearly flat, slowly draining land with a vegetative cover consisting primarily of pine flat wood uplands interspersed within cypress wetlands. Drought-induced fires in spring 1998 burned about 40% of the watershed and most of the burned area was logged in late-fall 1998. ET was measured using eddy correlation sensors placed on a tower 36.5-meter (m) high within an 18.3-m-high forest. About 27% of the 30-minute eddy correlation data were missing as a result of either inoperation of the sensors related to scaling of the hygrometer windows, collection of rainfall or dew on the sensors, or spurious turbulence created by the sensor mounting arm and the attached tower. These missing data generally occurred during periods (evening to early morning) when ET was relatively low. Linear relations between PAR and the fluxes of ET and sensible heat were used to estimate missing 30-minute values. Data were composited into daily values if the turbulent fluxes for more than 18 h of a given day were directly measured, rather than being estimated with the PAR-based relation. Daily values for which more than 6 h of data were missing were considered nonmeasured. This procedure resulted in 449 measurements of daily ET over the 2-year (730-day) period. An energy-budget variant of the standard eddy correlation method that accounts for the common underestimation of ET by the standard method was computed.

Following the fires, the daily measurements of ET were a composite of rates representative of burned and unburned areas of the watershed. The fraction of a given daily measurement derived from burned areas was estimated based on the diurnal pattern of wind direction and PAR for that day and on the transpiration station. The daily values of ET were used to calibrate a Priestley-Taylor model. The model was used to estimate ET for burned and unburned areas and to identify and quantify the environmental controls on ET.

The ET model successfully ($r^2 = 0.90$) reproduced daily measurements of site ET over a wide range of environmental conditions, giving credence to the ability of the model to estimate ET at the site. Estimation of ET from the watershed was based on an area-weighted composite of estimated values for burned and unburned areas. Annual ET from the watershed was 916 and 1,070 millimeters (mm) for 1998 and 1999, respectively, and averaged 993 mm. These values are comparable to those reported by previous researchers. ET has been estimated to average about 990 millimeters per year (mm/yr) over Volusia County [37] and to average about 890 mm/yr in the Tiger Bay watershed [7]. Bidlake and others [3] estimated annual cypress ET (970 mm) to be only 8.5% less than that of pine flat woods (1,060 mm) based on studies conducted in Sarasota and Pasco Counties, Florida. Liu [24] estimated average, annual ET of both cypress and pine flatwoods to be 1,080 mm based on a study in Alachua County, Florida.

The extensive burning and logging that occurred during the study produced a landscape that was not typical of forested areas of Florida. The estimated ET from unburned areas can be considered more representative of typical forest cover. Annual ET from unburned areas was 937 and 999 mm for 1998 and 1999, respectively, and averaged 968 mm. ET from burned areas for the 10-month period after the fire (July 1998–April 1999) averaged about 17% less than that from unburned areas and, for the following 8-month period (May–December 1999), averaged about 31% higher than from unburned areas. During the 554-day period after the fire, the average ET for burned areas (1,043 mm/yr) averaged 8.6% higher than that for unburned areas (960 mm/yr). Both actual and potential ET showed strong seasonal patterns and day-to-day variability. Actual ET from the watershed averaged only 72% of potential ET. ET declined from near potential rates in the wet conditions of January 1998 to less than 50% of potential ET after the fire and at the peak of the drought in June 1998. After the drought ended in early July 1998 and water levels returned to near land surface, ET increased sharply. The ET rate, however, was only about 60% of the potential rate in the burned areas, as compared to about 90% of the potential rate in the unburned areas. This discrepancy can be explained as a result of fire damage to vegetation. Beginning in spring 1999, ET from burned areas increased sharply relative to unburned areas, sometimes exceeding unburned ET by almost 100%. Possible explanations for the dramatic increase in ET from burned areas are not clear at this time, but may include phenological changes associated with maturation or seasonality of plants that emerged after the fire or successional changes in composition of plant community within burned areas.

Within the framework of the Priestley-Taylor model developed during this study, variations in daily ET were the result of variations in: surface cover, net radiation, PAR, air temperature, and water-table depth. Potential ET depended solely on net radiation and air temperature and increased as each of these variables increased. The extent to which potential ET was approached was determined by the Priestley-Taylor coefficient α. In this study, Priestley-Taylor α was a linear function of water-table depth and PAR. Unique parameters within the α function were estimated for each of four surface covers or time periods: unburned; burned, but unlogged; and both burned and logged (early postlogging and late postlogging). The ET model indicated that relative ET (the ratio of actual to potential ET) decreased as the depth to the water table increased. The rate of decline of relative ET with water-table depth was greater for the postlogging period than for the prelogging period, perhaps indicative of the replacement of many deeply rooted trees by shallow-rooted understory vegetation following the fires. Shallow-rooted plants would be less able to tap into deep soil moisture or the water table than deep-rooted trees. Relative ET decreased with increasing PAR, with the exception of the late postlogging period, which showed a slight increase in relative ET with increasing PAR.

A water budget for the watershed supported the validity of the estimates of ET produced with the energy-budget variant of the eddy correlation method. Independent estimates of average rates of rainfall (1,245 mm/yr), runoff (132 mm/yr), deep leakage (112 mm/yr), as well as ET (1,048 mm/yr) were compiled for a 570-day period over which the change in watershed storage was negligible. Water-budget closure was 47 mm/yr or 3.8% of rainfall, indicating good consistency between the estimated ET and estimates of the other terms of the water budget. Estimates of ET produced by the standard eddy correlation method were relatively inconsistent with the water budget (water-budget closure was 113 mm/yr or 9.1% of rainfall), indicating that the energy- budget variant is superior to the standard eddy correlation method. Specific yield was estimated based on estimated changes in watershed storage and water level. The change in watershed storage was estimated as a residual of the water budget. Specific yield values produced using ET estimated by the energy-bud- get variant of the eddy correlation method were reasonable and relatively consistent from year-to-year (0.19 to 0.27). However, specific yield values based on ET estimated by the standard eddy correlation method were unreasonable and inconsistent from year-to-year (–0.94 to 0.72). These results further support the premise that the energy – budget variant is more accurate than the standard eddy correlation method.

ET rates were about 74 and 77% of rainfall for 1998 and 1999, respectively, relatively constant considering the variability in surface cover and rainfall patterns between the 2 years. Potential ET was less consistent as an indicator of actual ET; ET was 67 and 77% of potential ET for years 1998 and 1999, respectively.

15.4 SUMMARY

Daily values of ET from a watershed in Volusia County, Florida, were estimated for a 2-year period (January 1998 through December 1999) by using an energy-budget variant of the eddy correlation method and a Priestley-Taylor model. The watershed consisted primarily of pine flat wood uplands interspersed within cypress wetlands. A drought-induced fire in spring 1998 burned about 40% of the watershed, most of which was subsequently logged. The model reproduced the 449 measured values of ET reasonably well (r^2= 0.90) over a wide range of seasonal and surface-cover conditions. Annual ET from the water- shed was estimated to be 916 millimeters (36 inches) for 1998 and 1,070 millimeters (42 inches) for 1999. ET declined from near potential rates in the wet conditions of January 1998 to less than 50% of potential ET after the fire and at the peak of the drought in June 1998. After the drought ended in early July 1998 and water levels returned to near land-surface, ET increased sharply; however, the ET rate was only about 60% of the potential rate in the burned areas, compared to about 90% of the potential rate in the unburned areas. This discrepancy can be explained as a result of fire damage to vegetation. Beginning in spring 1999, ET from burned areas increased sharply relative to unburned areas, sometimes exceeding unburned ET by almost 100 percent. Possible explanations for the dramatic increase in ET from burned areas could include phenological changes associated with maturation or seasonality of plants that emerged after the fire or successional changes in composition of plant community within burned areas.

Variations in daily ET are primarily the result of variations in surface cover, net radiation, PAR, air temperature, and water-table depth. A water budget for the watershed supports the validity of the daily measurements and estimates of ET. A water budget constructed using independent estimates of average rates of rainfall, runoff, and deep leakage, as well as ET, was consistent within 3.8 percent. An alternative water budget constructed using ET estimated by the standard eddy correlation method was consistent only within 9.1 percent. This result indicates that the standard eddy correlation method is not as accurate as the energy-budget variant.

KEYWORDS

- aerodynamic resistance
- anemometer
- aquifer
- atmospheric pressure
- big leaf assumption
- Bowen ratio
- Bowen ratio energy-budget variant (of eddy covariance method)
- cypress
- eddy covariance method
- energy-budget closure
- energy-budget variant
- evaporation
- evaporative fraction
- evapotranspiration
- evapotranspiration, potential

REFERENCES

1. Baldocchi, D. D.; Hicks, B. B.; Meyers, T. P.; Measuring biosphere-atmosphere exchanges of biologically related gases with micrometeorological methods: Ecology, 1988, 69(5), 1331–1340.
2. Baldwin, R.; Bush, C. L.; Hinton, R. B.; Huckle, H. F.; Nichols, P.; Watts, F. C.; Wolfe, J. A.; Soil Survey of Volusia County, Florida: US Soil Conservation Service, 1980, 207 p. and 106 pls.
3. Bidlake, W. R.; Woodham, W. M.; Lopez, M. A.; Evapotranspiration from areas of native vegetation in west-central Florida: US Geological Survey Open-File Report 93–415, 1993, 35 p.
4. Bowen, I. S.; The ratio of heat losses by conduction and by evaporation from any water surface: Physical Review, 2nd series, 1926, 27(6), 779–787.
5. Brutsaert, W.; Evaporation into the atmosphere-Theory, history, and applications: Boston, Kluwer Academic Publishers, 1982, 299 p.
6. Businger, J. A.; Yaglom, A. M.; 'Introduction to Obukhov's paper on "Turbulence in an atmosphere with a nonuniform temperature,"' Boundary-Layer Meteorology, 1971, 2, 3–6.
7. Camp, Dresser and McKee, Inc.; 1996, Volusia County, Florida-Tiger Bay water conservation and aquifer recharge evaluation-Phase I: Volusia County, Florida, Technical Report.
8. Campbell, G. S.; Norman, J. M.; An introduction to environmental biophysics: New York, Springer, 1998, 286 p.
9. Dyer, A. J. Measurements of evaporation and heat transfer in the lower atmosphere by an automatic eddy-correlation technique: Quarterly Journal of the Royal Meteorological Society, 1961, 87, 401–412.
10. Eichinger, W. E.; Parlange, M. B.; Stricker, H.; On the concept of equilibrium evaporation and the value of the Priestley-Taylor coefficient: Water Resources Research, 1996, 32(1), 161–164.
11. Eidenshink, J. C.; The 1990 conterminous U. S. AVHRR dataset: J. Photogramtry and Remote Sensing, 1992, 58, 809–813.
12. Fleagle, R. G.; Businger, J. A.; An introduction to atmospheric physics. New York, Academic Press, 1980, 432 p.
13. Flint, A. L.; Childs, S. W.; Use of the Priestley-Taylor evaporation equation for soil water limited conditions in a small forest clearcut: Agricultural and Forest Meteorology, 1991, 56, 247–260.
14. Garratt, J. R.; Surface influence upon vertical profiles in the atmospheric near-surface layer: Quarterly Journal of the Royal Meteorological Society, 1980, 106, 803–819.
15. German, E. R.; Regional evaluation of evapotranspiration in the Everglades: US Geological Survey Water-Resources Investigations Report 00–4217, 2000, 48 p.
16. Gillham, R. W.; The capillary fringe and its effect on water-table response: 1984, 67, 307–324.
17. Goulden, M. L.; Munger, J. W.; Fan, S-M, Daube, B. C.; Wofsy, S. C.; Measurements of carbon sequestration by long-term eddy covariance: methods and a critical evaluation of accuracy: Global Change Biology, 1996, 2, 169–182.

18. Johnson, A. I.; 1967, Specific yield-Compilation of specific yields for various materials: US Geological Survey Water-Supply Paper 1662-D, 74 p.
19. Kaimal, J. C.; Businger, J. A.; A continuous wave sonic anemometer-thermometer: J. Appl. Metero., 1963, 2, 156–164.
20. Kaimal, J. C.; Gaynor, J. E.; Another look at sonic thermometry: Boundary-layer meteorology, 1991, 56, 401–410.
21. Kimrey, J. O.; 1990, Potential for ground-water development in central Volusia County, Florida: US Geological Survey Water-Resources Investigations Report 90–4010, 31 p.
22. Knowles, L.; Jr.; 1996, Estimation of evapotranspiration in the Rainbow Springs and Silver Springs basins in north–central Florida: US Geological Survey Water-Resources Investigations Report 96–4024, 37 p.
23. Lee, X.; Black, T. A.; Atmospheric turbulence within and above a Douglas-fir stand. Part II: Eddy fluxes of sensible heat and water vapor: Boundary-layer meteorology, 1993, 64, 369–389.
24. Liu, S.; 1996, Evapotranspiration from cypress (Taxodium ascendens) wetlands and slash pine (Pinus elliottii) uplands in north-central Florida: Ph. D.; Dissertation, University of Florida, Gainesville, 258 p.
25. Lowe, P. R.; An approximating polynomial for the computation of saturation vapor pressure: J. Appl. Metero., 1977, 16(1), 100–103.
26. McDonald, M. G.; Harbaugh, A. W.; 1984, A modular three-dimensional finite-difference ground-water flow model: US Geological Survey Open-File Report 83–875, 528 p.
27. Monteith, J. L.; 1965, Evaporation and environment in The state and movement of water in living organisms, Symposium of the Society of Experimental Biology: San Diego, California, (Fogg, G. E.; ed.), Academic Press, New York, 205–234.
28. Monteith, J. L.; Unsworth, M. H.; 1990, Principles of environmental physics (2d ed.): London, Edward Arnold, 291 p.
29. Moore, C. J. Eddy flux measurements above a pine forest: Quarterly Journal of the Royal Meteorological Society, 1976, 102, 913–918.
30. National Oceanic and Atmospheric Administration, Climatological data-annual summary-Florida: 1998, 102(13), 21 p.
31. National Oceanic and Atmospheric Administration, Climatological data-annual summary-Florida: 1999, 103(13), 21 p.
32. Penman, H. L.; Natural evaporation from open water, bare soil, and grass: Proceedings of the Royal Society of London, Series, A.; 1948, 193, 120–146.
33. Phelps, G. G.; 1990, Geology, hydrology, and water quality of the surficial aquifer system in Volusia County, Florida: US Geological Survey Water-Resources Investigations Report 90–4069, 67 p.
34. Priestley, C. H. B.; Taylor, R. J. On the assessment of surface heat flux and evaporation using largescale parameters: Monthly Weather Review, 1972, 100, 81–92.
35. Reifsnyder, W. E.; Radiation geometry in the measurement and interpretation of radiation balance: Agricultural Meteorology, 1967, 4, 255–265.
36. Riekerk, H.; Korhnak, L. V.; The hydrology of cypress wetlands in Florida pine flatwoods: Wetlands. 2000, 20(3), 448–460.

37. Rutledge, A. T.; 1985, Ground-water hydrology of Volusia County, Florida with emphasis on occurrence and movement of brackish water: US Geological Survey Water-Resources Investigations Report 84–4206, 84 p.

38. Schotanus, P.; Nieuwstadt, F. T. M.; de Bruin, H. A. R.; Temperature measurement with a sonic anemometer and its application to heat and moisture fluxes. Boundary-Layer Meteorology, 1983, 50, 81–93.

39. Schuepp, P. H.; Leclerc, M. Y.; MacPherson, J. I.; Desjardins, R. L.; Footprint prediction of scalar fluxes from analytical solutions of the diffusion equation. Boundary-Layer Meteorology, 1990, 50, 355–373.

40. Simonds, E. P.; McPherson, B. F.; Bush, P.; 1980, Shallow ground-water conditions and vegetation classification, central Volusia County, Florida: US Geological Survey Water-Resources Investigations Report 80–752, 1 sheet.

41. Stannard, D. I.; Comparison of Penman-Monteith, Shuttleworth-Wallace, modified Priestley-Taylor evapotranspiration models for wildland vegetation in semiarid rangeland. Water Resources Research, 1993, 29(5), 1379–1392.

42. Stannard, D. I.; Interpretation of surface flux measurements in heterogeneous terrain during the Monsoon '90 experiment. Water Resources Research, 1994, 30(5), 1227–1239.

43. Stull, R. B.; An introduction to boundary layer meteorology. Kluwer Academic Publishers, Boston, 1988, 666 pp.

44. Sumner, D. M.; Evapotranspiration from successional vegetation in a deforested area of the Lake Wales Ridge, Florida. US Geological Survey Water-Resources Investigations Report 96–4244, 1996, 38 p.

45. Tanner, B. D.; Greene, J. P.; 1989, Measurement of sensible heat and water vapor fluxes using eddy correlation methods. Final report prepared for US Army Dugway Proving Grounds, Dugway, Utah, 17 p.

46. Tanner, B. D.; Swiatek, E.; Greene, J. P.; 1993, Density fluctuations and use of the krypton hygrometer in surface flux measurements: Management of irrigation and drainage systems, Irrigation and Drainage Division, American Society of Civil Engineers, July 21–23, 1993, Park City, Utah, 945–952.

47. Tanner, C. B.; Thurtell, G. W.; 1969, Anemoclinometer measurements of Reynolds stress and heat transport in the atmospheric boundary layer: United States Army Electronics Command, Atmospheric Sciences Laboratory, Fort Huachuca, Arizona, TR ECOM 66-G22-F, Reports Control Symbol OSD-1366, April 1969, 10 p.

48. The Orlando Sentinel, 1998, Special report--Florida ablaze: Sunday, July 12, 1998, 12.

49. Tibbals, C. H.; 1990, Hydrology of the Floridan aquifer system in east-central Florida: US Geological Survey Professional Paper 1403-E, 98 p.

50. Twine, T. E.; Kustas, W. P.; Norman, J. M.; Cook, D. R.; Houser, P. R.; Meyers, T. P.; Prueger, J. H.; Starks, P. J. Wesely, M. L.; Correcting eddy-covariance flux underestimates over a grassland: Agricultural and Forest Meteorology, 2000, 103, 279–300.

51. US Geological Survey, 1998a, Water resources data, Florida, water year 1998, v. 1A, north-east Florida surface water. US Geological Survey Water-Data Report FL-98–1A, 408 p.

52. US Geological Survey, 1998b, Conterminous U.S. AVHRR: US Geological Survey, National Mapping Division, EROS Data Center, 7 compact discs.

53. US Geological Survey, 1999a, Water resources data, Florida, water year 1999, v. 1A, north-east Florida surface water: US Geological Survey Water-Data Report FL–99–1A, 374 p.

54. US Geological Survey, 1999b, Conterminous U.S. AVHRR: US Geological Survey, National Mapping Division, EROS Data Center, 7 compact discs.

55. US Geological Survey, 2000, Water resources data, Florida, water year 2000, v. 1A, north-east Florida surface water: US Geological Survey Water-Data Report FL-00–1A, 388 p.

56. Volusia County Department of Geographic Information Services, 1996a, Vegetation, Daytona Beach, N. W.; prepared July 29, 1996, 1 sheet.

57. Volusia County Department of Geographic Information Services, 1996b, Vegetation-Daytona Beach, S. W.; prepared July 29, 1996, 1 sheet.

58. Webb, E. K.; Pearman, G. I.; Leuning, R.; Correction of flux measurements for density effects due to heat and water vapor transfer. Quarterly Journal of the Royal Meteorological Society, 1980, 106, 85–100.

59. Weeks, E. P.; Weaver, H. L.; Campbell, G. S.; Tanner, B. D.; Water use by saltcedar and by replacement vegetation in the Pecos River floodplain between Acme and Artesia, New Mexico. US Geological Survey Professional Paper 491-G, 1987, 37 pages.

APPENDIX I: DERIVATION OF EQ. (24) FOR W_B IN CHAPTER 15

The assumptions inherent in the weighting scheme used in Eqs. (23)–(25) can be seen through derivation of Eq. (24) for w_b. The latent heat flux measured by the eddy correlation sensors and derived from burned surface covers over a given day of 48 measurements is given by:

$$\lambda E_{bm} = \Sigma_i\, g_i\, \frac{1}{48} \Sigma_k\, \lambda E_{bk} \delta_i(\psi_k) \qquad (A1)$$

where, λE_{bm} is daily latent heat flux derived from burned surface covers and measured by the flux sensors, in watts per m²; g_i is fractional contribution of burned area within burn zone i to the measured latent heat flux when wind direction is from burn zone i; λE_{bk} is latent heat flux from burned surface covers for time step k, in watts per m²; $\delta_i(\psi k)$ is a binary function equal to 1 if ψk is within burn zone i and otherwise equals 0; and the index i is incremented from zone I to IV, and the index k is incremented from 1 to 48.

By definition, the expression in Eq. (A1) is equal to the second term of the right side of Eq. (23). Setting these two expressions equal and assuming that the high-resolution latent heat flux measurements for burned surfaces are directly proportional to PAR, and therefore, that the daily resolution latent heat flux for burned surfaces are directly proportional to average daily PAR:

$$w_b(\overline{aPAR}) = \frac{1}{48} \Sigma_i\, g_i \Sigma_k (aPAR_k)\, \delta_i \psi_k \qquad (A2)$$

where: w_b is the fraction of the measured latent heat flux originating from burned areas, dimensionless; and over-bars represent daily average values and the variable a is the constant of proportionality between latent heat flux and PAR.

Solving Eq. (A2), for w_b:

$$w_b = \frac{\frac{1}{48} \Sigma_i\, g_i \Sigma_k (aPAR_k) \delta_i(\psi_k)}{a\overline{PAR}} \qquad (A3)$$

$$w_b = \frac{\frac{1}{48} \Sigma_i\, g_i \Sigma_k (aPAR_k) \delta_i(\psi_k)}{a\frac{1}{n} \Sigma_k\, PAR_k} \qquad (A4)$$

$$w_b = \frac{\Sigma_i g_i \Sigma_k PAR_k \delta_i(\psi_k)}{\Sigma_k PAR_k} \tag{A5}$$

Eq. (A5) is identical to Eq. (24). The constant of proportionality a can change from day-to-day as environmental conditions (e.g., water level, air temperature, and green leaf density) change and, in fact, as shown in Eq. (A5), w_b is independent of the particular value of the constant. An equivalent expression, equal to $(1 - w_b)$, can be derived for the weight applied to daily latent heat flux from unburned surfaces. The constant of proportionality between unburned latent heat flux and PAR can be different than that between burned latent heat flux and PAR.

It is interesting to note that the use of measured high-resolution λE, rather than PAR, as a means of adjusting the weights for the combination of changing source area composition and diurnal variations in ET [Eq. (25)], produces excessive weighting towards zones with high-ET surface covers. This observation can be illustrated best by an example. Suppose, for a given day, the wind direction were from a lake (high ET) before solar noon and from a desert (near-zero ET) after solar noon. In this case, the appropriate weighting for each surface cover, within an equation of the form of Eq. (23), would be 0.5 and the average, measured ET for the day would be about one-half that of the lake. However, weighting by the fraction of ET measured from each zone would lead to a weight of near 1.0 for the lake zone and 0.0 for the desert zone, leading to a model for lake evaporation that would produce underestimates of true lake evaporation.

APPENDICES

APPENDIX A

CONVERSION SI AND NON-SI UNITS

To convert the Column 1 in the Column 2	Column 1 Unit	Column 2 Unit	To convert the Column 2 in the Column 1
Multiply by	SI	Non-SI	Multiply by

LINEAR

0.621 ——	kilometer, km (10^3 m)	miles, mi ——	1.609
1.094 ——	meter, m	yard, yd ——	0.914
3.28 ——	meter, m	feet, ft ——	0.304
3.94×10^{-2} —	millimeter, mm (10^{-3})	inch, in ——	25.4

SQUARES

2.47 ——	hectare, he	acre ——	0.405
2.47 ——	square kilometer, km^2	acre ——	4.05×10^{-3}
0.386 ——	square kilometer, km^2	square mile, mi^2 —	2.590
2.47×10^{-4} —	square meter, m^2	acre ——	4.05×10^{-3}
10.76 ——	square meter, m^2	square feet, ft^2 ——	9.29×10^{-2}
1.55×10^{-3} ——	mm^2	square inch, in^2 ——	645

CUBICS

9.73×10^{-3} —	cubic meter, m^3	inch-acre ——	102.8
35.3 ——	cubic meter, m^3	cubic-feet, ft^3 ——	2.83×10^{-2}
6.10×10^4 ——	cubic meter, m^3	cubic inch, in^3 ——	1.64×10^{-5}
2.84×10^{-2} ——	liter, L (10^{-3} m^3)	bushel, bu ——	35.24
1.057 ——	liter, L	liquid quarts, qt —	0.946
3.53×10^{-2} ——	liter, L	cubic feet, ft^3 ——	28.3

(Modified and reprinted with permission from: Megh R. Goyal, 2012. Appendices. Pages 317–332. In: *Management of Drip/Trickle or Micro Irrigation* edited by Megh R. Goyal. New Jersey, USA: Apple Academic Press Inc.)

0.265 ——— liter, L   gallon ————— 3.78
33.78 ——— liter, L   fluid ounce, oz —— 2.96×10^{-2}
2.11 ——— liter, L   fluid dot, dt ——— 0.473

WEIGHT

2.20×10^{-3} — gram, g (10^{-3} kg)   pound, ————— 454
3.52×10^{-2} — gram, g (10^{-3} kg)   ounce, oz ———— 28.4
2.205 ——— kilogram, kg   pound, lb ———— 0.454
10^{-2} ——— kilogram, kg   quintal (metric), q — 100
1.10×10^{-3} — kilogram, kg   ton (2000 lbs), ton — 907
1.102 ——— mega gram, mg   ton (US), ton ———— 0.907
1.102 ——— metric ton, t   ton (US), ton ———— 0.907

YIELD AND RATE

0.893 ——— kilogram per hectare   pound per acre ——— 1.12
7.77×10^{-2} — kilogram per cubic meter   pound per fanega ——— 12.87
1.49×10^{-2} — kilogram per hectare   pound per acre, 60 lb – 67.19
1.59×10^{-2} — kilogram per hectare   pound per acre, 56 lb – 62.71
1.86×10^{-2} — kilogram per hectare   pound per acre, 48 lb – 53.75
0.107 ——— liter per hectare   galloon per acre ——— 9.35
893 ——— ton per hectare   pound per acre ——— 1.12×10^{-3}
893 ——— mega gram per hectare   pound per acre ——— 1.12×10^{-3}
0.446—— ton per hectare   ton (2000 lb) per acre –2.24
2.24 ——— meter per second   mile per hour ——— 0.447

SPECIFIC SURFACE

10 ——— square meter   square centimeter
   per kilogram   per gram ——— 0.1
10^3 ——— square meter   square millimeter
   per kilogram   per gram ——— 10^{-3}

PRESSURE

9.90 ——— megapascal, MPa   atmosphere ——— 0.101
10 ——— megapascal   bar ————— 0.1
1.0 —— megagram per   gram per cubic —— 1.00
   cubic meter   centimeter

2.09 × 10⁻² — pascal, Pa$\qquad$pound per square feet— 47.9
1.45 × 10⁻⁴ — pascal, Pa$\qquad$pound per square inch— 6.90×10³

TEMPERATURE
1.00 (K-273)—Kelvin, K$\qquad$centigrade, °C — 1.00 (C+273)
(1.8 C + 32)—centigrade, °C$\qquad$Fahrenheit, °F — (F–32)/1.8

ENERGY
9.52 × 10⁻⁴ — Joule J$\qquad$BTU ———— 1.05 × 10³
0.239 —— Joule, J$\qquad$calories, cal —— 4.19
0.735 —— Joule, J$\qquad$feet-pound —— 1.36
2.387 × 10⁵ —— Joule per$\qquad$calories per square — 4.19 × 10⁴
   square meter$\qquad$ centimeter
10⁵ —— Newton, N$\qquad$dynes ———— 10⁻⁵

WATER REQUIREMENTS
9.73 × 10⁻³ — cubic meter$\qquad$inch acre ———— 102.8
9.81 × 10⁻³ — cubic meter per hour$\quad$cubic feet per second – 101.9
4.40 —— cubic meter per hour$\quad$galloon (US) per — 0.227
$\qquad$ minute
8.11 —— hectare-meter$\qquad$acre-feet ———— 0.123
97.28 —— hectare-meter$\qquad$acre-inch ———— 1.03 × 10⁻²
8.1 × 10⁻² — hectare centimeter$\quad$acre-feet ———— 12.33

CONCENTRATION
1 ———— centimol per kilogram$\quad$milliequivalents —— 1
$\qquad$ per 100 grams
0.1 —— gram per kilogram$\qquad$percents ———— 10
1 ———— milligram per kilogram$\quad$parts per million — 1

NUTRIENTS FOR PLANTS
2.29 —— P$\qquad$P₂O₅ ———— 0.437
1.20 —— K$\qquad$K₂O ———— 0.830
1.39 —— Ca$\qquad$CaO ———— 0.715
1.66 —— Mg$\qquad$MgO ———— 0.602

NUTRIENT EQUIVALENTS

Column A	Column B	Conversion A to B	Equivalent B to A
N	NH_3	1.216	0.822
	NO_3	4.429	0.226
	KNO_3	7.221	0.1385
	$Ca(NO_3)_2$	5.861	0.171
	$(NH_4)_2SO_4$	4.721	0.212
	NH_4NO_3	5.718	0.175
	$(NH_4)_2HPO_4$	4.718	0.212
P	P_2O_5	2.292	0.436
	PO_4	3.066	0.326
	KH_2PO_4	4.394	0.228
	$(NH_4)_2HPO_4$	4.255	0.235
	H_3PO_4	3.164	0.316
K	K_2O	1.205	0.83
	KNO_3	2.586	0.387
	KH_2PO_4	3.481	0.287
	Kcl	1.907	0.524
	K_2SO_4	2.229	0.449
Ca	CaO	1.399	0.715
	$Ca(NO_3)_2$	4.094	0.244
	$CaCl_2 \times 6H_2O$	5.467	0.183
	$CaSO_4 \times 2H_2O$	4.296	0.233
Mg	MgO	1.658	0.603
	$MgSO_4 \times 7H_2O$	1.014	0.0986
S	H_2SO_4	3.059	0.327
	$(NH_4)_2SO_4$	4.124	0.2425
	K_2SO_4	5.437	0.184
	$MgSO_4 \times 7H_2O$	7.689	0.13
	$CaSO_4 \times 2H_2O$	5.371	0.186

APPENDIX B

PIPE AND CONDUIT FLOW

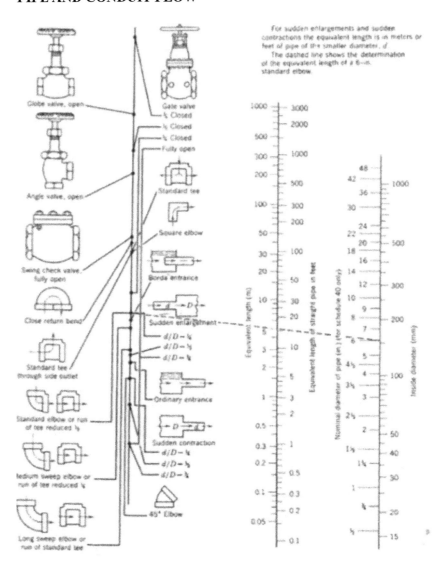

APPENDIX C

PERCENTAGE OF DAILY SUNSHINE HOURS: FOR NORTH AND SOUTH HEMISPHERES

Latitude	Jan	Feb	Mar	Apr	May	Jun	Jul	Aug	Sep	Oct	Nov	Dec
					NORTH							
0	8.50	7.66	8.49	8.21	8.50	8.22	8.50	8.49	8.21	8.50	8.22	8.50
5	8.32	7.57	8.47	3.29	8.65	8.41	8.67	8.60	8.23	8.42	8.07	8.30
10	8.13	7.47	8.45	8.37	8.81	8.60	8.86	8.71	8.25	8.34	7.91	8.10
15	7.94	7.36	8.43	8.44	8.98	8.80	9.05	8.83	8.28	8.20	7.75	7.88
20	7.74	7.25	8.41	8.52	9.15	9.00	9.25	8.96	8.30	8.18	7.58	7.66
25	7.53	7.14	8.39	8.61	9.33	9.23	9.45	9.09	8.32	8.09	7.40	7.52
30	7.30	7.03	8.38	8.71	9.53	9.49	9.67	9.22	8.33	7.99	7.19	7.15
32	7.20	6.97	8.37	8.76	9.62	9.59	9.77	9.27	8.34	7.95	7.11	7.05
34	7.10	6.91	8.36	8.80	9.72	9.70	9.88	9.33	8.36	7.90	7.02	6.92
36	6.99	6.85	8.35	8.85	9.82	9.82	9.99	9.40	8.37	7.85	6.92	6.79
38	6.87	6.79	8.34	8.90	9.92	9.95	10.1	9.47	3.38	7.80	6.82	6.66
40	6.76	6.72	8.33	8.95	10.0	10.1	10.2	9.54	8.39	7.75	6.72	7.52
42	6.63	6.65	8.31	9.00	10.1	10.2	10.4	9.62	8.40	7.69	6.62	6.37
44	6.49	6.58	8.30	9.06	10.3	10.4	10.5	9.70	8.41	7.63	6.49	6.21
46	6.34	6.50	8.29	9.12	10.4	10.5	10.6	9.79	8.42	7.57	6.36	6.04
48	6.17	6.41	8.27	9.18	10.5	10.7	10.8	9.89	8.44	7.51	6.23	5.86
50	5.98	6.30	8.24	9.24	10.7	10.9	11.0	10.0	8.35	7.45	6.10	5.64
52	5.77	6.19	8.21	9.29	10.9	11.1	11.2	10.1	8.49	7.39	5.93	5.43
54	5.55	6.08	8.18	9.36	11.0	11.4	11.4	10.3	8.51	7.20	5.74	5.18
56	5.30	5.95	8.15	9.45	11.2	11.7	11.6	10.4	8.53	7.21	5.54	4.89
58	5.01	5.81	8.12	9.55	11.5	12.0	12.0	10.6	8.55	7.10	4.31	4.56
60	4.67	5.65	8.08	9.65	11.7	12.4	12.3	10.7	8.57	6.98	5.04	4.22
					SOUTH							
0	8.50	7.66	8.49	8.21	8.50	8.22	8.50	8.49	8.21	8.50	8.22	8.50
5	8.68	7.76	8.51	8.15	8.34	8.05	8.33	8.38	8.19	8.56	8.37	8.68
10	8.86	7.87	8.53	8.09	8.18	7.86	8.14	8.27	8.17	8.62	8.53	8.88
15	9.05	7.98	8.55	8.02	8.02	7.65	7.95	8.15	8.15	8.68	8.70	9.10
20	9.24	8.09	8.57	7.94	7.85	7.43	7.76	8.03	8.13	8.76	8.87	9.33

Latitude	Jan	Feb	Mar	Apr	May	Jun	Jul	Aug	Sep	Oct	Nov	Dec
25	9.46	8.21	8.60	7.74	7.66	7.20	7.54	7.90	8.11	8.86	9.04	9.58
30	9.70	8.33	8.62	7.73	7.45	6.96	7.31	7.76	8.07	8.97	9.24	9.85
32	9.81	8.39	8.63	7.69	7.36	6.85	7.21	7.70	8.06	9.01	9.33	9.96
34	9.92	8.45	8.64	7.64	7.27	6.74	7.10	7.63	8.05	9.06	9.42	10.1
36	10.0	8.51	8.65	7.59	7.18	6.62	6.99	7.56	8.04	9.11	9.35	10.2
38	10.2	8.57	8.66	7.54	7.08	6.50	6.87	7.49	8.03	9.16	9.61	10.3
40	10.3	8.63	8.67	7.49	6.97	6.37	6.76	7.41	8.02	9.21	9.71	10.5
42	10.4	8.70	8.68	7.44	6.85	6.23	6.64	7.33	8.01	9.26	9.8	10.6
44	10.5	8.78	8.69	7.38	6.73	6.08	6.51	7.25	7.99	9.31	9.94	10.8
46	10.7	8.86	8.90	7.32	6.61	5.92	6.37	7.16	7.96	9.37	10.1	11.0

APPENDIX D

PSYCHOMETRIC CONSTANT (γ) FOR DIFFERENT ALTITUDES (Z)

$$\gamma = 10^{-3} \, [(C_p.P) \div (\varepsilon.\lambda)] = (0.00163) \times [P \div \lambda]$$

γ, psychrometric constant [kPa C^{-1}]

$c_{p,}$ specific heat of moist air = 1.013 [kJ kg^{-1} °C^{-1}]

P, atmospheric pressure [kPa].

ε, ratio molecular weight of water vapor/dry air = 0.622

λ, latent heat of vaporization [MJ kg^{-1}]
= 2.45 MJ kg^{-1} at 20 °C.

Z (m)	γ kPa/°C	z (m)	γ kPa/°C	z (m)	γ kPa/°C	z (m)	γ kPa/°C
0	0.067	1000	0.060	2000	0.053	3000	0.047
100	0.067	1100	0.059	2100	0.052	3100	0.046
200	0.066	1200	0.058	2200	0.052	3200	0.046
300	0.065	1300	0.058	2300	0.051	3300	0.045
400	0.064	1400	0.057	2400	0.051	3400	0.045
500	0.064	1500	0.056	2500	0.050	3500	0.044
600	0.063	1600	0.056	2600	0.049	3600	0.043
700	0.062	1700	0.055	2700	0.049	3700	0.043
800	0.061	1800	0.054	2800	0.048	3800	0.042
900	0.061	1900	0.054	2900	0.047	3900	0.042
1000	0.060	2000	0.053	3000	0.047	4000	0.041

APPENDIX E

SATURATION VAPOR PRESSURE [e$_s$] FOR DIFFERENT TEMPERATURES (T)

Vapor pressure function = e$_s$ = [0.6108] × exp{[17.27 □ T]/[T + 237.3]}

T °C	e$_s$ kPa	T °C	e$_s$ kPa	T °C	e$_s$ kPa	T °C	e$_s$ kPa
1.0	0.657	13.0	1.498	25.0	3.168	37.0	6.275
1.5	0.681	13.5	1.547	25.5	3.263	37.5	6.448
2.0	0.706	14.0	1.599	26.0	3.361	38.0	6.625
2.5	0.731	14.5	1.651	26.5	3.462	38.5	6.806
3.0	0.758	15.0	1.705	27.0	3.565	39.0	6.991
3.5	0.785	15.5	1.761	27.5	3.671	39.5	7.181
4.0	0.813	16.0	1.818	28.0	3.780	40.0	7.376
4.5	0.842	16.5	1.877	28.5	3.891	40.5	7.574
5.0	0.872	17.0	1.938	29.0	4.006	41.0	7.778
5.5	0.903	17.5	2.000	29.5	4.123	41.5	7.986
6.0	0.935	18.0	2.064	30.0	4.243	42.0	8.199
6.5	0.968	18.5	2.130	30.5	4.366	42.5	8.417
7.0	1.002	19.0	2.197	31.0	4.493	43.0	8.640
7.5	1.037	19.5	2.267	31.5	4.622	43.5	8.867
8.0	1.073	20.0	2.338	32.0	4.755	44.0	9.101
8.5	1.110	20.5	2.412	32.5	4.891	44.5	9.339
9.0	1.148	21.0	2.487	33.0	5.030	45.0	9.582
9.5	1.187	21.5	2.564	33.5	5.173	45.5	9.832
10.0	1.228	22.0	2.644	34.0	5.319	46.0	10.086
10.5	1.270	22.5	2.726	34.5	5.469	46.5	10.347
11.0	1.313	23.0	2.809	35.0	5.623	47.0	10.613
11.5	1.357	23.5	2.896	35.5	5.780	47.5	10.885
12.0	1.403	24.0	2.984	36.0	5.941	48.0	11.163
12.5	1.449	24.5	3.075	36.5	6.106	48.5	11.447

APPENDIX F

SLOPE OF VAPOR PRESSURE CURVE (Δ) FOR DIFFERENT TEMPERATURES (T)

$$\Delta = [4098.\ e°(T)] \div [T + 237.3]^2$$
$$= 2504\{\exp[(17.27T) \div (T + 237.2)]\} \div [T + 237.3]^2$$

T °C	Δ kPa/°C	T °C	Δ kPa/°C	T °C	Δ kPa/°C	T °C	Δ kPa/°C
1.0	0.047	13.0	0.098	25.0	0.189	37.0	0.342
1.5	0.049	13.5	0.101	25.5	0.194	37.5	0.350
2.0	0.050	14.0	0.104	26.0	0.199	38.0	0.358
2.5	0.052	14.5	0.107	26.5	0.204	38.5	0.367
3.0	0.054	15.0	0.110	27.0	0.209	39.0	0.375
3.5	0.055	15.5	0.113	27.5	0.215	39.5	0.384
4.0	0.057	16.0	0.116	28.0	0.220	40.0	0.393
4.5	0.059	16.5	0.119	28.5	0.226	40.5	0.402
5.0	0.061	17.0	0.123	29.0	0.231	41.0	0.412
5.5	0.063	17.5	0.126	29.5	0.237	41.5	0.421
6.0	0.065	18.0	0.130	30.0	0.243	42.0	0.431
6.5	0.067	18.5	0.133	30.5	0.249	42.5	0.441
7.0	0.069	19.0	0.137	31.0	0.256	43.0	0.451
7.5	0.071	19.5	0.141	31.5	0.262	43.5	0.461
8.0	0.073	20.0	0.145	32.0	0.269	44.0	0.471
8.5	0.075	20.5	0.149	32.5	0.275	44.5	0.482
9.0	0.078	21.0	0.153	33.0	0.282	45.0	0.493
9.5	0.080	21.5	0.157	33.5	0.289	45.5	0.504
10.0	0.082	22.0	0.161	34.0	0.296	46.0	0.515
10.5	0.085	22.5	0.165	34.5	0.303	46.5	0.526
11.0	0.087	23.0	0.170	35.0	0.311	47.0	0.538
11.5	0.090	23.5	0.174	35.5	0.318	47.5	0.550
12.0	0.092	24.0	0.179	36.0	0.326	48.0	0.562
12.5	0.095	24.5	0.184	36.5	0.334	48.5	0.574

APPENDIX G

NUMBER OF THE DAY IN THE YEAR (JULIAN DAY)

Day	Jan	Feb	Mar	Apr	May	Jun	Jul	Aug	Sep	Oct	Nov	Dec
1	1	32	60	91	121	152	182	213	244	274	305	335
2	2	33	61	92	122	153	183	214	245	275	306	336
3	3	34	62	93	123	154	184	215	246	276	307	337
4	4	35	63	94	124	155	185	216	247	277	308	338
5	5	36	64	95	125	156	186	217	248	278	309	339
6	6	37	65	96	126	157	187	218	249	279	310	340
7	7	38	66	97	127	158	188	219	250	280	311	341
8	8	39	67	98	128	159	189	220	251	281	312	342
9	9	40	68	99	129	160	190	221	252	282	313	343
10	10	41	69	100	130	161	191	222	253	283	314	344
11	11	42	70	101	131	162	192	223	254	284	315	345
12	12	43	71	102	132	163	193	224	255	285	316	346
13	13	44	72	103	133	164	194	225	256	286	317	347
14	14	45	73	104	134	165	195	226	257	287	318	348
15	15	46	74	105	135	166	196	227	258	288	319	349
16	16	47	75	106	136	167	197	228	259	289	320	350
17	17	48	76	107	137	168	198	229	260	290	321	351
18	18	49	77	108	138	169	199	230	261	291	322	352
19	19	50	78	109	139	170	200	231	262	292	323	353
20	20	51	79	110	140	171	201	232	263	293	324	354
21	21	52	80	111	141	172	202	233	264	294	325	355
22	22	53	81	112	142	173	203	234	265	295	326	356
23	23	54	82	113	143	174	204	235	266	296	327	357
24	24	55	83	114	144	175	205	236	267	297	328	358
25	25	56	84	115	145	176	206	237	268	298	329	359
26	26	57	85	116	146	177	207	238	269	299	330	360
27	27	58	86	117	147	178	208	239	270	300	331	361
28	28	59	87	118	148	179	209	240	271	301	332	362
29	29	(60)	88	119	149	180	210	241	272	302	333	363
30	30	—	89	120	150	181	211	242	273	303	334	364
31	31	—	90	—	151	—	212	243	—	304	—	365

APPENDIX H

STEFAN-BOLTZMANN LAW AT DIFFERENT TEMPERATURES (T)

$$[\sigma'(T_K)^4] = [4.903 \times 10^{-9}], \text{ MJ K}^{-4} \text{ m}^{-2} \text{ day}^{-1}$$

where: $T_K = \{T[°C] + 273.16\}$

T	$\sigma*(T_K)^4$	T	$\sigma*(T_K)^4$	T	$\sigma*(T_K)^4$
		Units			
°C	MJ m^{-2} d^{-1}	°C	MJ m^{-2} d^{-1}	°C	MJ m^{-2} d^{-1}
1.0	27.70	17.0	34.75	33.0	43.08
1.5	27.90	17.5	34.99	33.5	43.36
2.0	28.11	18.0	35.24	34.0	43.64
2.5	28.31	18.5	35.48	34.5	43.93
3.0	28.52	19.0	35.72	35.0	44.21
3.5	28.72	19.5	35.97	35.5	44.50
4.0	28.93	20.0	36.21	36.0	44.79
4.5	29.14	20.5	36.46	36.5	45.08
5.0	29.35	21.0	36.71	37.0	45.37
5.5	29.56	21.5	36.96	37.5	45.67
6.0	29.78	22.0	37.21	38.0	45.96
6.5	29.99	22.5	37.47	38.5	46.26
7.0	30.21	23.0	37.72	39.0	46.56
7.5	30.42	23.5	37.98	39.5	46.85
8.0	30.64	24.0	38.23	40.0	47.15
8.5	30.86	24.5	38.49	40.5	47.46
9.0	31.08	25.0	38.75	41.0	47.76
9.5	31.30	25.5	39.01	41.5	48.06
10.0	31.52	26.0	39.27	42.0	48.37
10.5	31.74	26.5	39.53	42.5	48.68
11.0	31.97	27.0	39.80	43.0	48.99

T	$\sigma^*(T_K)^4$	T	$\sigma^*(T_K)^4$	T	$\sigma^*(T_K)^4$
		Units			
°C	MJ m^{-2} d^{-1}	°C	MJ m^{-2} d^{-1}	°C	MJ m^{-2} d^{-1}
11.5	32.19	27.5	40.06	43.5	49.30
12.0	32.42	28.0	40.33	44.0	49.61
12.5	32.65	28.5	40.60	44.5	49.92
13.0	32.88	29.0	40.87	45.0	50.24
13.5	33.11	29.5	41.14	45.5	50.56
14.0	33.34	30.0	41.41	46.0	50.87
14.5	33.57	30.5	41.69	46.5	51.19
15.0	33.81	31.0	41.96	47.0	51.51
15.5	34.04	31.5	42.24	47.5	51.84
16.0	34.28	32.0	42.52	48.0	52.16
16.5	34,52	32.5	42.80	48.5	52.49

APPENDIX I

THERMODYNAMIC PROPERTIES OF AIR AND WATER

1. Latent Heat of Vaporization (λ)

$$\lambda = [2.501-(2.361 \times 10^{-3})\ T]$$

where: λ = latent heat of vaporization [MJ kg^{-1}]; and T = air temperature [°C].

The value of the latent heat varies only slightly over normal temperature ranges. A single value may be taken (for ambient temperature = 20 °C): $\lambda = 2.45$ MJ kg^{-1}.

2. Atmospheric Pressure (P)

$$P = P_o\ [\{T_{Ko}-\alpha(Z-Z_o)\} \div \{T_{Ko}\}]^{(g/(\alpha.R))}$$

where: P, atmospheric pressure at elevation z [kPa]

P_o, atmospheric pressure at sea level = 101.3 [kPa]

z, elevation [m]

z_o, elevation at reference level [m]

g, gravitational acceleration = 9.807 [m s^{-2}]

R, specific gas constant == 287 [J kg^{-1} K^{-1}]

α, constant lapse rate for moist air = 0.0065 [K m^{-1}]

T_{Ko}, reference temperature [K] at elevation z_o = 273.16 + T

T, means air temperature for the time period of calculation [°C]

When assuming P_o = 101.3 [kPa] at z_o = 0, and T_{Ko} = 293 [K] for T = 20 [°C], above equation reduces to:

$$P = 101.3[(293-0.0065Z)\ (293)]^{5.26}$$

3. Atmospheric Density (ρ)

$$\rho = [1000P] \div [T_{Kv}\ R] = [3.486P] \div [T_{Kv}], \text{ and } T_{Kv} = T_K[1-0.378(e_a)/P]^{-1}$$

where: ρ, atmospheric density [kg m^{-3}]

R, specific gas constant = 287 [J kg$^{-1\,K-1}$]

T_{Kv}, virtual temperature [K]

T_K, absolute temperature [K]: $T_K = 273.16 + T$ [°C]

e_a, actual vapor pressure [kPa]

T, mean daily temperature for 24-hour calculation time steps.

For average conditions (e_a in the range 1–5 kPa and P between 80–100 kPa), T_{Kv} can be substituted by: $T_{Kv} \approx 1.01\,(T + 273)$

4. Saturation Vapor Pressure function (e_s)

$$e_s = [0.6108] \times \exp\{[17.27 \times T]/[T + 237.3]\}$$

where: e_s, saturation vapor pressure function [kPa]

T, air temperature [°C]

5. Slope Vapor Pressure Curve (Δ)

$$\Delta = [4098.\ e°(T)] \div [T + 237.3]^2$$
$$= 2504\{\exp[(17.27T) \div (T + 237.2)]\} \div [T + 237.3]^2$$

where: Δ, slope vapor pressure curve [kPa C^{-1}]

T, air temperature [°C]

e°(T), saturation vapor pressure at temperature T [kPa]

In 24-hour calculations, Δ is calculated using mean daily air temperature. In hourly calculations T refers to the hourly mean, T_{hr}.

6. Psychrometric Constant (γ)

$$\gamma = 10^{-3}\ [(C_p.P) \div (\epsilon.\lambda)] = (0.00163) \times [P \div \lambda]$$

where: γ, psychrometric constant [kPa C^{-1}]

c_p, specific heat of moist air = 1.013 [kJ kg$^{-1\,°C-1}$]

P, atmospheric pressure [kPa]: equations 2 or 4

ϵ, ratio molecular weight of water vapor/dry air = 0.622

λ, latent heat of vaporization [MJ kg^{-1}]

7. Dew Point Temperature (T_{dew})

When data is not available, T_{dew} can be computed from e_a by:

$$T_{dew} = [\{116.91 + 237.3 Log_e(e_a)\} \div \{16.78 - Log_e(e_a)\}]$$

where: T_{dew}, dew point temperature [°C]

e_a, actual vapor pressure [kPa]

For the case of measurements with the Assmann psychrometer, T_{dew} can be calculated from:

$$T_{dew} = (112 + 0.9 T_{wet})[e_a \div (e° T_{wet})]^{0.125} - [112 - 0.1 T_{wet}]$$

8. Short Wave Radiation on a Clear-Sky Day (R_{so})

The calculation of R_{so} is required for computing net long wave radiation and for checking calibration of pyranometers and integrity of R_{so} data. A good approximation for R_{so} for daily and hourly periods is:

$$R_{so} = (0.75 + 2 \times 10^{-5} z) R_a$$

where: z, station elevation [m]

R_a, extraterrestrial radiation [MJ m^{-2} d^{-1}]

Equation is valid for station elevations less than 6000 m having low air turbidity. The equation was developed by linearizing Beer's radiation extinction law as a function of station elevation and assuming that the average angle of the sun above the horizon is about 50°.

For areas of high turbidity caused by pollution or airborne dust or for regions where the sun angle is significantly less than 50° so that the path length of radiation through the atmosphere is increased, an adoption of Beer's law can be employed where P is used to represent atmospheric mass:

$$R_{so} = (R_a) \exp[(-0.0018P) \div (K_t \sin(\Phi))]$$

where: K_t, turbidity coefficient, $0 < K_t < 1.0$ where $K_t = 1.0$ for clean air and $K_t = 1.0$ for extremely turbid, dusty or polluted air.

P, atmospheric pressure [kPa]

Φ, angle of the sun above the horizon [rad]

R_a, extraterrestrial radiation [MJ m^{-2} d^{-1}]

For hourly or shorter periods, Φ is calculated as:

$$\sin \Phi = \sin \varphi \sin \delta + \cos \varphi \cos \delta \cos \omega$$

where: φ, latitude [rad]
   δ, solar declination [rad] (Eq. (24) in Chapter 3)
   ω, solar time angle at midpoint of hourly or shorter period [rad]
For 24-hour periods, the mean daily sun angle, weighted according to R_a, can be approximated as:

$$\sin(\Phi_{24}) = \sin[0.85 + 0.3 \, \varphi \sin\{(2\pi J/365)-1.39\}-0.42 \, \varphi^2]$$

where: Φ_{24}, average Φ during the daylight period, weighted according to R_a [rad]
   φ, latitude [rad]
   J, day in the year
The Φ_{24} variable is used to represent the average sun angle during daylight hours and has been weighted to represent integrated 24-hour transmission effects on 24-hour R_{so} by the atmosphere. Φ_{24} should be limited to >0. In some situations, the estimation for R_{so} can be improved by modifying to consider the effects of water vapor on short wave absorption, so that: $R_{so} = (K_B + K_D) R_a$ where:

$$K_B = 0.98\exp[\{(-0.00146P) \div (K_t \sin \Phi)\}-0.091\{w/\sin \Phi\}^{0.25}]$$

where: K_B, the clearness index for direct beam radiation
   K_D, the corresponding index for diffuse beam radiation
   $K_D = 0.35-0.33 \, K_B$ for $K_B > 0.15$
   $K_D = 0.18 + 0.82 \, K_B$ for $K_B < 0.15$
   R_a, extraterrestrial radiation [MJ m^{-2} d^{-1}]
   K_t, turbidity coefficient, $0 < K_t < 1.0$, where $K_t = 1.0$ for clean air
and $K_t = 1.0$ for extremely turbid, dusty or polluted air.
   P, atmospheric pressure [kPa]
   Φ, angle of the sun above the horizon [rad]
   W, perceptible water in the atmosphere [mm] $= 0.14 \, e_a \, P + 2.1$
   e_a, actual vapor pressure [kPa]
   P, atmospheric pressure [kPa]

INDEX